ROCKEFELLER

ROCKEFELLER
Controlling the Game

JACOB NORDANGÅRD

Skyhorse Publishing

Skyhorse Publishing books may be purchased in bulk at special discounts for sales promotion, corporate gifts, fund-raising, or educational purposes. Special editions can also be created to specifications. For details, contact the Special Sales Department, Skyhorse Publishing, 307 West 36th Street, 11th Floor, New York, NY 10018 or info@skyhorsepublishing.com.

Skyhorse® and Skyhorse Publishing® are registered trademarks of Skyhorse Publishing, Inc.®, a Delaware corporation.

Visit our website at www.skyhorsepublishing.com.

Please follow our publisher Tony Lyons on Instagram @tonylyonsisuncertain

10 9 8 7 6 5 4 3 2 1

Library of Congress Cataloging-in-Publication Data is available on file.

Cover design by David Ter-Avanesyan
Cover image from Getty Images

Print ISBN: 978-1-5107-8021-7
Ebook ISBN: 978-1-5107-8022-4

Printed in the United States of America

CONTENTS

Preface

THIS BOOK IS the result of some questions that arose when working on my 2012 dissertation, *Ordo Ab Chao: The Political History of Biofuels in the European Union*.

My interest in energy and environmental issues, as well as reflections on our increasingly technological society, began in 1998 when I took a social science course on Cybernetics and Transhumanism, and was introduced to Jeremy Rifkin's 1981 book, *Entropy*. His grim conclusions had a strong impact on my view of the world.

Shortly thereafter, I joined the Swedish Green Party and became active on the city Planning Board in Norrköping. At that time, I could hardly imagine how closely related the ideas of cybernetics were to Rifkin's theories.

In 2004, I stumbled upon the Peak Oil theory and became very concerned over what would happen if basic functions of society could no longer be maintained due to energy shortages. This led me back to the university to study these areas more in depth. It surprised me that the climate issue at that time so dominated the debate while few spoke of Peak Oil.

My surprise became even greater when, in the spring of 2009, I was working on a background chapter on the origin and history of climate change and found that the powerful Rockefeller oil and finance family had been deeply involved in bringing the climate issue onto the international political agenda! This seemed highly contradictory, given their great influence over both the oil industry and the economic globalization of recent decades.

I continued to dig deeper for my climate history chapter contribution to the 2013 book *Domedagsklockan* (*The Doomsday Clock*), where new connections were found and investigated. However, my analysis ended with the founding of the United Nations Panel on Climate Change (IPCC) and the first UN Rio Summit in 1992. This meant that there was much more material to research about the decades that followed.

After writing the autobiographical *An Inconvenient Journey* in 2015, about what happens when you challenge powerful interests connected to the university and the difficulties getting one's research results out when they contradict general presuppositions, I wanted to continue researching the Rockefeller family's involvement in world politics and find answers to several remaining questions.

Why had the Rockefeller family funded and influenced climate research since the 1950s and helped shape climate policy since the 1980s? And why did Rockefeller Brothers Fund announce in 2014 that they would divest from all of their fossil energy holdings? Why attack the very industry on which their immense wealth was founded? What was their motive in their own words—and how did it all begin?

My digging resulted in a series of articles in which I followed the family from the founding of Standard Oil and the Rockefeller Foundation, up to the aftermath of the Paris Agreement, with the declaration of the Fourth Industrial Revolution in January 2016. The series was then expanded upon, resulting in this book. This in-depth research project that took almost two years to complete was made possible through the large quantity of material made available on the internet. The Rockefeller family's foundations' own reports and annual reviews formed the basis for further research. I also read biographies and articles about the family to get a better picture of the most prominent members and to learn what ambitions drove each of them.

My main focus has been the Rockefeller family's involvement in climate research and politics, but the actions and motives of some of their allies are also mentioned, as well as the family's influence on the development of modern medicine, family planning, agriculture, art, architecture, behavioral science, information technology, and politics.

It should be pointed out that I do not see my research as an absolute or complete account of the history of climate change research and politics. This story only gives *one* perspective on how this issue has come to grow to the proportions it has today. However, the Rockefeller family has undoubtedly been one of the most influential global players through its top position in American business, close contacts with the White House, and as one of the world's leading private research funders. With their immense financial power they have, in collaboration with other influential business partners, been able to anchor the climate issue both scientifically and politically.

Finally, I want to thank Hans Holmén and Staffan Wennberg for their manuscript reading, comments, and financial support for the project, and readers who have encouraged and helped fund my writings. And last but not least my wife Inger for her editing, translation, and invaluable support during the work.

—*Jacob Nordangård, PhD*

Taft oil well blow-out in Kern County, California, owned by Standard Oil, c. 1920.

Prologue

The Rockefeller Family Fund is proud to announce its intent to divest from fossil fuels. While the global community works to eliminate the use of fossil fuels, it makes little sense—financially or ethically—to continue holding investments in these companies.[1]
—The Rockefeller Family Fund, 2016

IN MARCH 2016, the nonprofit Rockefeller Family Fund (RFF) announced with much fanfare that they would be divesting from all holdings in fossil fuels, including the Rockefeller family's old crown jewel, ExxonMobil. The Paris Agreement had given a clear signal: if humanity and the ecosystem were to survive the coming decades, fossil fuels would have to stay in the ground.

At the same time, RFF accused ExxonMobil of having misled the public and of having spread doubts about the theory of anthropogenic climate change. An Exxon spokesman said,

"It's not surprising that they're divesting from the company since they're already funding a conspiracy against us."[2]

Less than two years earlier, during the great Climate March in New York, the bigger foundation, the Rockefeller Brothers Fund, also announced that due to their fight against climate change they would start divesting from coal and oil sands. The Rockefeller family, having for nearly a decade tried to pressure their old family company Exxon into changing their position on climate change,[3] now accused Exxon of having known about climate change since the 1980s and attempting to hide its implications from the public.

Around the same time, Exxon became subject of a real indictment by the state prosecutors in New York and California for having lied to their shareholders and the public about the seriousness of climate change.[4] The indictment was initiated by the Rockefeller-funded Climate Accountability Institute.[5]

The oil industry, now joined in with social activists against neoliberalism and globalisation like Naomi Klein, and climate activists like Bill McKibben—both members of the Rockefeller-funded 350.org which organized the People's Climate March in 2014 and the Global Climate March in 2015.

It was a very odd situation where the old oil barons attacked the very business upon which their power and wealth had been built. The industry which had, for good and ill, enabled the twentieth-century industrial development, the agricultural revolution, the pharmaceutical industry, and mass motoring. The family that had made us dependent on oil was now taking a leading position against it, by declaring the burning of fossil fuels as immoral, destructive, and sinful, with mankind as the sinner.

Had they suddenly changed their position on moral grounds?

Perhaps not. Despite its eagerness to divest, the Rockefeller Brothers Fund would keep its shares in Exxon in order to "be able to continue exerting pressure," while their largest foundation, the Rockefeller Foundation, totally opposed any divestments in fossil energy. The family still had strong ties to its old company.

Looking closer at the Rockefeller family's actions from the founding of Standard Oil to their climate activism today, their accusations against the flagship company appears to rather be part of a larger scheme pursued for more than a century—a plan to consolidate power on a worldwide scale and creating a world with a more effective global governance and a new economic system, Smart Globalization.[6] A technocratic world where the interdependent parts relinquish their sovereignty to serve a wider community under a planetary-wide institutional management.

This vision rings of Thomas Hobbes's *Leviathan* and the old dream of the Omega Point where the world (man, economy, and ecology) is united into in a technologically interconnected and synchronised whole—a cybernetic world organism. It was now time for the Great Transformation.[7]

Behind the philanthropic façade one finds a desire to manage both the population and the planet's natural resources by using Hegelian dialectics to reach the desired synthesis—a sustainable Utopia and a new economic system. All to save the world from the great climate catastrophe.

In the words of French priest and paleontologist Pierre Teilhard de Chardin, *The age of nations is past. The task before us now, if we would not perish, is to build the earth.*

John D. Rockefeller Sr. and Jr.

No longer can any man live to himself alone, nor any nation. The world has become a unit. Crop failure in South America is felt in Europe. A panic in London or New York creates financial depression throughout the world. Industrial difficulties in any one country have their influence in all countries.

—John D. Rockefeller Jr.

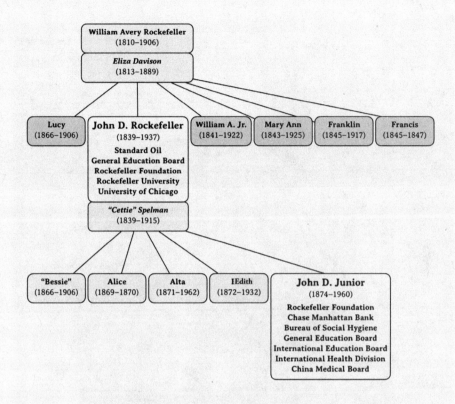

William Avery Rockefeller
(1810–1906)

Eliza Davison
(1813–1889)

Lucy
(1866–1906)

John D. Rockefeller
(1839–1937)

Standard Oil
General Education Board
Rockefeller Foundation
Rockefeller University
University of Chicago

William A. Jr.
(1841–1922)

Mary Ann
(1843–1925)

Franklin
(1845–1917)

Francis
(1845–1847)

"Cettie" Spelman
(1839–1915)

"Bessie"
(1866–1906)

Alice
(1869–1870)

Alta
(1871–1962)

IEdith
(1872–1932)

John D. Junior
(1874–1960)

Rockefeller Foundation
Chase Manhattan Bank
Bureau of Social Hygiene
General Education Board
International Education Board
International Health Division
China Medical Board

JOHN DAVISON ROCKEFELLER was born in 1839 as the second child of William Avery Rockefeller and Eliza Davison. He married Laura Spelman and had five children with her. John founded Standard Oil in 1870. This would make him one of the richest people in the history of the world. His wealth was used for philanthropic endeavours such as the founding of the University of Chicago, the Rockefeller University, the General Education Board and the Rockefeller Foundation. John D. died in 1937 of atherosclerosis at ninety-seven years old.

John D. Rockefeller Jr. was born in 1874 as the youngest child and only son of John D. Rockefeller and Eliza Davison. He married Abigail "Abby" Aldrich and had six children with her. John became chairman of the Rockefeller Foundation, supported internationalist initiatives such as the League of Nations and the United Nations, eugenics, ecumenics, and conservation. He built Rockefeller Center and offered the family estate for the building of the Museum of Modern Art in New York. John D. Jr. died in 1960 at eighty-six years old.

Rockefeller Center, New York, inaugurated in May 1933, with a statue of Atlas.

Chapter One

Controlling the Game

Good leadership consists of showing average people how to do the work of superior people.

—John D. Rockefeller

BACKGROUND

When the Rockefeller family is involved in something, one can be certain that is very carefully planned. Their extraordinary global influence has been achieved by the ability to work towards specific goals over a very long timescale.

Standard Oil

The Rockefeller fortune was created by developing and controlling the oil industry in the United States. It all started with Standard Oil of Ohio, founded in 1870 in Cleveland, Ohio, by brothers John D. and William Avery Rockefeller Jr. Initially, the oil was only used to produce kerosene—a revolutionary product at that time that helped light up homes across the world. John D. Rockefeller had a special gift for anticipating public demand.

Due to his aversion towards wasting resources, he made use of a kerosine waste product that the rest of the petroleum industry simply dumped: gasoline. By being so proactive, a fuel was developed ready to power the nascent auto industry. The investment in gasoline also helped Standard Oil survive the electrification of private homes which dramatically decreased kerosene demand. By 1910, gasoline was outselling kerosene.[1]

William Avery Rockefeller Jr. (1841–1922).   John D. Rockefeller (1839–1937).

In the early 1900s scientists also discovered that petroleum (rock oil) could be used or modified into other chemical products such as pesticides, ink, paint, paraffin wax, beauty and hygiene products, synthetic fibers, pharmaceuticals, and various forms of plastic. Bakelite was the first plastic material to be developed, as early as 1907.

The Rockefeller brothers' business plan was to control the whole production chain, from refining to the finished product, while constantly looking for ways to make production more efficient. With renowned ruthlessness, they created a monopoly on the refining and transportation of oil, the Standard Oil Trust.

John D., the most ambitious of the brothers, saw "competition as a sin" and would use any means to outsmart their competitors. State after state fell into their hands as competing petroleum companies were either acquired or eliminated. In the end, Standard Oil dominated the US market.

In 1882 their power was consolidated into a nine-man-strong board of directors, with John as chairman. The board appointed the directors and officers of all subordinate companies. Thus, all the interdependent parts acted as one disciplined unit. This cartel was the first of its kind. At its peak, the Standard Oil Trust controlled 90 percent of the petroleum market.

In 1885, Standard Oil's assets were gathered in the impressive new headquarters at 26 Broadway, New York, which, with its proximity to Wall Street, soon became a global hub. William became the representative of Standard Oil Company in New York until 1911. The Rockefeller family institutions also began redrawing the map and character of New York City.

Standard Oil also expanded worldwide and became one of the first and largest multinational corporations. After only a few decades, oil became the lifeblood of a world economy in which Standard Oil played a central part. Oil revolutionized travel and transportation, while at the same time causing huge pollution problems and fuelling the destructive wars of the twentieth century. It caused geopolitical conflicts right from the start and those who controlled it could easily be intoxicated by the power it gave.

John took the "survival of the fittest" business logic of capitalism to mean that in the end only the strongest company would emerge as lone victor at the top. To reach this goal, Standard's products had to be the best on the market and were developed and refined to perfection. The name *Standard* reflected not only a monopolist mindset but also the ambition to provide a product with a reliable standardized quality all over the world.

The other essential component was being able to outmaneuver all competition and becoming a master at manipulation. When competitors could not be beaten, they would be joined.

An early example was the 1880s oil price war between the Rockefellers, the Rothschild family, the Nobel brothers, and newly formed oil companies such as Royal Dutch and Shell (later merged to form Royal Dutch Shell). The powerful Rothschild family had entered the oil business and given loans to the Nobel brothers for oil drilling in Russia. John D. naturally saw this as a serious threat to the world dominance of Standard. The price war forced him to invite the head of the Rothschild family, Baron

Alphonse de Rothschild, to a meeting in New York in 1892. The negotiations resulted in a truce where the competing families instead formed
an alliance.[2] These close ties would remain and influence the social, economic, and political development of the twentieth century.

In the late 1800s, the Rockefeller brothers started expanding their
business ventures. William founded Amalgamated Copper with Henry
H. Rogers. John D. invested in the mining industry and Colorado Fuel
and Iron, partnering with the unscrupulous industrialist Jay Gould. The
brothers both sensed that power and wealth was based on controlling natural resources.

Devil Bill

The art of winning by any means, fair or foul, had been passed down
from their father, William Rockefeller, a notorious fraudster and bigamist
who went by the nickname "Devil Bill." Bill was tall, broad-chested and
handsome and had a special charm for both attracting women and parting
gullible people from their hard-earned cash.

Bill made a living both from legitimate business ventures such as lumber, and as a travelling salesman and charlatan. In his twenties he started
posing as a deaf and dumb peddler of trinkets.[3] Later he travelled under

Eliza Davison Rockefeller
(1813–1889).

William "Bill" Rockefeller (1810–1906).

the name Dr. William Levingston, offering "herbal remedies" to unsuspecting country folk.[4]

Despite his dubious business methods, he was still meticulous about bookkeeping, paying bills, respecting contracts, and avoiding alcohol (likely deterred by his alcoholic father).[5]

When marrying Eliza Davison, who came with a five hundred dollar dowry, Bill brought his former sweetheart, the beautiful but poor Nancy Brown, as housekeeper and had children alternately with his wife and with his mistress (until Nancy grew quarrelsome and was sent back to her family, later marrying another man).[6]

Bill was constantly travelling, now and then returning home, extravagantly dressed with his pockets full of money which he spent generously as compensation for his absence, and entertaining his children with captivating tales. Between these short showers of abundance, John D. and his mother struggled to make ends meet.

On one of Bill's business trips he later met and married a younger woman, Margaret Allen, under his alias Dr. Levingston, while continuing to be married to Eliza.[7] Bill's double life and shady background remained an embarrassing family secret.

John D. Rockefeller

Even in his youth, John D. was introverted and solemn, with piercing eyes and high moral standards. Like his parents he was a teetotaller, and minded his health. Early in life he got involved with the Baptist church and was much influenced by its teachings. He did not, however, take lightly to those who led a sinful life with gambling or drunkenness and was viewed as a bit of a killjoy.

When he started earning his own money, at age sixteen, all his personal economic activity, both incomes and expenses, were meticulously entered into his red *Ledger A*. It was his most precious possession and would for the rest of his life be treated almost as a sacred relic.[8] This bookkeeping habit was later transferred to his own children who were taught the art of financial management from an early age.

Starting his career, first as a diligent bookkeeping assistant and later as partner in a commissions firm before going into the oil business, John's industriousness and serious manners earned him the trust and respect of the older businessmen in the community. People started calling him "Mr. Rockefeller" while still in his teens. He was punctual, orderly, and loved to work from early morning to late at night.

From his father, Bill, John D. had learned the art of bookkeeping, writing contracts, evaluating the quality of goods, and unconventional negotiation techniques—skills which turned out very useful in his professional life.

However, he also came to harbor a well-concealed resentment against his father who had boasted to a neighbor, *"I trade with my boys and I skin 'em and I just beat 'em every time I can. I want to make 'em sharp."* When Bill knew that John was in dire straights, he would demand immediate repayment of a loan he had given earlier, keep the money for a while and then hand it back to his son.[9]

Even when John was a young boy, Bill had played cruel games with him, with the explicit purpose of teaching his son not to trust *anyone* completely, not even his own father. Such tough love taught John to be a ruthless and efficient businessman who would not accept defeat. His father had given him the clear message that the business world was fierce and ruthless and that in order to succeed, any means were acceptable.[10]

From his conscientious, stern, and deeply religious mother, John got both his strategical and mathematical thinking and his Christian faith. He internalised her constant admonitions to not let success go to his head. These disparate influences made John D. very successful, but also gave him somewhat of a dual personality, a Dr. Jekyll and Mr. Hyde.

Like his father, John D. loved his money more than anything but was not inclined to spend it willy-nilly. His rapidly increasing fortune was to grow and not be scattered on excesses, parties, fine clothing, or luxuries. Every purchase and investment was to be carefully calculated and nothing go to waste. Always attending to the most minute detail in optimising and streamlining every part of his growing business empire, by 1880, he had reached the top of the list of the richest men in the United States.

PHILANTHROPY

From a young age, John D. was a strong believer in giving a certain percentage of his earnings to charity (carefully entered into *Ledger A*, of course). As soon as he earned an income he would make donations to various charities, both Baptist, Methodist, Catholic, and abolitionist.

Inspired by Baptist ethics, and by the prosperity theology not uncommon among businessmen of his time, he saw his wealth as a gift from God, given to the most worthy and hardworking for stewardship.[11] With his immense resources he wanted to make the world a better place.

Those who refused to participate in his plans and become part of his growing empire were, however, seen as sinners and not deserving of any sympathy. Charity also became a way of redeeming his often ruthless and predatory business methods.

In 1896, the same year that Svante Arrhenius (1859–1927) became the first scientist to calculate the levels of carbon dioxide in the atmosphere and its possible influence on the greenhouse effect, John D. Rockefeller Sr. retired from the management of Standard Oil and handed the leadership of the company over to his closest man, John Dustin Archbold (1848–1916).

His son, John D. Jr., was also placed in more responsible positions within the company. Senior had new plans. Just like steel magnate Andrew Carnegie (1835–1919), he started investing in philanthropy and social reform.

There was now enough capital to start reshaping the world and provide it with a more efficient management.

Only a few decades later, Rockefeller philanthropy would start financing the research field emerging from the theories of Svante Arrhenius.

EDUCATION AND RESEARCH

Early on in his career, John D. Rockefeller realised the benefits of gaining control over education and the production of knowledge. With the aid of his financial advisor, Baptist pastor Frederick Gates (1853–1929), who was hired to run the philanthropic endeavours, the founding or financing of a long list of significant institutions began.

Through the generations, the Rockefeller family came to actively support seventy-five top colleges and universities, including Harvard, Yale, Brown, Columbia, Cornell, Massachusetts Institute of Technology (MIT), Princeton, Tufts, University of California Berkley, and University of Chicago. This gave a significant degree of influence over the field of education and research, mainly in the United States but also overseas (e.g. through Central Philippines University and the London School of Economics).[12]

University of Chicago

In 1890, John D. Rockefeller and the American Baptist Education Society co-founded the University of Chicago (on land donated by department store owner Marshall Field). It was Rockefeller's million dollar donation that turned the college into a university, and he continued to finance its operation. As the university required more and more funding, Rockefeller asked for representatives on the board of directors in order to get more oversight and control of how donations were used.

In 1896, Frederick Gates was elected board member, joined the following year by his son, John D. Jr.[13] Rockefeller's great grandson, David

University of Chicago, inaugurated in 1891.

Rockefeller, became a board member in 1947, and would have close ties to the university throughout his life as advisor to several of its principals.[14]

Rockefeller had high ambitions for the university right from the start and appointed the young innovative William Rainey Harper as its first president.[15] Success soon followed and the University of Chicago became one of the world's most distinguished universities, with a significant impact on the scientific, political, cultural, and social development of the twentieth century.

As of October 2019, one hundred Nobel laureates have been affiliated with the university. Several disciplines were developed there of benefit to Rockefeller's power base, including economy, sociology, and behavioural sciences. It also spawned neoclassical economics, known as the Chicago school of economics, and part of the Manhattan Project.

Several decades later, in 1940, Swedish Carl-Gustaf Rossby was elected chair to the Department of Meteorology at the University of Chicago, which would emerge as a leading center for studying the impact of carbon dioxide on the greenhouse effect.

Despite the enormous success of Standard Oil, John D. Sr. viewed the University of Chicago as the best investment of his life.

Negative Publicity

Soon, however, there were problems for Rockefeller Sr., which forced him to return to the bustle of Standard Oil.

In an investigative article series, "The History of the Standard Oil Company" published between 1902 and 1904, journalist Ida M. Tarbell had revealed the company's shady business practices. The series, later published as a book, reached a large audience.[16] It inspired several antitrust acts and in 1906 led to Standard Oil being accused of conspiring to prevent free trade.

Ida M. Tarbell (1857–1944), journalist.

Already in 1890 the Sherman Antitrust Act had been adopted to counteract monopolies. Following a court order in 1892, Rockefeller had still managed to avoid the breaking up of Standard Oil by creating the holding company, Standard Oil Company of New Jersey.

In 1911, the law finally caught up with Rockefeller, resulting in Standard Oil being divided into thirty-four smaller companies. The most powerful were Standard Oil Company of New Jersey, using the brands Jersey Standard and ESSO (S.O., which later became Exxon) and Standard Oil Company of New York (which became Mobil) and had its head office in Rockefeller Center. The damage, however, turned out to be minimal as Rockefeller had been well prepared for this eventuality. At the time of the breakup, Rockefeller owned 25 percent of the shares and retained the same percentage of shares in all the new companies. In a short period of time, the original value of his stock increased fivefold, while he retained control over all the companies.

THE STANDARD OIL COMPANY	
Standard Oil of New Jersey (SONJ) → **ESSO** + **Humble Oil** = **Exxon**	**ExxonMobil**
Standard Oil of New York (SOCONY) + Vacuum Oil Co. = **Mobil**	
Standard Oil of California (SOCAL) + Texaco	**Chevron**
Standard Oil of Kentucky (KYSO)	
Standard Oil of Ohio (SOHIO)	BP
Standard Oil of Indiana (Stanolind Oil & Gas) → **AMOCO**	
The Ohio Oil Company → **Marathon**	Marathon Petroleum
Bold = Rockefeller-controlled petroleum companies	

Many years later, these companies would slowly begin merging again. Today, the remnants of Standard Oil can be found in ExxonMobil and Chevron, while BP acquired most of the rest (a few were acquired by Shell and Unilever). These companies still retain the rights to the name *Standard*.

Rockefeller Foundation

In order to improve his tarnished reputation as a ruthless industrialist—and avoid taxation—John D. Rockefeller in 1913 founded the Rockefeller

Foundation. It had been carefully prepared for several years and was a development of the General Education Board, based on the ideas of Frederick Gates. The plan was to run the Rockefeller Foundation with the same efficiency as Standard Oil.

Some of the sharpest minds in the country were invited to the board of directors (see Appendix A). The foundation still has close ties to the political and business elites of the US. The capital was based on shares in the family's oil companies. During its first year, Rockefeller transferred $100 million to the foundation.[17] By 1929, $300 million in share capital had been transferred from Standard Oil. The foundation became another pillar of the empire.

After the creation of the foundation, Rockefeller withdrew from his commitments. He remained a nominal board member until 1923 but did not participate in any meetings. His position was instead left to his son, John Jr., first as president and from 1917 to 1939 as chairman.

John D. Rockefeller Jr.

Like his father, John Jr. was reserved and disciplined and the responsibilities would at times weigh heavily on his shoulders, always working in the shadow of his legendary father. With time, however, he matured into his position as custodian of his family's immense fortune and its growing economic, philanthropic, and political influence.

Nelson W. Aldrich

The Rockefeller family grew even more powerful in 1901 when John Jr. married socialite Abby Aldrich Rockefeller, daughter of Republican Senator Nelson W. Aldrich. The marriage can be seen as an early example of a public-private partnership, as it combined the powers of the richest family in the United States with that of one of its most influential politicians at the time. Senator Aldrich's connections in the political and economic establishment was of considerable value to the Rockefellers, and would later inspire the political aspirations of the second son to John and Abby, Nelson Rockefeller. Aldrich, who was also a prominent Freemason, was known as "the general manager of the nation." He defended the interests of Big Business, introduced the income tax (1909) and laid the foundation for the Federal Reserve System through the Aldrich Plan (1911).[18] Aldrich was as despised as Rockefeller Sr. due to his business-friendly policies, allegations of corruption, and investments in and support for the ruthless Belgian King Léopold II's colony in Congo. He died of a stroke in 1915.

Abby's more outgoing and relaxed attitude towards life, however, brought an element of liveliness and progressiveness into the strict Baptist family and promptly declared her refusal to conform to any restraints.[19]

Rockefeller Center

One of John D. Junior's major achievements was the planning of Rockefeller Center; twenty-one buildings for offices, culture, and commerce (including Radio City Music Hall, RCA Building, Time Life Building) built between 1929 and 1940. The center became a visible symbol of his father's achievements—the great benefactor, the brilliant businessman, and an icon for his heirs to admire.

During the period 1933 to 2015, the family head office was located in the legendary Room 5600 on the fifty-sixth floor of the central building, 30 Rockefeller Plaza (RCA Building, later GE Building). This was the control center of the empire. John D. Jr. had a private elevator installed that would take him from his office straight down to the vault in the basement.[20]

The Ludlow Massacre

In 1914 something happened that would greatly affect the family's reputation. Via holdings in the Colorado Fuel & Iron Company, John Jr. was the chief owner of a coal mine in Ludlow, Colorado, where workers were on strike for better working conditions.

The company refused to agree to their demands, resulting in a massacre where twenty-one workers, women, and children where shot to death with machine guns when the company guards and the state national guard attacked the tent camp, and many more deaths in the ensuing skirmishes.[21]

This tragedy weighed heavily on John Jr.

Margaret Sanger wrote a scathing article against the company owners in her magazine *The Woman Rebel,* where she described them as worse than cannibals.

> Cannibals, you see, are uncivilized, primitive folk, low in the scale of human intelligence. Their tastes are not so fastidious, so refined, so Christian, as those of our great American coal operators, who have subsidized the State of Colorado, and treat the President of the United States as an office boy—these leering, bloody hyenas of the human race who smear themselves with the stinking honey of Charity to attract those foul flies of religion who spread pollution throughout the land.[22]

Public hatred against the super-capitalists was now at its peak.

Public Relations

To clean up the Rockefeller family image, press agent Ivy Lee was recruited. John D. Sr. had an old habit of handing out dimes to adults and nickels to children—with an admonition to save them and work hard.[23]

Ivy Lee made sure these small acts of generosity were captured on photos and newsreels. After a few years of effective PR campaigns, the image of the family began to change into one of generous philanthropists.

Some years later, several major nationwide periodicals, including *Time* and *Newsweek*, were acquired by Senior and his banking colleague, J. P. Morgan, to be used as their own propaganda channels. John D. Sr., who had been quite offended by Tarbell's articles, wanted to prevent any major newspaper from writing anything unflattering about him ever again.[24]

Propaganda Techniques

The Rockefeller Foundation also made substantial donations for the study of propaganda techniques and how to influence people's behaviour and decision making, both through politics and media.

Harold Lasswell (1902–78) from the University of Chicago proposed that people needed to gradually get used to new ideas and measures.

Would-be propagandists were recommended to introduce and slowly nurture these by well-developed long-term campaigns. Emotionally loaded symbols should be created in order to generate specific responses, to be used for large-scale mass action. These propaganda tools would then be offered to a technocratic elite of scientists.[25]

Another suggested method of influencing was to cooperate with established authorities and institutions, pointing out solvable problems and encouraging them to assume responsibility for their long-term solutions. This has proven a successful concept. Many international organizations have been created in this way (e.g., CGIAR, see "Agriculture," chapter 2).

The Rockefeller Foundation also funded the study and development of psychological warfare and management strategies. These strategies were later used on a large scale to further the family's own long-term goals.

MEDICINE

The stated purpose of the Rockefeller Foundation was "promoting the well-being of humanity throughout the world." Medicine became an early focus for the foundation's charity.

Rockefeller Institute for Medical Research

In 1901, John D. Sr., with assistance from Frederick Gates, founded the Rockefeller Institute for Medical Research (which in 1965 became Rockefeller University). John D. Jr.'s advisor Simon Flexner became its first director of laboratories.[26] Flexner was the man behind the *Flexner Report* (1910), financed by Carnegie Corporation and resulting in a standardised science-based medical education.[27] Alternative medical treatments were marginalised or outlawed, and replaced by the emerging pharmaceutical industry with patentable chemical compounds, often made from oil derivatives—in which the Rockefellers and some of their fellow philanthropists happened to have large financial stakes. A very profitable strategy, later to be used in other areas.

General Education Board

In 1904, the Rockefeller Foundation founded the General Education Board. Again, Simon Flexner was involved as a board member. These medical institutions would become pillars of the Rockefeller empire and gave the family unprecedented influence over education, health, and medicine in the United States.

The Rockefeller Foundation also had an international outlook and in 1913 founded the precursor to the WHO, International Health Division (initially focused on curing hookworm, then malaria, tuberculosis, and yellow fever).

In 1923 the accompanying International Education Board (IEB) was founded. In China the foundation established the China Medical Board. The relationships with Chinese authorities were very good, until the revolution severed contacts with the West. When China began opening up again in the 1970s, the Rockefeller family were among the first to establish contacts with the Communist regime. This laid the foundation for China's part in world economy after the fall of Communism in Europe.

EUGENICS

Like many of their peers in the upper classes of the early 1900s, the Rockefellers were strongly influenced by the Malthusian ideals of population control and the improvement of social and racial hygiene. The American eugenics movement had its roots in the biological determinist ideas of Sir Francis Galton from the 1880s.

The ultimate goal was to create a genetically improved super-human through selective breeding. This would entail appointed experts (rather than the individual) choosing suitable mating partners, in order to prevent degeneration of future generations.[28]

Others, such as Margret Sanger, founder of Planned Parenthood Federation of America and friend of Abby Rockefeller, felt the choice should be left to the women, but instead advocated birth control and sterilisation, offered especially to those with mental health problems or physical defects.

Mandatory sterilization programs were initiated in several US states, the first as early as 1907 in Indiana, only decades later spreading overseas (Denmark 1929, Germany 1933, Sweden 1934–77, Finland 1935, China 1978).

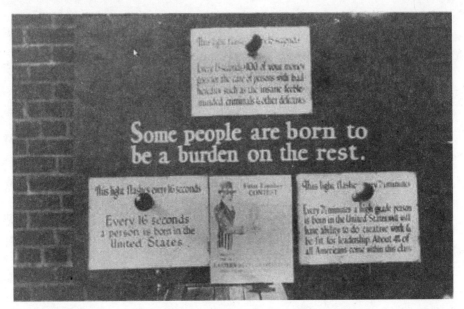

"U.S. eugenics poster advocating for the removal of genetic 'defectives' such as the insane, 'feeble-minded' and criminals, and supporting the selective breeding of 'high-grade' individuals, c.1926" (Image and caption from Eugenics in the United States, Wikipedia).

Bureau of Social Hygiene

In 1911, John D. Rockefeller Jr. helped found the Bureau of Social Hygiene (BSH) to address social problems in New York such as prostitution, corruption, drug use, and juvenile delinquency, and to create better sanitary conditions.

The bureau funded sexual education and research into psychological, physiological, and sociological aspects of the "sex instinct," abstinence, masturbation, contraception, venereal disease, and sexual relationships.[29]

Sensitive studies which BSH was unwilling to openly fund, such as the request for $10,000 from Margaret Sanger's American Birth Control League to study contraceptives in 1924, John D. Jr. would support privately.[30] The hostile attacks from Sanger a decade earlier were by then clearly forgiven by their mutual interest in curtailing reproduction.

From 1928, John D. III became a board member of the BSH. Just like his father, he developed a lifelong interest in population matters and family planning.

BSH ceased its operations in 1934 and research into sexual and repro-ductive matters was instead financed by the Rockefeller Foundation. They funded such sensitive projects as the groundbreaking but controversial 1948 study *Sexual Behavior in the Human Male* by Alfred Kinsey, which "transformed American society by challenging American perceptions and attitudes toward sex," according to the Rockefeller Foundation Archives.[31]

Kaiser Wilhelm Institute

Through the 1930s, the Rockefeller Foundation also helped finance the infamous Kaiser Wilhelm Institute of Anthropology, Human Heredity, and Eugenics, founded in 1927 in Berlin, Germany. After World War II, when eugenics had become disreputable for well-known reasons, the Rockefeller family still kept pursuing the cause under different names, such as population control, family planning, genetics, and transhuman-ism (see also chapter 12).

BANKING

In 1911, the same year Standard Oil was split up, the Rockefeller busi-ness ventures expanded to include banking. Through the acquisition of the Equitable Trust Company, all accounts from the former Standard Oil companies and the Rockefeller philanthropic ventures could be gathered in their own bank. It soon grew to become the eighth largest in the USA.

Chase Manhattan Bank

In 1929, the Equitable Trust Company was merged with Chase Manhattan (later Chase Bank) and became the bastion and financial tool of the Rockefeller clan. It was closely linked with the Standard Oil Company, especially Exxon, and would become one of the world's most influential banks.[32]

The Board of Directors of Chase included several representatives from the Rockefeller oil companies. Winthrop Aldrich, brother-in-law of John D. Rockefeller Jr., was elected chairman of the bank. Later, John D. Jr.'s son David became chief executive (1960–80), and remained chairman until 1981. In 2000, Chase merged with Rothschild's bank J. P. Morgan and grew even more powerful.

National City Bank

The other brother, William Avery Rockefeller Jr., helped build up the competing National City Bank of New York (later Citigroup), which two generations later would be led by his grandson, James Stillman Rockefeller.

Long-term Strategic Planning

By using foundations and charities the Rockefellers and their allies have been able to avoid the short-term goals of politics and business and promote any long-term development they wish to see, by methodical step-by-step action, while also appearing as generous and benevolent philanthropists. Through strategic donations, their philanthropies have become an invisible hand which has almost imperceptibly wielded a considerable influence on the turn of events in the United States and the world.

Right from the onset, the family has been very careful in choosing what to support, while meddling in almost every area of human activity; energy, politics, religion, banking, media, education, medicine, agriculture, technology, futurism, population control, and conservation—areas which, during the next century, would be merged under the major global threat of climate change, requiring the ultimate solution: a sustainable Utopia with global control over natural and human resources.[33]

John D. Rockefeller III
RBF chairman 1941–1956

The road to happiness lies in two simple principles: find what it is that interests you and that you can do well, and when you find it, put your whole soul into it—every bit of energy and ambition and natural ability you have.

—John D. Rockefeller III

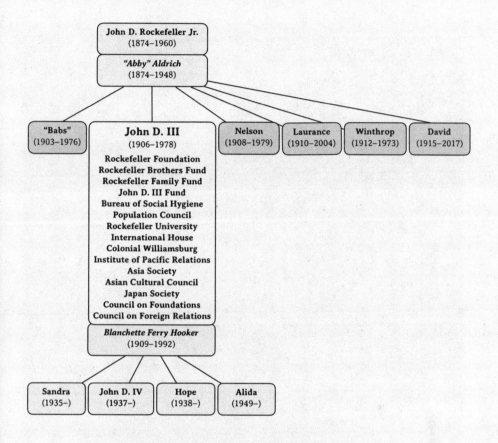

John D. Rockefeller Jr.
(1874–1960)

"Abby" Aldrich
(1874–1948)

"Babs"
(1903–1976)

John D. III
(1906–1978)

Rockefeller Foundation
Rockefeller Brothers Fund
Rockefeller Family Fund
John D. III Fund
Bureau of Social Hygiene
Population Council
Rockefeller University
International House
Colonial Williamsburg
Institute of Pacific Relations
Asia Society
Asian Cultural Council
Japan Society
Council on Foundations
Council on Foreign Relations

Blanchette Ferry Hooker
(1909–1992)

Nelson
(1908–1979)

Laurance
(1910–2004)

Winthrop
(1912–1973)

David
(1915–2017)

Sandra
(1935–)

John D. IV
(1937–)

Hope
(1938–)

Alida
(1949–)

JOHN D. ROCKEFELLER III was born in 1906 as second child and eldest son of John D. Rockefeller Jr. and Abby Aldrich. He married Blanchette Ferry Hooker in 1932 and had four children with her: Sandra, John D. "Jay" IV, Hope, and Alida. John developed a lifelong interest in population issues in Asia and was the chairman of the Rockefeller Foundation for two decades. He founded organizations such as the Population Council, the Asia Society, and the Asian Cultural Council. He played a leading role in the Council on Foundations, the Institute for Pacific Relations, and the Japan Society. He was a board member of the International House of New York, Colonial Williamsburg, the Rockefeller University, and the China Medical Board. John was co-founder of the Rockefeller Brothers Fund (RBF) and its first chairman. He was also a member of the Council on Foreign Relations (CFR). John died in a car accident on July 10, 1978, at seventy-two years old.

The United Nations Headquarters in New York, designed by Le Corbusier and Oscar Niemeyer under Harrison & Abramovitz, built 1948–52, on land donated by John D. Rockefeller Jr.

Chapter Two

The Brothers

We cannot escape, and indeed should welcome, the task which history has imposed on us. This is the task of helping to shape a new world order in all its dimensions—spiritual, economic, political, social.

—Special Studies Project: The Mid-Century Challenge to U.S. Foreign Policy, 1959[1]

ROCKEFELLER BROTHERS FUND

The 1940s marked the start of a new era with the children of John D. Rockefeller Jr. entering the stage. They had now received the education and experience necessary to develop the legacy of their father and grandfather.

In 1940, the brothers John D. III, Nelson, Laurance, Winthrop, and David founded the Rockefeller Brothers Fund (RBF) after a series of meetings initiated by Nelson, where they had discussed problems and mutual interests. The stated purpose of RBF was to "advance social change that contributes to a more just, sustainable, and peaceful world."[2] Through this foundation the brothers would be able to hone their philanthropic skills.

All five brothers became members of the first board of directors, with the eldest, John D. III, as chairman. Their sister, Abigail "Babs" Rockefeller Mauzé (1903–1976), did not join until 1954.[3] The two most philanthropically minded brothers, John D. III and Laurance, were initially the most active in RBF, while Nelson and Winthrop went into politics (as the governors of New York and Arkansas, respectively) and David focused on banking and finance.

In 1951, modernist architect Wallace Harrison was invited to join the board, together with the chairman of the National Academy of Science, Detlev Bronk. Through the latter, the family gained a valuable link to the scientific community, which would be masterfully used as Bronk became a central player in the climate research policies of the 1950s (see chapter 3).

While the Rockefeller Foundation mainly funded research, RBF was more focused on activism and politics. RBF thus became a powerful tool both for building strategies and for financing activities advantageous to the Rockefeller family's ambitions for the world.

The Rockefeller Foundation and RBF complemented each other and were closely linked, as John D. III was also chairman of the Rockefeller Foundation (1952–1971). Both foundations used large-scale planning "for the survival of humanity and the planet" in order to change both mankind and the world in a fundamental way.

The family's influence over American politics, philanthropy, and finance was unprecedented. After World War II the world lay practically at their feet.

Central themes have from the onset been population control, conservation, and resource management, as well as the promotion of an international political structure for handling these issues. This was made possible and became especially obvious when John D. Rockefeller Jr. donated $58 million to RBF's Special Program in 1951. The *1951–53 Annual Report* specified:

> Accordingly, the trustees decided that the program of the Fund should be expanded to include the support or possibly in some instances the direct operation of experimental or new undertakings in areas of special interest to the trustees, which fall generally into the broad fields of human relations, international relations, and development of human and natural resources.[4]

CONSERVATION

Conservation was an early priority for RBF, especially in relation to the effects of humans on the natural environment. The Rockefeller family's foundations would play a leading role in the emergence of the green movement, mainly through Laurance, chairman of RBF for more than twenty years, who was the most passionate about conservation of the brothers. He was nicknamed Mr. Conservation.

In 1935, Laurance became board member of the New York Zoological Society (later Wildlife Conservation Society) where he met Fairfield Osborn[1] who became his mentor and close friend.

In 1948, Osborn wrote *Our Plundered Planet* which, together with William Vogt's *Road to Survival*, published the same year, laid the foundation for the modern conservation and population discourse. Osborn felt that a sustainable use of natural resources could only be attained in an international context.[5]

Conservation Foundation

In 1948, Laurance Rockefeller and Fairfield Osborn founded the Conservation Foundation with funding from RBF, the Rockefeller Foundation, and the Ford Foundation (which had close ties to the Rockefeller charities). The board of directors came from the Rockefeller network and included William Vogt, Samuel Ordway, and Nobel Prize laureate Sir John Boyd-Orr from the UN Food and Agriculture Organization (FAO).

Conservation Foundation was dominated by Neo-Malthusian ideas about the planet's carrying capacity in relation to the number of people and the resources available. Its purpose was to halt degradation of the natural environment and reinstate a balance between man and nature.

There was also an early interest in man's impact on the climate. Through Laurance, several Conservation Foundation members would come to play key roles in bringing these concerns to the political arena in both the United States and the rest of the world. Contacts with the

---

1   Fairfield Osborn (Henry Fairfield Osborn Jr., 1887–1969) was the son of palaeontologist and eugenicist Henry Fairfield Osborn (1857–1935), a disciple of British Darwinist Thomas Huxley. Fairfield was the cousin of the eugenicist Frederick Osborn (1889–1981) of the Population Council and the American Eugenics Society. The cousins were heirs to the J. P. Morgan banking family and railway magnate Cornelius Vanderbilt. It was a rich man's club with close connections between the boardrooms of their foundations.

White House were well established, not least through Laurance becoming advisor to several US presidents. The Conservation Foundation became pivotal in the creation of the US Environmental Protection Agency (EPA) in the early 1970s.

Laurance's commitment to conservation went beyond the Conservation Foundation. In 1958 he founded the American Conservation Association and the year after, Resources for the Future. All in all, he was connected to over fifty environmental organizations, including the National Geographic Society and the World Wildlife Fund (WWF).[6]

International Union for Conservation of Nature

Conservation became an international concern. In 1948, Julian Huxley, general secretary of UNESCO, founded the International Union for Conservation of Nature (IUCN).[7]

Huxley was the grandson of anthropologist Thomas Huxley and brother of the author Aldous Huxley. Julian had been secretary of London Zoological Society from 1935 to 1942 and was chairman of both the British Eugenics Society and the British Humanist Society. In 1957 he coined the term *transhumanism*.

Huxley's advocacy of internationalism, eugenics, population control, and evolutionary humanism (a secular humanist religion) coincided with the goals of the Rockefeller family and the Conservation Foundation.[8] He became a valuable ally.

Up until the 1970s, conservation was largely a concern for an Anglo-American elite. The Rockefeller family became an important link between the two countries, while also financing both IUCN and UNESCO.

POPULATION

After World War II, the population issue, like conservation, became even more pressing. This subject became John D. III's special interest. John had struggled in his role as the eldest brother and often found himself in the shadow of his extroverted brother Nelson.

Even though John made significant contributions in philanthropy and in the relations with Asia (through the Asia Society and other NGOs), it was his involvement in the population matter that became his lasting legacy and he became known as Mr. Population.

The Population Council

In 1952, the Conference on Population Problems was held, at the initiative of John D. III, Lewis Strauss (RBF's financial advisor), and Detlev Bronk (chairman of the conference). Thirty handpicked proponents for population control were invited.[9] The conference resulted in the founding of the Population Council six months later. Its mission was to develop a global plan for keeping the world's population growth in check.

In the 1940s and early '50s, several methods had been proposed as solutions to the population issue: increased social and economic equality, better distribution of the world's population by international migration, and fertility control.

The Population Council chose the latter and would work primarily in developing countries with social studies and experiments aimed at lowering fertility by family planning and sterilisations. This was done in collaboration with the pharmaceutical industry both dominated by and closely linked to the Rockefellers. After its founding, the Population Council moved into a building on the campus of Rockefeller Institute (1953–68 headed by Detlev Bronk).

The Population Council Board of Directors (with John III as chairman), included Bronk, Strauss, and Frederick Osborn (from the American Eugenics Society)[10] who had expressed clearly elitist ideas about who should inhabit the planet in the future. Frederick's cousin Fairfield Osborn from the Conservation Foundation, who stated, "[W]e need the greatest number of births among genetically superior individuals" in a 1956 *Eugenics Review* article, was also a member of the Population Council.

The old ideals had been rebranded but the goals were the same.

Meanwhile, initiatives were taken to sway public opinion in support of more drastic measures.[11]

AGRICULTURE

Closely related to the population issue, was agriculture. It was a major area
of philanthropy for the Rockefeller Foundation. Their agricultural mod-
ernisation programs were initiated during WWII and spawned the green
revolution of postwar agriculture. In the foundation's own words:

> Today it is nearly impossible to imagine the global transformation
> of agriculture without the Rockefeller Foundation (RF). From the
> 1940s through the 1960s, it founded permanent research facilities in
> Mexico, the Philippines, Colombia, and Nigeria. These centers bred
> higher-yield grains, reduced crops' susceptibility to disease, improved
> fertilizers, and instructed farmers in efficient sowing and irrigation
> techniques.[12]

This revolution, initially focusing on seed hybridisation and more efficient
fertilisation and irrigation systems, meant traditional farming methods
would be replaced by more large-scale, energy-intensive industrial agriculture. This significantly increased crop yields and relieved hunger, but also
made agriculture highly dependent on petroleum products and synthetic
fertilisers and pesticides, creating huge profits for leading chemical companies such as Dow Chemicals, BASF, Monsanto, DuPont, and Bayer
AG.

In the 1970s, the Rockefeller Foundation started creating global institutions for coordinating international agricultural research, such as the
Consultative Group for International Agricultural Research (CGIAR),
founded in 1971 in partnership with other philanthropic foundations, governments, and international institutions (such as the Ford Foundation, the
OPEC fund, the European Commission, UNDP, and the World Bank).

The Rockefeller Foundation, under the leadership of John D. III, was
also deeply involved in the biotech revolution with genetically modified
organisms (GMO) to relieve world hunger. This fundamental transformation of agriculture may be founded in a genuine will to create food security for the world, but also gave influence over the world system through
controlling life itself and changing it at the most fundamental level—an

old alchemical dream. The crises highlighted by RF and RBF also became sources of revenue and influence for the Rockefellers and their allies in the agricultural, chemical and later biotech industries. Through the technologies developed to save the world, the power of associated multinational corporations would also keep growing.

The enduring legacy of the RF is a changed world agriculture regime, characterized by scientific methods, global information exchange, and the treatment of food production as a business enterprise.[13]

RELIGION

Religion was always a strong motivating force for the Rockefeller family.

Ecumenism

A Baptist like his parents, John D. Jr. favoured an ecumenical approach, uniting all Protestant churches. He donated large sums to the Interchurch World Movement, the Federal Council of Churches, the Institute of Social and Religious Research, and to the Cathedral of St. John the Divine in New York. In 1930, with Baptist pastor Harry Emerson Fosdick Jr. founded the ecumenical Riverside Church in New York, built in neo-gothic style and open to all denominations with faith in Christ.

In 1958, he also donated towards the building of Interchurch Center, 475 Riverside Drive, opposite Riverside Church, near Columbia University and St. John the Divine. This was an ecumenical center, housing a large number of religious organizations, nicknamed the God Box. Tellingly, both the Rockefeller Brothers Fund and the Rockefeller Family Fund share office space in the God Box.

The interest in religion was also shared by Junior's son Laurance and later Nelson's son, Steven Rockefeller, who became professor of religion.

Evolutionary Humanism

A major inspiration for the Rockefeller family was the French Jesuit priest, palaeontologist, and geologist Pierre Teilhard de Chardin. According to Teilhard's teachings, mankind was evolving into a point of cultural convergence, called the noosphere, and would eventually reach a mystical state called the Omega Point—the end goal of humanity which also included the Second Coming of Christ.[14]

Teilhard's transhumanist teachings, combining Christianity with Darwinism, suited the Rockefellers perfectly and provided a deterministic foundation for their utopian vision for the world and gave it a divine blessing.

Even the secular humanist Sir Julian Huxley, friend of "Pére Teilhard," had similar views, summarised in his vision for UNESCO.

> [T]he general philosophy of Unesco should, it seems, be a scientific world humanism, global in extent and evolutionary in background.[15]

When Teilhard de Chardin's 1955 book, *The Phenomenon of Man*, was translated into English in 1959, Huxley wrote in its introduction:

> The incipient development of mankind into a single psychosocial unit, with a single noosystem or common pool of thought, is providing the evolutionary process with the rudiments of a head. It remains for our descendants to organise this noosystem more adequately . . . Accordingly, we should endeavour to equip it with the mechanisms necessary for the proper fulfilment of its task—the psychosocial equivalents of sense organs, effector organs, and a co-ordinating central nervous system with dominant brain.[16]

Pierre Teilhard de Chardin (1881–1955).

Teilhard's visions of the future were also closely connected to the development of cybernetic technology. The first step towards the development of the noosphere is represented by the internet and will, according to the vision, be followed by the development of a technologically improved Super Human, with a united global consciousness and unified nations under a world government. This vision of a global society was something the Rockefeller family wanted to help turn into reality.

During the 1970s, Laurance, RBF, the Rockefeller Foundation, and Lilly Endowment funded the Lindisfarne Association, where discussions were held on how to make Teilhard de Chardin's visions of a planetary culture and ethics a reality.[17] This had a major impact on the emerging New Age movement, which Laurance also helped fund.

CYBERNETICS

As early as 1946, Laurance Rockefeller and his brothers had launched Rockefeller Brothers Inc. for investing in emerging technologies, especially information technology.

In 1969 Rockefeller Brothers Inc. was turned into Venrock, a major financier of Silicon Valley and the computer revolution. Leading tech companies such as Apple and Intel Corporation both received their founding financing from Venrock.[18] In the 1970s Intel developed the 8080 processor for the first personal computer, Altair 8800.

Through Venrock's research director, Warren Weaver, the Rockefeller Foundation had supported both mathematician Norbert Wiener who, in 1948, founded the research field cybernetics, and the first conference on artificial intelligence, held at Dartmouth College in 1956.[19]

While investing in early computer technology, Laurance also supported a number of New Age gurus and organizations in which the technological revolution was sold in spiritual trappings. The seemingly disparate areas of religion and technology would decades later be merged with the threat of climate change to get everyone on board for the Great Transformation.

INTERNATIONALISM

The internationalism advocated by both Pierre Teilhard de Chardin and Julian Huxley has had a strong influence on most of the Rockefeller family's activities from the outset. Domination over the American system was

only a first step. Now, these ideas would be spread globally. By supporting the founding of international organizations, the family's goals and
initiatives could more easily be disseminated across the world. For the
Rockefellers, it was also about furthering their business interests in international trade. Some of the brightest minds available were enlisted to
achieve these goals.

Council of Foreign Relations

In 1921, one of the Rockefeller family's most influential centers of power
was founded, the elite think tank Council on Foreign Relations (CFR),
sister organization of the Royal Institute of International Affairs (Chatham
House), founded in London in 1919.

Initially, CFR was funded by leading banker families J. P. Morgan,
Lehman, Schiff, and Rockefeller (John D. Rockefeller Jr. financed both its
foundation and its first headquarters in New York).[20] After the death of J.
P. Morgan Jr. in 1953, the think tank was run by the Rockefeller family.
David Rockefeller, who had become board member in 1949, was chairman
from 1979 to 1985, after which he retained an honorary chairmanship.

Several other board members from the Rockefeller Foundation and
RBF have been members of the CFR.[2] Many key players involved in
the climate threat discourse have also been members, including Carroll
L. Wilson, Walter Orr Roberts, Edward Teller, Frederick Seitz, and Al
Gore.

CFR, which advocates the same internationalism as the Rockefeller
family, soon became a major power base—especially in influencing foreign policy by its members holding key positions in presidential administrations ever since its inception, including national security advisors such
as Henry Kissinger and Zbigniew Brzezinzki.

That David Rockefeller left CFR $25 million in his will is an indica-
tion of how much he valued the organization.

---

2   New members are nominated by sitting members and must be recommended by at least three.
     There are also three levels of corporate memberships. The top level (at a $100,000 fee) includes
     ExxonMobil, Chevron, and JP Morgan Chase from the Rockefeller sphere.

Two other fora of great value to the family in their internationalist ambitions have been the Trilateral Commission (founded in 1973 by David Rockefeller and Zbigniew Brzezinski) and the Bilderberg group (initiated in 1954 by Prince Bernhard of the Netherlands).

The League of Nations

The League of Nations (1920–46), precursor to the United Nations, was the first worldwide intergovernmental organization dedicated to maintaining world peace. John D. Rockefeller Jr. was a major donor and had close ties to the organization.[21]

The project, however, did not succeed. The United States never joined. The Soviet Union was expelled after attacking Finland. Germany, Italy, Spain, Japan, and other nations withdrew. It also failed to prevent the Second World War.

United Nations

In 1945, the Rockefeller family (via Council on Foreign Relations), together with the British élite (via Royal Institute of International Affairs), were involved in the launching of the United Nations as replacement for the LN.

During World War II (1939–45), the Rockefeller Foundation and Carnegie Corporation had funded CFR's War and Peace Studies project, in which recommendations were made that led to the founding of the United Nations, the International Monetary Fund (IMF), and the World Bank.[22] Vice-chairman of the project was Allen Dulles.[3] His brother, John Foster Dulles, helped draft the preamble to the UN Charter together with David Rockefeller in the Carnegie Endowment for International Peace.

At the United Nations Conference on International Organization held in San Francisco from April 25 to June 26, 1945, which launched the United Nations as an organization, Nelson Rockefeller was part of the US delegation. According to his son Steven,

---

3   Allen Dulles was a board member of Rockefeller Foundation and chairman from 1950 to 1952. In 1953 he became director of the CIA, the same year his brother, John Foster Dulles, became secretary of state in the Eisenhower administration.

My father, who served in government positions under four presidents and as governor of New York, had a passionate interest in international affairs. He and my grandfather [John. D. Jr.] played an active part in helping to establish the United Nations in New York City, and my family has been a supporter of the UN ever since.[23]

In late 1945, when the newly founded United Nations needed location for its headquarters, John D. Rockefeller Jr. and his sons offered their Pocantico estate in upstate New York. As this was too suburban to be suitable, John D. Jr. instructed his son Nelson to purchase a seventeen-acre site on the East River in New York City, walking distance from Rockefeller Center.

Nelson and his favorite architect, Wallace Harrison, negotiated the purchase and got all the necessary city, state, and federal permits and waivers—a process accomplished in only thirty-six hours.[24]

The property, worth $8.5 million, was then donated to the UN by John D. Jr.[25] Nelson then commissioned Wallace Harrison and Max Abramovitz to lead the project, and they chose the designs of Le Corbusier and Oscar Niemeyer for the modernist landmark.[26]

In a speech in 2012 to commemorate the eighty-fifth anniversary of the John D. Rockefeller Jr. donation to the League of Nations library, UN General Secretary Ban Ki-moon expressed his gratitude towards the family:

I personally want to thank the Rockefeller family for my own office—and the entire United Nations campus on the East Side of Manhattan. When Rockefeller's donation of the land was announced in the General Assembly in 1945, the Hall was filled with loud applause. The United States Ambassador cheered Mr. Rockefeller's 'magnificent benevolence.' I am deeply grateful to the esteemed members of the Rockefeller family and the Rockefeller Foundation for continuing the noble tradition of supporting international organizations devoted to peace.[27]

The Rockefeller brothers had grand visions for the UN as carrier of internationalism. Right from the start, ways of strengthening the organization's influence were sought. The UN was the embryo for a world government and a first step towards the dream of Pierre Teilhard de Chardin's utopian Kingdom of Peace.

In their internationalist aspirations, the family cooperated with other philanthropic foundations such as the Ford Foundation, the Carnegie Endowment for International Peace, and the Guggenheim Foundation, often sharing board members.[28]

CULTURE

Part of the international project, and a crucial tool for reshaping the world, was the new modernist movement from Europe, aimed at reforming all the fine arts in order to create a new international progressive culture.

Museum of Modern Art

In 1929, the Museum of Modern Art (MoMA) in New York was founded by John D. Jr.'s wife, Abby Aldrich Rockefeller, and her friends Lillie P. Bliss and Mary Quinn Sullivan. Abby invited industrialist A. Conger Goodyear to become its first president, with herself as treasurer and driving force.[29]

MoMA's first exhibition in 1929, in a rented venue, displayed works of van Gogh, Gaugin, Cézanne, and Seurat, followed a decade later by the retrospective Picasso exhibition, which gave MoMA international prominence.[30]

John Jr. was at first opposed to the museum and Abby had to find funding elsewhere, but eventually he gave in and in 1939 donated the family's property on 53rd Street for the building of MoMA and became one of its major donors.[31]

In 1939, their son Nelson became chairman and financier of "mommy's museum" until 1958, when he became governor of New York and left the position to his younger brother David, who was initially not too keen on modern art but was eventually converted and turned into one of the world's greatest art collectors. After his death in 2017, his collection sold for $830 million, of which $125 million was bequeathed to MoMA.[32]

The Rockefeller family's ties to the museum have remained strong. RBF started funding it in 1947. David Rockefeller Jr. and Sharon Percy Rockefeller (the wife of Senator Jay Rockefeller) are still board members.

MoMA's board of directors has over the decades included Walt Disney, department store owner Marshall Field, Henry Luce, Beardsley Ruml (New York Federal Reserve), and other luminaries from the New York high society, including members of the Ford, Goodwin, Guggenheim, Payson, Phillips, Sachs, Vanderbilt, and Warburg families.

According to curator Philip Johnson, however, there was no question of who was in charge: "It's a Rockefeller institution; it's a democracy of one."[33]

The Abby Aldrich Rockefeller Sculpture Garden, Museum
of Modern Art, New York, built on the family estate of Abby
and John D. Rockefeller Jr.

The Covert CIA Cultural War

In 1950, the Central Intelligence Agency (CIA), led by Ivy League alumni with a liberal education, set up the International Organizations Division for "combating communism" by infiltrating leftist cultural circles abroad with the goal of swaying them towards American rather than Soviet sympathies. For this purpose, organizations such as Congress for Cultural Freedom (CCF) were founded in Europe, Asia, Africa, North America, Latin America, and Australia. CCF published dozens of cultural magazines in these countries and enrolled many leading intellectuals, including Arthur Koestler and Bertrand Russell.

Individualistic abstract expressionism, by artists such as Jackson Pollock, Mark Rothko, and Willem de Kooning, was chosen as cultural weapon against Stalinism, and for crafting an image of America as the progressive cultural leader of the postwar world.

During the cold war, when conservative congressmen, the communist-hunting Federal Bureau of Investigation (FBI), and an unimpressed general public made avant-garde art by progressive artists with a history of radical leftism less than appreciated in the US, the CIA set up an unofficial Arts Council, with museum directors, philanthropists, entrepreneurs, art critics, and magazine editors, for exporting art and covertly funnelling funds for cultural projects overseas.

Many prominent foundations, including the Ford Foundation, the Whitney Museum, and even the German Marshall Plan, were used by the CIA for this purpose. MoMA (led by Nelson Rockefeller) was a key player and started exporting avant-garde art to Europe.[34]

The board of MoMA also included William S. Paley (founder of CBS broadcasting), Julius Fleischmann (president of CIA's fake Farfield Foundation), and former Office of Strategic Services (OSS) member John Hay Whitney (of the Whitney Museum) as chairman. Another driving force was former OSS agent and architect behind CIA, John McCloy, who also had a prominent position in the Ford Foundation and was chairman of Chase Manhattan.[35]

At this time, the Rockefellers also launched an ambitious program for corporate investment in avant-garde art, starting with their own Chase

Manhattan Bank. It was given a yearly art budget of $500,000 resulting
in a collection of more than thirteen thousand abstract works, which must
have helped raise the value and status of the genre considerably.

Architecture

The dashing young architect Philip Johnson, who came from an affluent background, personally founded and funded MoMA's Department of
Architecture, and became its curator.

In February 1932, "Modern Architecture: International Exhibition"
opened at MoMA, displaying works by architects such as Walter Gropius,
Le Corbusier, Ludwig Mies van der Rohe, J. J. P. Oud, Frank Lloyd
Wright, and Alvar Aalto.[36] The minimalist rectilinear shapes in concrete,
steel, and glass were presented as the "International Style" and would after
World War II come to dominate architecture across the world, replacing
both traditional building methods and regional styles.

After an interlude from 1936 to 1939 as an overseas reporter, where
he became enamored with national socialism at a Hitler Youth rally,
Johnson returned to MoMA,[37] continuing to promote primarily modernist architecture.

In 1988, Johnson curated another groundbreaking exhibition,
"Deconstructivist Architecture"—again defining and naming this new,
intentionally unsettling and destructive style which has come to dominate
landmark architecture ever since—awarding instant fame and status to its
architects.[38]

Back to the days of the first exhibition, the modernist movement and
the International Style suited the Rockefeller brothers' internationalist
aspirations like a glove. It also inspired radically new zoning laws and
urban planning models, leading not only to a boxy skyline of rectangular
high-rise slivers, but to extensive sprawl and automobile dependency—
which also happened to be highly profitable for the oil and auto industries.

The Rockefeller family had an enormous influence, not only on the
promotion of the International Style but—more directly—on New York
itself by commissioning a long list of prominent buildings, including the
Rockefeller University (John D. Sr.); Rockefeller Center, MoMA, the

Memorial Sloan Kettering Cancer Center, Riverside Church, and The Cloisters (John D. Jr.); Lincoln Center for the Performing Arts (John. D. III); Empire State Plaza in Albany, New York (where Nelson was governor); the World Trade Center and One Chase Manhattan Plaza (David).[39]

"What David Rockefeller wanted built got built," noted an observant *New York Times* journalist.[40]

Robert Moses

Grand building projects in New York were often developed in cooperation with the infamous public official, Coordinator of Construction Robert Moses, through the so-called urban renewal project, which meant that the city could expropriate the properties desired by a developer (e.g., a Rockefeller), and have the tenants and small businesses evicted from the area. The building of Lincoln Center alone resulted in the removal of forty thousand tenants.

According to David Rockefeller, Moses was "authoritarian and ruthless."[41] This suited the Rockefeller brothers perfectly—as long as it coincided with their own plans. Their mutually beneficial relationship, however, started going sour when Moses challenged Nelson Rockefeller.

In 1960, the year after Nelson was elected governor of New York, he wanted Moses, then seventy-two, to leave his position as chairman of the State Council of Parks to Nelson's brother Laurance. Moses had no plans of stepping down from any of his influential posts and tried the threatening-to-resign act that had worked so well before. This time it didn't.[42]

Having provided New York City with well-needed highways, bridges, parks, playgrounds, and housing projects since the 1920s, Moses's plans for another highway through the popular Greenwich Village in 1955 did not go down well with the citizens of New York. Jane Jacobs successfully rallied up public support against the project and Moses had to back off.

Surprisingly, in 1958 Jacobs had received funding from Rockefeller Foundation to write her groundbreaking book, *The Death and Life of American Cities*, published in 1961, which further damaged Moses's reputation. The book criticized exactly the type of ruthless urban renewal that the Rockefeller family had been a major part of, and indirectly inspired all over the United States and the world by promoting modernist city planning with its widespread urban sprawl and automobile dependency. However, the main blame landed squarely on Moses.[43]

By the end of the 1960s, the conflict with Nelson came to a head. Moses still retained a very powerful position in the Triborough Bridge and Tunnel Authority which Nelson wanted to incorporate into his own empire. Nelson was known as the most ruthless man in politics. Moses didn't have a chance. Former *New York Times* Albany bureau chief William Farrell once said, "Nelson is a true democrat. He has contempt for everyone, regardless of race, color, creed, religion or anything else."[44]

Robert Moses (1888–1981) with a model of Battery Bridge.

A community of c. 7,000 was erased to make space for Nelson Rockefeller's Empire State Plaza, Albany, New York, designed by Wallace Harrison and built 1965–76.

Aspen Institute

In 1949, the Aspen Institute of Humanistic Studies, an international think tank, cultural center, and research institute was founded in Aspen, Colorado, by industrialist and philanthropist Walter Paepcke, CEO of Container Corporation and board member of the University of Chicago.[45]

After Paepcke's wife, Elizabeth, had discovered the run-down and almost derelict former mining town Aspen in the scenic setting by the Aspen Mountain, Paepcke had a vision of redeveloping it and turning it into a fashionable ski resort and an American cultural center for art, design, literature, and music (with an annual music festival) in a setting of breathtaking scenery and avant-garde architecture and design. Paepcke's friend, Bauhaus architect, graphic designer, and co-visionary Herbert Bayer (who had made ad designs for Paepcke's company), was to make a master plan and design many of Aspen's buildings and art pieces.[46]

At the initiative of professor Giuseppe Borgese, Robert M. Hutchins, president of the University of Chicago, suggested to Paepcke that a bicentennial celebration of the birth of Johann Wolfgang von Goethe be organized at Aspen, "to honor the great German author as a universal thinker who could bring the world together" and help heal the scars after the war.[47]

The Seminar Building (later renamed the David H. Koch Building) at the Aspen Institute, designed by Herbert Bayer and built in 1953.

The twenty-day event (including a music festival), held in the summer of 1949, was a success and attracted more than two thousand participants through luminaries such as Albert Schweizer and Arthur Rubinstein. It resulted in the formal founding of the Aspen Institute and marked the start of a yearly summer music festival (in the tent designed by Eero Saarinen, using the offseason ski resort huts for lodging).

The new cultural center was met with enthusiasm in the press:

U.S. CULTURE MOVES WEST WHEN IT'S TIME TO THINK
IN THE ROCKIES BRAIN SPA[48]

The center soon developed into Paepcke's vision of spreading "eternal truths" as ethical guidance for business leaders and offering a platform for leaders in government, education, culture, religion, business, media, and science, to discuss fundamental problems of society and Western civilisation. As magazine publisher Henry Luce Jr. (*Time, Life*) suggested, to save the nation's soil and soul, you had to "start with the men at the top."[49] The Aspen Institute's influence on contemporary thinking cannot be overstated. Several UN conferences were held there, and this was where the climate issue was first discussed in the early sixties by solar physicist Walter Orr Roberts, director of the National Center for Atmospheric Research in Boulder.

Besides private donations and seminar fees, the institute would be funded by RBF, the Rockefeller Foundation, Carnegie Corporation, the Ford Foundation, the German Marshall Fund, the Kettering Foundation, the Lilly Endowment, Exxon, Chase Manhattan Bank, Coca-Cola, and IBM. Oil tycoon Robert O. Anderson (founder of ARCO and board member of Chase Manhattan) got involved in the Aspen Institute at an early stage and became its president in 1957. Coal industrialist David Koch funded the David H. Koch Building and the Aspen Music Festival campus and later became a board member.

Through other substantial donations the Aspen Institute continued to grow, with more research departments, conferences, and programs. Campuses were established in other states and eventually also in other countries.

The Great Books Program

When founding the Aspen Institute, Paepcke had been inspired by Mortimer Adler's Great Books Program at the University of Chicago. Professor Adler was the principal coworker of president Robert M.

Hutchins. Both dreamt of reforming American higher education: teaching the liberal arts from high school and turning learning into a lifelong personal enterprise.[50]

The Great Books curriculum featured a list of 100 to 150 Western classics, from Homer and the Ancient philosophers to Machiavelli, Shakespeare, Swift, Goethe, Hegel, Comte, Galton, Milton, Thoreau, Marx, Freud, Sartre, Pavlov, Trotsky, as well as some of the books of inspiration to those aspiring to reshape the world, e.g., Plato's *The Republic*, More's *Utopia*, Bacon's *New Atlantis*, Hobbes's *Leviathan*, and Malthus's *An Essay on the Principle of Population*.[51]

Hutchins and Adler would also hold Great Books seminars at Aspen Institute. Like the many other cultural events and workshops at Aspen Institute, the Great Books curriculum and the book series with selected classics had a profound impact on the mid-century cultural life in the US. Versions of the Great Books curriculum are still available at more than a hundred institutions of higher learning in the United States, Canada, and Europe.

World Federalist Movement

Internationalism as a solution to all the world's ills dominated both Aspen and the University of Chicago. Hutchins, who later went on to head the Ford Foundation, advocated a world federation and world citizenship, and was one of the leaders of the World Federalist Movement, founded in 1947. Together with Professor Borgese, he had written a draft for a World Constitution.[4] Hutchins was also involved in the creation of *Bulletin of the Atomic Scientists*.

The Doomsday Clock

In 1945, shortly after the bombings of Hiroshima and Nagasaki ended World War II, physicists Eugene Rabinowitch and Hyman Goldsmith launched the academic journal *Bulletin of the Atomic Scientists*.

---

4   Borgese was married to the daughter of Thomas Mann, professor Elisabeth Mann Borghese, an oceanic expert, who later became one of the founders of the Club of Rome.

Rabinowitch and Goldsmith had both been involved in the Manhattan Project at the University of Chicago Metal Laboratory, where the first chain reaction, on December 2, 1942, signalled the start of the Atomic Era.[52] *Bulletin* writers have included luminaries such as Albert Einstein, Robert Oppenheimer, Bertrand Russell, Lord Boyd Orr, Stuart Chase, and Max Born. In order to manage the threat they themselves had helped create, the *Bulletin of the Atomic Scientists* repeatedly advocated for the United Nations developing into a world government. Editor Rabinowitch clarified how to gain public support for this notion in the December 1947 issue.

> If the world government cause is to triumph it will need more than sympathetic endorsement by the majority. People must be made to feel that their own security, freedom, and prosperity, yes their very own survival, depend on the creation in our time of a world rule of law. They must be made to believe that the establishment of a World Government is more urgent than the maintenance of a high domestic standard and as, if not more, practical than the pursuit of a deceptive security by full military preparedness.

One of the leading voices of the *Bulletin* was Hungarian physicist Edward Teller, also working at the University of Chicago Institute for Nuclear Studies. In 1948, he joined in the advocacy for a world government to handle the threat of a thermonuclear war—before going on to develop the hydrogen bomb in 1952.[53]

In 1947, the *Bulletin* began using the symbolic Doomsday Clock to illustrate how close world was to a nuclear holocaust. The clock was initially set to seven minutes to midnight.

In 1949, when the Soviet Union launched its first atomic bomb testing program, the clock was set to four minutes to midnight. Its arms have since been adjusted backwards and forwards depending on the geopolitical status of the world. In 1991, after the fall of the Soviet Union and the end of the cold war, the clock was set back to seventeen minutes to midnight. In 2007, climate change was added as a threat and the hands of the Doomsday Clock have since been moved ever closer to midnight.[54]

The strategies for increasing global warming awareness were developed in close collaboration between the *Bulletin of the Atomic Scientists* and the Rockefeller family institutions. Besides the journal's conception at the Rockefeller-founded University of Chicago, the *Bulletin* got direct financial support from the Rockefellers. Its Board of Sponsors included Detlev Bronk and his successor at Rockefeller University and National Academy of Sciences, Frederick Seitz.

Cover of the first issue, June 1947 (designed by Martyl Langsdorf).

The issue of global warming would not gain political traction until several decades later, however. First, the elite had to be persuaded—at institutions such as Aspen Institute—and then it needed to become popularised and brought down to the grassroots level.

Nelson Rockefeller
RBF chairman 1956–58

*I am a great believer in planning—economic, social, political,
military, total world planning.*
—Nelson Rockefeller, *Playboy* magazine, October 1975

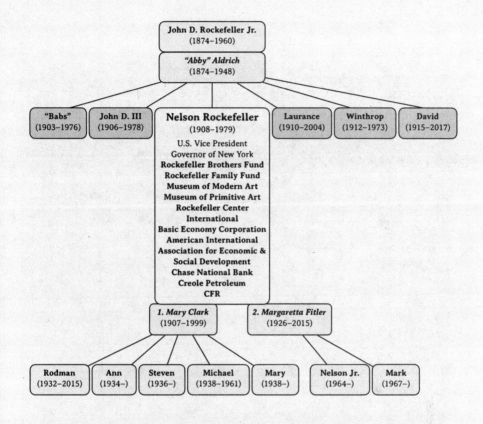

John D. Rockefeller Jr.
(1874–1960)

"Abby" Aldrich
(1874–1948)

| "Babs"
(1903–1976) | John D. III
(1906–1978) | **Nelson Rockefeller**
(1908–1979) | Laurance
(1910–2004) | Winthrop
(1912–1973) | David
(1915–2017) |

U.S. Vice President
Governor of New York
Rockefeller Brothers Fund
Rockefeller Family Fund
Museum of Modern Art
Museum of Primitive Art
Rockefeller Center
International
Basic Economy Corporation
American International
Association for Economic &
Social Development
Chase National Bank
Creole Petroleum
CFR

1. Mary Clark
(1907–1999)

2. Margaretta Fitler
(1926–2015)

| Rodman
(1932–2015) | Ann
(1934–) | Steven
(1936–) | Michael
(1938–1961) | Mary
(1938–) | Nelson Jr.
(1964–) | Mark
(1967–) |

NELSON ALDRICH ROCKEFELLER was born in 1908 as the third and next eldest son of John D. Rockefeller Jr. and Abby Aldrich. He married Mary Clark and they had five children: Rodman, Ann, Steven, Michael, and Mary. In 1964, Nelson married Margaretta "Happy" Fitler and had sons Nelson Jr. and Mark. Taking after his maternal grandfather Nelson Aldrich, Nelson had great political ambitions and made a political career as governor of New York and vice president of the United States. He also worked at Chase Manhattan Bank and Creole Petroleum, and had a life-long involvement with the Museum of Modern Art in New York. Nelson also had a strong interest in Latin America and founded the American International Association for Economic and Social Development (AIA) and the International Basic Economy Corporation (IBEC). He was co-founder of the RBF and served as chairman between 1956 and 1958. Nelson died of cardiac arrest on January 26, 1979, at age seventy.

THE DELUGE.

The Deluge, xylography by Gustave Doré, from the illustrated Bible 1865.

Chapter Three

The Carbon Dioxide Theory

If it becomes possible to interfere actively in the big processes with the atmosphere, the results are likely to transcend national boundaries. The problems that will then arise must be handled on an international basis. They may well be insoluble if the development leading up to weather control has been carried out by uncorrelated national efforts.

—The Rockefeller Brothers Fund, *Prospect for America: The Rockefeller Panel Reports*[1]

THE BIRTH OF THE CO_2 THEORY

The theory of anthropogenic global warming was conceived in the late 1800s, when the Swedish chemist and physicist Svante Arrhenius (1859–1927) presented his hypothesis that a doubling of carbon dioxide concentrations in the atmosphere would lead to a global increase in temperature of five to six degrees Celsius.

Unlike leading advocates of later decades, however, Arrhenius felt that this warming would be a good thing as it would help avoid a new Ice Age, and benefit vegetation and crop yield. At this time, there was very little interest in carbon dioxide emissions within the scientific community.

In 1938, the British engineer and inventor G. S. Callendar expanded on Arrhenius's ideas and proposed that the human burning of fossil fuels might have caused the observed warming since 1880. His findings were initially met with skepticism.

In 1941, during World War II, German meteorologist Hermann Flohn (1912–97) picked up Callendar's theory and developed it further in his article "Die Tätigkeit des Menschen als Klimafaktor" while serving in the Luftwaffe High Command and participating in the planning of Operation Barbarossa. After the war, Flohn (later dubbed "one of the world's greatest climatologists") became head of the Institute of Meteorology of Bonn University and would in the 1970s be teaching the leaders of the United Nations' newly initiated Environmental Program, UNEP, about man's impact on the climate.

The Military Origins of Climate Science

In the US, the CO2 theory at first had no impact whatsoever on the scientific community. After 1945, however, when the US military sought a deeper understanding of the forces of weather, there was a marked increase in research grants, making funds available for studies and the development of climate modelling tools. Callendar's claims could now be tested properly.

The threats of nuclear war and climate change were closely linked, with the former leading to the latter. Research into the possible effects of nuclear weapons on the climate became a priority for both the military and for the Rockefeller Foundation.

In 1946, US President Harry S. Truman (1884–1972) founded the Office of Naval Research (ONR) and the Atomic Energy Commission (AEC), both of which became leading organizations in the financing of studies on carbon dioxide impact on climate during the 1950s, following the recommendations in the report *Science—The Endless Frontier* (1945) by the president's science advisor, mathematician Vannevar Bush.[2] The report advocated a deeper involvement and funding from the government, resulting also in the founding of the National Science Foundation five years later.

The ONR had close connections to the Rockefeller family. Bush, who was involved in the creation of the ONR, had developed a calculating machine called the Rockefeller Differential Analyzer with funding from the Rockefeller Foundation.[3] It was used by the military and a precursor to

Frederick Seitz, Detlev Bronk, and Paul Weiss, with Ava Pauling in the background.

the emerging microcomputer revolution which, during the 1950s, would enable calculations and prognoses on man's impact on the climate.

The ONR's first Scientific Advisory Group, led by Warren Weaver (director of the science department at the Rockefeller Foundation) included ten prominent scientists, businessman Lewis Strauss (financial advisor to the Rockefeller Brothers Fund from 1950 to 1953), and Detlev Bronk (principal of Johns Hopkins University—which had close associations with US military on various projects).[4] Both Strauss and Bronk were also members of the Atomic Energy Commission, Strauss as chairman (1953–58) and Bronk as member of its Advisory Committee.

Detlev Bronk

Through his positions in key institutions, Detlev Bronk would have a significant influence in the world of academia. From 1950 to 1962 he was chairman of National Academy of Sciences (NAS), member (and 1952 chairman) of the American Association for the Advancement of Sciences (AAAS), chairman of the executive committee of National Science Foundation (NSF), and member of the President's Science Advisory Board (under presidents Truman, Eisenhower, and John F. Kennedy).[5]

Bronk's extensive network in the scientific community made him indispensable to the Rockefeller family. The NAS, for example, was part of the International Council of Scientific Unions (ICSU) which played a key role in giving the climate issue international attention.

In 1951, Bronk was invited to become a board member of both Rockefeller Foundation and Rockefeller Brothers Fund. Two years later, he left his presidency at Johns Hopkins to become principal of the Rockefeller Institute for Medical Research (later Rockefeller University), on the recommendation of its chairman, David Rockefeller.

Significant decisions on the direction of climate research were also made at the Woods Hole Oceanographic Institute (founded by Rockefeller Foundation in 1930) where Bronk was a board member (1950–69) and Laurance Rockefeller a lifelong board member from 1957. Such key positions gave Bronk and the Rockefeller brothers a unique insight into, and influence over, the institutions which laid the foundation for the emerging interest in the theory of carbon dioxide's impact on climate.

Bronk shared the Rockefeller family's views on future challenges, including the constantly looming threat of overpopulation. He had been instrumental in the founding of John D. Rockefeller III's Population Council.[6]

Roger Revelle

At the population conference which resulted in the founding of Population Council, one of the attendees was Roger Revelle from the Scripps Institution of Oceanography. He shared the same concerns for overpopulation and became a key figure in the establishment of the carbon dioxide theory during the 1950s.[7] He faithfully served on many of the committees initiated by Bronk as chairman of NAS. Revelle later described his relation to Bronk with the words, "I was kind of a pet of Det Bronk's."[8]

Revelle had a highly successful career and was deeply involved in international research at both ICSU and UNESCO. As a young oceanographer he had been crucial in the development of the Department of Geophysics at the Office for Naval Research.[9] In 1951, Revelle became head of the Scripps Institution of Oceanography which, during the 1950s, was largely funded by the Office of Naval Research and the Rockefeller Foundation.

Carl-Gustaf Rossby

Roger Revelle established a contact with Sweden through a friend, Swedish meteorologist Carl-Gustaf Rossby. The two pioneers had met in 1936 and their work in the field was groundbreaking.

Rossby became a well-connected institution builder, and was supervisor of a number of doctoral students who would pass on his legacy.[10] He also had a direct link to the originator of the CO_2 theory, Svante Arrhenius, who had been his mentor.

Through Rossby, Sweden after World War II became the United States' most important ally in studying human impact on the climate. After earning his licentiate degree in mathematical physics at Stockholm University in 1926, Rossby moved overseas to work for the US Weather Bureau. With support from the Guggenheim Foundation, Rossby had a stellar career.[11]

Two years later, he established the MIT Meteorological Program, where he stayed until 1939. In 1940, Rossby organised and became head of the newly established Department of Meteorology at the University of Chicago, where the Chicago School of Meteorology would be developed.[12]

Already in 1931, he had become a research fellow at the newly founded Woods Hole Oceanographic Institute, and came to be a crucial member of the research group.[13] Rossby soon advanced to become one of the superstars of meteorology. In 1943, he became a member of National Academy of Sciences (NAS).

During and after the Second World War, Rossby, who had become an American citizen, was also involved with the US Department of Defense at the Pentagon and their interest in weather modification. He was, among other things, advisor to the US Secretary of War, and part of the Joint Research and Development Board (where Vannevar Bush was chairman).[14]

In 1946, Rossby was persuaded back to Sweden to found the Meteorological Institution at Stockholm University (MISU). Support for this initiative was given by the University of Chicago and the US Weather Bureau (where Rossby's former graduate student from MIT, Harry Wexler, was research director). Funding came from the US Office of Naval Research (ONR) and there was also support from the Swedish government, through the Minister of Education (later prime minister), Tage Erlander.

Rossby established a link between the two nations and laid the foundation for the close ties which would remain and be further deepened in the field of climate research. The joint project also included the US military, who had military staff were posted for advanced training in Stockholm until the mid-1950s.[15]

Other international links were also created by Rossby. The pioneer Hermann Flohn, who became head of the West German Weather Service in 1958, and who later came to play a prominent role in the emerging climate agenda, was invited

Carl-Gustaf Rossby (1898–1957) on the cover of *Time* magazine, December 1956.

to Stockholm by Rossby.[16] Stockholm became, in effect, a US base in Europe, and the nexus of a growing international network.

MISU, however, did not engage in field studies but based its research mainly on theories of atmospheric dynamics. Rossby became known for not having made a single observation throughout his career.[17]

In 1949, Rossby started the scientific journal *Tellus* for the Swedish Geophysical Society in Stockholm, which would grow in importance over the next decade. Stockholm was established as a center for atmospheric research. The carbon dioxide theory became an early priority.[18]

In 1954, the year after physicist Gilbert Plass (1920–2004) from Johns Hopkins University had said in *Time* magazine that increasing carbon dioxide levels could lead to rising global temperature, it was decided at a conference in Stockholm that trace gases in the atmosphere should be researched more thoroughly, and that a worldwide network of monitoring stations for carbon dioxide was to be established.[19] All as preparation for the International Geophysical Year 1957–58.

Growing Blanket of Carbon Dioxide Raises Earth's Temperature

Earth's ground temperature is rising 1½ degrees a century as a result of carbon dioxide discharged from the burning of about 2,000,000,000 tons of coal and oil yearly. According to Dr. Gilbert N. Plass of the Johns Hopkins University, this discharge augments a blanket of gas around the world which is raising the temperature in the same manner glass heats a greenhouse. By 2080, he predicts the air's carbon-dioxide content will double, resulting in an average-temperature rise of at least four percent. If most of man's industrial growth were over a period of several thousand years, instead of being crowded within the last century, oceans would have absorbed most of the excess carbon dioxide. But because of the slow circulation of the seas, they have had little effect in reducing the amount of the gas as man's smoke-making abilities have multiplied over the past hundred years.

AUGUST 1953   119

Excerpt from Gilbert Plass's article in *Time* magazine 1953.

In 1955, Rossby founded the International Meteorological Institute (IMI) to facilitate international collaboration in meteorology. Rossby wanted the IMI to be recognised by UNESCO and sought financial support from the Rockefeller Foundation for its formation. Through his good relations with former minister for foreign affairs Richard Sandler and Prime Minister Tage Erlander, the institute had direct support from the Swedish government. Sandler became chairman.

The close connections between the government and the institute continued and were further deepened over the years. The political dimension was interwoven from the start. This also resulted in a close collaboration with Rossby's earlier doctoral student, Jule Charney (1917–81), and John von Neumann at the Institute for Advanced Study. Rossby was offered a position there but declined.[20]

Institute for Advanced Study

The Institute for Advanced Study (IAS) had been founded in 1930 by Abraham Flexner (member of the Council on Foreign Relations and brother of Simon Flexner).[21]

Institute for Advanced Study, Princeton University, New Jersey.

There were other close ties between the IAS and the Rockefeller family. Lewis Strauss (RBF's financial advisor and chairman of the Atomic Energy Commission) was a board member of IAS. Walter W. Stewart (chairman of the Rockefeller Foundation, 1941–45) was also affiliated with the institute.

The plans for the institute had been drawn up by Tom Jones from the British Round Table Group. It was intended as an American equivalent of All Souls College in Oxford.

IAS was headed by atomic physicist J. Robert Oppenheimer (1904–67) who had been the scientific director of the Manhattan Project and later principal advisor to the Atomic Energy Commission.

At the IAS, the science of climate forecasting was developed, using data modelling which was the first of its kind. In order to avoid climate disasters, human intervention for balancing the weather system was advocated, though not through reducing emission via political agreements, as it is the case today, but by weather modification.

John von Neumann

Professor John von Neumann of the Institute for Advanced Study, who had studied mathematics at the University of Göttingen through a grant from the Rockefeller Foundation, was a leading advocate for using climate modification to tackle the effects of possible climate change.

John von Neumann (1903–1957).

Weather modification technology had been patented by the Serbian inventor Nikola Tesla (1856–1943). One patent was later developed further by the US military and the defense contractor Advanced Power Technologies, Inc., founded by Robert O. Anderson's ARCO.[22]

Von Neumann, a key figure in the development of game theory, held the view that most natural phenomena could be expressed in mathematical terms and suggested that whoever could provide accurate predictions about the future would also control the world.[23]

This prospect was also of great interest to the US military and would soon also became a priority for the ruling elite and for the many global planners, systems theorists, and futurists on their payroll. The idea later came to particularly influence the philosophy of the Club of Rome (see chapter 4).

Von Neumann had also been involved in the Manhattan Project, a virtual playground for climate theorists. It was von Neuman who connected the threat of global nuclear war with the threat of climate change. In 1955 he made the following prophecy:

> Intervention in atmospheric and climatic matters . . . will unfold on a scale difficult to imagine at present . . . this will merge each nation's affairs with those of every other, more thoroughly than the threat of a nuclear or any other war would have done.[24]

Study on Radioactivity

In 1954, the Rockefeller Foundation decided to make an independent study on radioactivity and appointed Detlev Bronk to put together a research group (Biological Effects on Atomic Radiation).[25]

The executive committee included RF chairman John D. Rockefeller III, Detlev Bronk, and Wallace Harrison (board members of the RBF), with Dean Rusk (US secretary of state, 1960–69) as president. The collaboration also included the Department of Defense, the Atomic Energy Commission (headed by Lewis Strauss), and the Medical Research Council of Great Britain.[26]

This resulted in six subcommittees formed in 1955 under the National Academy of Sciences, with a hundred specialists in the areas of pathology, genetics, meteorology, oceanography, agriculture, and waste. Roger Revelle was appointed chairman of the oceanography group while Harry Wexler chaired meteorology. Tests included what effect nuclear explosions had on the weather and on the oceans.

The study, published in June 1956, revealed that radioactive radiation would stay in the upper atmosphere for years, and that this was true of almost anything released into the atmosphere (including carbon dioxide). No direct effects, however, could be found during the first decade of the Atomic Age.[27] The Rockefeller Foundation concluded in its *1956 Annual Report* that the panels had highlighted new problems, revealed in the initial analysis.[28] These would now be studied closer.

The Great Geophysical Experiment

In April 1956, Detlev Bronk, as chairman of the National Academy of Sciences, had already appointed a national meteorologic committee that would "consider and recommend means for increasing understanding and control of the atmosphere." The committee included, among others, Lloyd Berkner as chairman, Carl-Gustaf Rossby, Jule Charney, John von Neumann, and Edward Teller, with Roger Revelle as advisor.

Their efforts led to the founding in 1960 of the National Center for Atmospheric Research (NCAR) in Boulder under the leadership of solar scientist Walter Orr Roberts.[29] Together with the Aspen Institute, and with support from both RBF and the Rockefeller Foundation, NCAR would be instrumental in bringing the climate issue to the next level during the 1960s and '70s.

Revelle continued studying the phenomenon and connected it to the theory of pioneers Plass and Callendar of the heating effect of carbon dioxide. Early the same year he testified before a congressional hearing about how the human-induced increase of carbon dioxide concentrations in the atmosphere in effect was a huge geophysical experiment.

In connection with the upcoming International Geophysical Year (IGY), Revelle then sought federal funding for studying how this CO_2 increase was affecting global temperatures.[30] The application was granted, and Wexler, who was also head of the research department at the US Weather Bureau, secured the funding.

Scripps thereby became a central node in the IGY atmospheric carbon dioxide program. Charles Keeling (1928–2005) from the California Institute of Technology was recruited for the project in July 1956.

The Tellus Article

In September 1956, Revelle and chemist Hans Suess wrote the article "Carbon Dioxide Exchange between Atmosphere and Ocean and the Question of an Increase of Atmospheric CO2 during the Past Decades." Rossby suggested it be submitted to *Tellus* (where Rossby's former doctoral student Bert Bolin was editor). It was published in February 1957.[31]

The previous year, the article "The Carbon Dioxide Theory of Climatic Change" by Gilbert Plass had also been published in *Tellus*.[32] Both articles were funded by the Office of Naval Research and used to establish the climate issue both in the scientific community and politically.

The media had also started reporting on the phenomenon. In October 1956, the *New York Times* wrote that human carbon dioxide emissions could lead to a warmer climate.[33]

The December issue of *Time* magazine had Rossby on the cover, with a warning in the featured article of the consequences of human impact on the environment.[34] This became his last media appearance. The following year he passed away prematurely from a heart attack, while his friend, John von Neumann, died from cancer that same year.

In his last essay, Rossby addressed the problem of carbon dioxide emissions from the burning of fossil fuels: "It has been pointed out frequently that mankind now is performing a unique experiment of impressive planetary dimensions by now consuming during a few hundreds of years all the fossil fuel deposited during millions of years"[35]

The International Geophysical Year 1957–58

The International Geophysical Year (IGY) had been proposed in 1950 by physicist Lloyd Berkner of the National Academy of Sciences (NAS). It became a major international research project with participation from nations across the world. The goal was to identify important geophysical phenomena in the Earth system, including the climate  as a small component.[36] This resulted in a scientific satellite program.

In Sweden, the project was coordinated by the Royal Swedish Academy of Sciences and organised by Rossby and his assistant and former doctoral student, Bert Bolin (1925–2007).

In the United States, the NAS, under Detlev Bronk, became the central node for the preparations. In 1953, Bronk appointed the US National Committee for IGY with members such as Lloyd Berkner, Roger Revelle, and Harry Wexler.[37] Funding for the American participation came from the National Science Foundation.

Official host of the project was the Paris-based International Council of Scientific Unions (ICSU), closely related to UNESCO. As the chairman of ICSU, Lloyd Berkner could now oversee the project and would have a growing impact on climate science. In 1969, he founded the Scientific Committee on Problems of the Environment (SCOPE).

The International Geophysical Year resulted in the first breakthrough for the carbon dioxide theory. Measurements of CO_2 levels in the atmosphere were initiated at Mauna Loa, Hawaii.

The early proponents of the theory were clearly not especially alarmed over climate change. They appear rather as careerists sensing opportunities for establishing scientific institutions, making a name for themselves and gathering their own followings by being early proponents of the theory. They were curious scientists eager for more resources for new exciting projects and experiments, in collaboration with collegues abroad.

Behind them, however, were other players who saw how the theory could be exploited politically and who knew how to take advantage of some of the scientists' egos and eagerness for scientific recognition.

The Rossby Memorial Volume

In 1959, a memorial volume was compiled by Bert Bolin in memory of Rossby, who had passed away two years earlier.[38] It also included the article "Changes in the Carbon Dioxide Content of the Atmosphere and Sea Due to Fossil Fuel Combustion" by Bert Bolin and Erik Eriksson.

This was its first publication issued by Rockefeller Institute Press and quite an out-of-place piece for a medical institute. The publishing manager was not amused.[39] Its publication can be seen as an indication of the enormous influence Detlev Bronk and the Rockefeller family had on early climate science.

The Swedish-American Climate Project

After Rossby's death in 1957, Bert Bolin assumed leadership of MISU and IMI and followed in the footsteps of his mentor. Under Bolin, IMI came to play a central role in climate research before the founding of IPCC (of which he became chairman in 1988).

The successful collaboration on climate research would mark the beginning of a long Swedish-American affair, with a closely interconnected network of scientists. It was also a military project, in which Sweden became the "neutral ground" and would play a leading part.

In December 1957, Edward Teller warned that the polar ice caps would melt due to increased carbon dioxide levels (though four decades later he would reject his earlier opinion).[40]

On February 12, 1958, the new threat was presented to the general public in the propaganda film *The Unchained Goddess*, produced by Frank Capra for the telephone company AT&T/Bell Laboratories. The film was shown in schools and other institutions. One of the scientific advisors was Warren Weaver from Rockefeller Foundation.[41]

After the climate issue had been raised in institutions which had come about with funding from the Rockefeller charities—supported by the National Academy of Sciences, the Office of Naval Research, and the US Atomic Energy Commission (1946–75)—the Rockefeller family foundations increasingly took over as it got more political implications. A new political climate was on the agenda and RBF would be leading the way.

Special Studies Project

This would be obvious in RBF's Special Studies Project. In the early 1950s, the extroverted and goal-oriented Nelson Rockefeller had been working for President Eisenhower and in 1954 became Special Assistant to the President for Foreign Affairs. His ideas, however, were not well received by the head of the CIA, Allen Dulles. In 1955, Nelson resigned from his position in the White House and turned to a forum where he would meet less resistance.

The following year, Nelson replaced his older brother, John D. III, as chairman of the Rockefeller Brothers Fund, and launched the RBF Special Studies Project. It could be seen as a preparation for Nelson's political career and his aspirations to become president of the United States.[42]

The project ran from 1956 to 1961 and resulted in a number of sub-reports, which analyzed the problems and opportunities that the United States would be facing in the coming decades. It was a very ambitious program aiming to establish guiding principles for governing institutions on how to deal with the challenges ahead. It included eight panels for different areas:

- Panel I—International Objectives and Strategies
- Panel II—International Security Objectives and Strategy

- Panel III—International Economic and Social Objectives and Strategy
- Panel IV—US Economic and Social Policy
- Panel V—US Utilization of Human Resources
- Panel VI—US Democratic Process
- Panel VII—The Moral Framework of National Purpose

An impressive list of experts were brought on board for the panels, including Dean Rusk (secretary of state, 1961–69), Detlev Bronk (NAS), and Edward Teller (father of the hydrogen bomb), with family protégé Henry Kissinger as project leader and organizer.

Several of the experts were part of Nelson's personal "portable brain trust"—which had irked President Eisenhower, who remarked that Nelson was "too used to borrowing brains instead of using his own."[43]

Kissinger, a professor at Harvard University, became one of the family's most valuable assets as national security advisor and later secretary of state.

Kissinger felt that modern governments were ill equipped to handle complex problems, tending to become victims to circumstances instead of actively shaping the world of tomorrow.[44] A new political architecture would be required, better able to offer long-term governance.

Henry Kissinger (1923–2023), US Secretary of State 1973–77.

The Special Studies Project became a part of this long-term leadership and would be followed by the founding of new institutions for global cooperation. The conclusions in the project reports would serve both party platforms before the presidential election in 1960.

Conservation, overpopulation, and the overtaxing of the planet's resources were areas identified as requiring special measures. These ideas would have a major impact on the political agenda in the coming decades. Nelson, too, mentioned environment as an area that could motivate international cooperation.

This concern would later be merged with the threat of climate change. The Special Studies Project gave a hint of the future of the climate agenda. Several panel members would be involved in its implementation.

The panel "International Security Objectives and Strategy" included Edward Teller, Detlev Bronk, and publisher Henry Luce. Teller became one of the first to warn the polar icecaps would melt due to increased levels of carbon dioxide (forty years later, however, he would change his opinion on this).[45]

The panel chairman, Laurance Rockefeller, and Carroll L. Wilson, professor of management at MIT, would introduce the climate issue to both US presidential administrations and to the United Nations in the 1960s and '70s, while Henry Luce gave it media coverage.[46]

Global Issues

The panel "International Economic and Social Objectives and Strategy" also included David Rockefeller. According to its 1958 analysis, the current world order lacked both a central purpose and international coordination. The scientific community was, however, seen as an example to the contrary, as transnational cooperation was essential for scientific progress.

The United States should therefore, as it had during the International Geophysical Year, continue its efforts to initiate international agreements for the benefit of both international exchange and scientific progress. These efforts could then be expanded to include other areas. Oceanography, meteorology, and global health were seen as particularly suitable as these problems transcended national borders (see this chapter's epigraph).[47]

Echoing John von Neumann, it was concluded that human impact on the atmosphere would lead to problems which could only be solved by international cooperation. The forces of weather needed to be controlled.

Three decades later these ideas would lead to the founding of the United Nations Panel on Climate Change (IPCC). Under the leadership of David Rockefeller and Henry Kissinger, RBF would be deeply involved in this process. As a philanthropy with nearly unlimited resources it was possible to work patiently over a long time frame.

Laurance Rockefeller
RBF chairman 1957–79

Individually, people are finding that a simpler lifestyle provides greater satisfaction than relentless pursuit of materialism.

—Laurance Rockefeller[48]

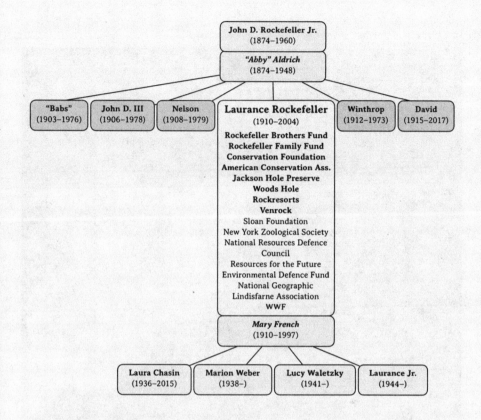

LAURANCE SPELMAN ROCKEFELLER was born in 1910 as the fourth child and third son of John D. Rockefeller Jr. and Abby Aldrich. He married Mary French and had four children: Laura, Marion, Lucy, and Laurance "Larry" Jr.

Laurance developed a lifelong interest in risk capitalism, conservation, and religion. He founded Rockefeller Brothers Inc., the American Conservation Association, and Rockresorts, and also funded the Lindisfarne Association, the Fund for the Enhancement of the Human Spirit, and the Foundation for Conscious Evolution. He was board member, vice president, and vice executive of the Conservation Foundation; board member of the National Resources Defense Council, National Geographic Society, Woods Hole Oceanographic Institute, Resources for the Future, and the Sloan Foundation; and a member of the Environmental Defense Fund and the WWF. He co-founded the RBF and served as chairman from 1958 to 1979. Laurance passed away in 2004 at ninety-six years old.

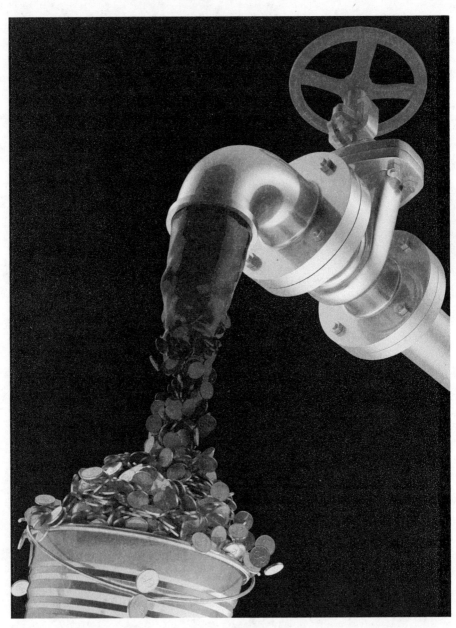

Oil Money.

Chapter Four

Future Shock

Man is now degrading his environment at a terrifying rate. The cumulative effects of advancing technology, massive industrial-ization, urban concentration, and population growth have all combined . . . not only to create imminent danger to the quality of human life, but even to pose threats to life itself.
—Rockefeller Foundation, *1969 Annual Report*

DECADE OF ENVIRONMENTALISM

While Nelson Rockefeller tried, unsuccessfully, to get himself nominated as a presidential candidate before the 1960, 1964, and 1968 elections, the environmental concerns had an international breakthrough. Nelson's brothers Laurance and John worked hard to get the issues of conservation and population on the political agenda.

World Wildlife Fund

In 1961, Julian Huxley founded the World Wildlife Fund (WWF) to raise funds for International Union for Conservation of Nature (IUCN). Prince Bernhard of the Netherlands, founder of the Bilderberg group, was elected chairman. Godfrey A. Rockefeller (1924–2010), great grandson of John D. Rockefeller's brother William, became president and turned it into a leading organization in international conservation.[1]

Royal Dutch Shell became one of WWF's first funders. In 1966, Shell's president, John Loudon, became a board member of WWF and in 1977 succeeded Prins Bernhard as chairman. Loudon also helped David

Rockefeller lead and recruit for the International Advisory Committee of Chase Manhattan.[2]

The champions for conservation were thus some of the most prominent oil magnates, industrialists, and bankers. They would eventually also come to have a significant influence on the climate change discourse.

The First Conference on Carbon Dioxide

On March 12, 1963, the Conservation Foundation (which in 1985 would become part of the WWF) organized the first international conference specifically focused on the effects of carbon dioxide on the climate. The chairman of the conference was Conservation Foundation vice president Fraser Darling, with Conservation Foundation president Samuel Ordway as one of the observers. Attending the conference were three of the scientists who, during the previous decade, had helped develop the theory: Gilbert Plass, Charles Keeling (from Scripps), and Erik Eriksson (Bert Bolin's colleague from the IMI at Stockholm University).

The results from Keeling's CO_2 measurements at Mauna Loa could now confirm a rising trend. It was discussed how this could result in a temperature increase that could have negative consequences. A scientific committee to investigate the consequences was proposed, under the National Academy of Sciences. The close collaboration between Scripps Institution and the Department of Meteorology of Stockholm University was considered appropriate for the research that had been done so far. However, the Conservation Foundation proposed that a more formal international organization would be required if more nations were to participate in the measurements.[3]

This in turn required a much wider acceptance of the theory. For this purpose the World Meteorological Organization founded World Weather Watch (WWW) in April 1963, based on an idea by Harry Wexler (1911–1962).

That same year, RBF donated $50,000 to the Conservation Foundation. Laurance Rockefeller, chairman of RBF, was at this time also the vice president of Conservation Foundation.[4]

Conservation in the White House

After the assassination of President John F. Kennedy on November 22, 1963, Vice President Lyndon B. Johnson became president. Exactly six months later, Johnson declared protection of the environment and its beauty to be one of the three pillars in his reform package "The Great Society."[5] This was good news for leading conservationist Laurance Rockefeller, who had worked for this behind the scenes for a long time.

Lyndon B. Johnson (1908–1973), Democrat, US president 1963–69.

In August 1964, work on the report, *Restoring the Quality of Our Environment*, was initiated. It was produced under the direction of John W. Tukey from Bell Laboratories, with members of the President's Science Advisory Committee. Around fifty scientists, in eleven different sub-panels, were to look into various forms of environmental pollution and their possible consequences.

Roger Revelle, who had been advisor to the secretary of the interior in the Kennedy Administration, Stewart Udall, became chairman of the subpanel "Atmospheric Carbon Dioxide." The Revelle panel also included Charles Keeling and Wallace Broecker (who would coin the term *global warming* in 1975). The report, published November 5, 1965, concluded,

> By the year 2000 the increase in CO2 will be close to 25%. This may be sufficient to produce measurable and perhaps marked changes in climate.[6]

President Johnson had held a speech in February 1964 about CO2's impact on the climate where he stated: "This generation has altered the composition of the atmosphere on a global scale through radioactive materials and a steady increase in carbon dioxide from the burning of fossil fuels."

This was the first time a president or head of state discussed the climate issue. Johnson's speech writer, Richard Goodwin, with the help of Stewart Udall and Laurance Rockefeller, had carefully prepared the speech.[7]

As one of the most influential men in the country in the area of conservation, Laurance became deeply involved in US politics and the close contacts with the White House would remain, almost regardless of who was in power. Both Laurance and his brothers had a revolving door to the very center of political power.

In 1958, the same year that Laurance became president of the RBF, Laurance was appointed by President Eisenhower to lead the Outdoor Recreation Resource Review Commission.[8] He then went on as advisor to John F. Kennedy. After Lyndon B. Johnson became president, the amiable relationship was reinforced, especially through Laurance's close friendship with the First Lady, "Lady Bird" Johnson, who shared Rockefeller's passion for conservation.

His brother, Nelson Rockefeller, however, in 1964 lost the nomination as the Republican presidential candidate to Barry Goldwater, but managed, through his faithful squire Laurance, to still hold some influence over Democrat environmental policy.

Nelson's failure can partly be explained by the divorce from his wife Mary Clark, to marry a member of his office staff, Margaretta "Happy" Fitler. This did not go down well with the general public, particularly not with his female voters. An honorable man did not leave his wife and children—especially not when aspiring to become president of the United States. The charming Nelson was a womanizer and would throughout his career be involved in a number of extramarital affairs, thus not living up to the high moral standards set by his paternal grandfather. His brothers were painfully reminded of the legacy of Devil Bill and all except Laurence boycotted his wedding.

Nelson tried once more to get nominated before the next election, but his divorce was the last straw for his presidential ambitions.

There were, however, other means of reaching this goal . . .

Growing Environmental Awareness

In 1966 Laurance was again appointed to now head the Citizens' Advisory Committee on Recreation and Natural Beauty. This project would be developed further under President Nixon, when Laurance became chairman of the Citizens Advisory Committee on Environmental Quality.[9] After successfully having influenced several presidential administrations, it was now time to bring conservation and population awareness to a general public.

Already, Rachel Carson's book *Silent Spring* (1962) had sparked a public debate on the insecticide DDT and environmental pollution around the world. In the 1950s, the Conservation Foundation had, with support from the Rockefeller Foundation, researched pesticides such as DDT and its effects on animals. Carson, who died of a heart attack just two years after the book was published, used the results in her book.

Center for Population Studies

In 1964, the Center for Population Studies at Harvard University was founded, made possible by a donation of $600,000 from the Rockefeller Foundation.[1]

The Center was headed by Roger Revelle, possibly as a reward for his accomplishments.[10] Here, he would educate the later vice president, Al Gore, about climate and population issues.

Besides tuition, the Center for Population Studies was to focus on projects on how to solve the population problem through various forms of population control.[11] The climate issue was slowly being merged with the fear of an unchecked population growth, in relation to the planet's resources. A smaller world population meant lower CO_2 emissions.

---

1   Harvard had close ties to the Rockefeller Family. Laurance's brother David was member of Harvard's Board of Overseers from 1954 to 1966 and chairman from 1966 to 1968. Henry Kissinger was also involved with the Harvard Center for International Affairs.

World Future Society

In October 1966, the World Future Society (WFS) was founded by Edward Cornish, a Washington journalist driven by a strong fear of thermonuclear war and a passionate interest in predicting the future. Cornish himself became its (reluctant) president.

From a modest start as a simple newsletter in February 1966, the World Future Society soon became the largest and most influential community of futurists in the world, including luminaries such as Barbara Marx Hubbard, Maurice Strong, and Robert McNamara (secretary of defense, 1961–68 and president of the World Bank, 1968–81) on the board of directors.

Other notable members and contributors have included Buckminster Fuller, Herman Kahn (founder of the Hudson Institute), Gene Roddenberry (originator of *Star Trek*), Ray Kurzweil (chief engineer at Google), Carl Sagan (science fiction author), and Neil deGrasse Tyson (astrophysicist).

Buckminster Fuller (1893–1983), architect, designer, system theorist, inventor, author, and futurist (photo by Staffan Wennberg, 1969).

Through its conferences, writings, and extensive network of influential people in science, culture, media, business, and politics, the World Future Society would become a major player in setting the global agenda for the decades to come, offering technocratic solutions to environmental problems (see chapters 8, 11, and 12).

In 1967, inspired by the French futurist Bertrand de Jouvenel's journal *Futuribles*, Cornish also launched and became editor of *The Futurist*.[12]

Major Social Trends

The October 1968 issue of *The Futurist* featured a list of thirty-one coming social trends based on *The Next 500 Years* (1967) by Burnham Putnam Beckwith. Some of the main points:

- Population growth "will increasingly be offset by rational individual and social control over reproduction."
- Social control. "The steady growth of scientific knowledge, especially in the social sciences, makes government control over social trends ever more feasible."
- Rationalization of all social policies. "This trend will be reinforced by eugenic reform that will increase innate human intelligence. This trend . . . justifies predicting that men will eventually adopt any social policy or reform which can now be shown to be rational or scientific under probable future social conditions."
- Spread of birth control.
- Eugenic progress "will soon begin to appear in the most advanced countries and will spread and improve steadily during the remainder of the next five centuries."
- Urbanization "will transform most backward states and continue in most advanced states until their population is over 95% urban."
- Industralization, automation, specialization, professionalization, increase in scale of production, and growth of monopoly.
- Centralization of control. "The invention of nuclear weapons has recently made world conquest and rule possible and a world government inevitable."
- Collectivisation, economic specialization, and interdependence. "Rapid increase in the efficiency of large-scale management and government."

- Meritocracy—"rule and management by the most able."
- Feminism, due to "industrialization, urbanization, birth control, the growth of international peace and order, education, and the invention of new household machines."
- Paternalism. "There will be an increase in legislation which protects relatively incompetent minorities against exploitation or their own poor judgement. . . . The inevitable progress of applied science will enable experts to give more and more valuable advice concerning many personal decisions now left to individuals."
- Humanitarianism.
- Cultural homogenization. "There will be a standardization of all human beliefs and activities—political, economic, educational, sexual, artistic, scientific, and recreational—due to the increase in travel, migration, communication, and freight transportation, which will prevent isolation of human groups. The most significant homogenization trend will be the further Europeanization of all non-European countries."

Planning the Stockholm Conference

In 1967, British diplomat and mathematician William Penney (who had created the first British nuclear bomb) proposed an environmental conference arranged by the United Nations.

However, it was Swedish diplomat Sverker Åström (Sweden's Permanent Representative to the UN from 1964) who made the official request in the General Assembly.[13] The United States representatives were sympathetic to the proposal—which is hardly surprising as it was the result of diplomatic negotiations between Sweden and themselves.[14] There were also the joint research projects on population and climate.

The background paper, written by Swedish lobbyist Hans Palmstierna, included concerns over what rising CO_2 emissions might lead to. In 1968, Palmstierna helped spread environmental awareness to the general public through his bestselling book *Plundring, svält, förgiftning (Looting, Starvation, Poisoning)*, where the carbon dioxide issue was mentioned as one of several looming threats.[15]

The choice of hosting country for the conference was likely no coincidence, as American initiatives were often met with mistrust from many developing nations, and in light of the Vietnam war having escalated when Lyndon B. Johnson assumed office as president (with Robert McNamara as secretary of defense). Stockholm became an ideal location, given Sweden's good standing in the Third World.

Part of the planning and agenda for the Stockholm Conference had been developed at the Aspen Institute and the International Institute for Environmental Affairs (IIEA), founded in 1971 to assist the conference secretariat. The Aspen contribution specifically stressed the need for international governance to handle the environmental problems.[16] The board of IIEA included Robert O. Anderson (ARCO, Aspen Institute), Maurice Strong (Aspen Institute), and Robert McNamara (newly appointed head of the World Bank after his post as secretary of defense).

The IIEA was later renamed the International Institute for Environment and Development (IIED) when the institute relocated from Washington, DC, to London, under the direction of Dame Barbara Ward.

The Club of Rome

The following year, 1968, the powerful environmental think tank the Club of Rome was created by Aurelio Peccei (president of Olivetti, executive at Fiat, and founder of Alitalia), and Alexander King (Director General for Scientific Affairs of OECD), with contributions from Fiat president Gianni Agnelli (1921–2003), grandson of Fiat founder Giovanni Agnelli.

Agnelli, seen as the most powerful capitalist in Italy, was a close ally of David Rockefeller's and was part of the International Advisory Committee of Chase Manhattan Bank. This network would grow to become a major player in the international arena, with David as coordinator.

In 1965, Peccei had attracted the interest of US Secretary of State Dean Rusk (former chairman of the Rockefeller Foundation and panelist in RBF's Special Studies Project) when delivering a speech to the private investment organization ADELA (Atlantic Community Development

Group for Latin America) in Buenos Aires.[2] Peccei had talked about global crises such as population explosion, environmental degradation, the gap between North and South, and the need for a new industrial electronic revolution, with long-term planning at a global level.[17]

These concerns were also of interest to Jermen Gvishiani, vice president of the State Committee for Science and Technology in the Soviet Union. An international collaboration between East and West was initiated, that was to lay the foundation for the Club of Rome and for the institutes IIASA (International Institute for Applied Systems Analysis) in Laxenburg, Austria, and IFIAS (International Federation of Institutes of Advanced Study) in Stockholm.[18]

Astrophysicist, systems theorist, and futurist Erich Jantsch[19] further developed Peccei's ideas ("Framework for Initiating System-Wide Planning on a World Scale") during a conference at the Academia dei Lincei in Rome in April 1968.[20] The meeting was described as a failure and no consensus could be reached, but some of the participants (Peccei, King, Jantsch, Hasan Özbekhan) decided to go ahead and assumed the name Club of Rome.

A few months later (from October 27 to November 2, 1968), the Working Symposium on Long-Range Forecasting and Planning was held at the Rockefeller Foundation's conference center Villa Serbelloni, in Bellagio, organised by the OECD and funded by the Rockefeller Foundation.[21] Several Club of Rome founders attended, as well as some of its future spokespersons, such as René Dubos (Rockefeller University) and Jay Forrester (MIT).[22] The conference resulted in the *Bellagio Declaration on Planning* (1969), highlighting the need for international and holistic planning to handle technological, economic, political, and social stresses.[23]

In 1970, after a rather disorganised start, the Club of Rome got a more solid foundation. Its stated purpose was "to promote understanding of the global challenges facing humanity and to propose solutions through scientific analysis, communication and advocacy"—with the creation of a planetary civilisation as the ultimate goal.

---

2   ADELA included industrialists such as Agnelli, Marcus Wallenberg Sr., and Henry Ford II.

The Rockefeller Foundation conference center in Bellagio, Como, Italy.

The philosophy was largely based on cybernetics and systems theory, originating from Ludwig von Bertalanffy's view of the world as a closed system with interdependent parts which, according to the Club of Rome, was in need of a central governance—a technocratic system with a futurist dream of being able to predict and handle all upcoming challenges. The systems theorists also used John von Neumann's ideas about controlling the world by making mathematical calculations about future development. This was the basis of the computer modelling later used as political motivation for a global transformation into a new system.

The Club of Rome had close connections to the Rockefeller family and the financial elite. Max Kohnstamm, member of the Club of Rome inner circle, was also involved in the Bilderberg group steering committee with David Rockefeller and Gianni Agnelli.[24] Detlev Bronk from Rockefeller University was recruited in 1969.[25] The executive board included Nelson Rockefeller's personal friend Carroll Louis Wilson from the Sloan School of Management at MIT.

Limits to Growth
In 1972, the Club of Rome published *Limits to Growth*, with pessimistic Neo-Malthusian data projections about future development. The computer models used in the book, on commission from Carroll L. Wilson, had been developed at the MIT Sloan School of Management (a private business school founded by GM CEO Alfred P. Sloan, headed by the Rockefeller family's financial advisor, William F. Pounds). The book suggested two solutions: a system transformation into Global Governance and zero growth.[26]

The Climate Conferences SCEP and SMIC
During the prelude to the upcoming Stockholm Conference, Carroll L. Wilson was the one who put the climate change on the international agenda, by organizing two conferences, Study of Critical Environmental Problems (SCEP) in Williamstown, Massachusetts, in 1970, and Study of Man's Impact on Climate (SMIC) in Wijk, Sweden, the following year.

The conference reports are often cited as the origins of public interest in the emerging climate discourse. According to climate researcher and author of the climate chapter of the SMIC report William Kellogg (NCAR), they were "required reading" for all participants at the Stockholm Conference and inspired a whole generation of climate scientists.[27]

The conferences were initiated by the Woods Hole Oceanographic Institute (where Wilson was a board member with Detlev Bronk), and MIT.[28] Funding came from the National Science Foundation, Ford Foundation, Alfred P. Sloan Foundation, and American Conservation Association (ACA). The influence of the Rockefeller family was significant through Laurance Rockefeller being chairman of ACA and board member of the Sloan Foundation, besides Wilson being a close associate. The SCEP report was directly financed by the Rockefeller Foundation.

For the conference in Williamstown (SCEP), Wilson had gathered forty scientists and experts from a number of disciplines "to raise the level of informed public and scientific discussion and action on global environmental problems." The research group wanted to find leverage points where relatively small human environmental stressors might have a significant and global impact on climate change.[29]

The second conference was held in Svante Arrhenius's birthplace, Wijk, outside Uppsala, Sweden, in collaboration with the Royal Academy of Sciences (KVA) and the Royal Academy of Engineering (IVA).[30] The Swedish-American connections remained strong. This conference was intended to serve as an authoritative source of virtually all issues connected with climate change and related areas, and had been sought for by Maurice Strong, who was secretary general of the Stockholm Conference.

There was still at this time some disagreement on whether the main problem was aerosols (with soot causing cooling) or carbon dioxide (having a warming effect). It was concluded that in any case *humans* influenced the climate system through the burning of fossil fuels.

The reports did not gain much attention during the Stockholm Conference but would soon prove very useful as basis for research programs such as ICSU–SCOPE and the Global Atmospheric Research Program (GARP).

Big Oil and Auto Fund the Green Movement

At the end of the 1960s more active measures were taken to spread environmental awareness to the general public.

In 1967, the Environmental Defense Fund was funded by George Woodwell from Conservation Foundation to inform about the dangers of DDT.

Three years later, the closely related National Resources Defence Council (NRDC) was founded by, among others, environmental lawyer James Gustave Speth, with Woodwell on the board of directors and funded by the Ford Foundation. Laurance Rockefeller was member of both organizations. His son Larry, who would later succeed him as chairman of the American Conservation Association, would work for NRDC for twenty-five years.

Woodwell, Speth, and the RBF would in the 1980s come to play crucial roles in making the climate a global political concern. Speth (a Rhodes Scholar), would in 1982 also found the World Resources Institute

and thereafter be appointed to the board of
directors of RBF.

In 1969, the Rockefeller Foundation and
Ford Foundation launched their environmental program, Quality of the Environment.

Around the same time, the British had
convinced the Council of European Union
to declare 1970 the European Conservation
Year.

Through the Rockefeller family's efforts,
conservation was also included in the agenda
of the new US president, Richard Nixon.

Russell E. Train from the Conservation
Foundation became chairman of the Council
on Environmental Quality, while Laurance
Rockefeller continued as presidential advisor.[31]

Richard Nixon (1913–1994),
Republican, US president 1969–
74, at inauguration ceremony,
January 1969 (photo by Staffan
Wennberg).

Shortly thereafter, several new climate research institutes were formed.
In 1971, the Climate Research Unit (of later "Climategate" fame) at the
University of East Anglia in England was founded by Graham Sutton
and Lord Zuckerman, and funded by BP, Shell Oil, and the Rockefeller
Foundation.

Friends of the Earth

In 1969, Friends of the Earth (FoE) was launched, as a more radical branch
of the nature and wildlife conservation movement (previously dominated
by the upper class concerned mainly with the creation of national parks).

Funding came from oil magnate Robert O. Anderson from the Aspen
Institute. David Brower from the Sierra Club was recruited as chairman.
Aurelio Peccei (chairman of the Club of Rome and founder of Alitalia)
was also a board member.[32] It was all quite contradictory.

Friends of the Earth now attracted young people and radicals who
fought the establishment and resented the big finance dynasties' undue
influence over politics and business, without understanding who was
really pulling the strings. The aim of the organization was to change the

economic world order towards one based on solidarity and ecological sustainability.

During the 1970s, nuclear power and nuclear waste became a target of Friends of the Earth activism. It was seen as a competitor by the oil industry, which can explain some of its support for radical environmentalism. Such unholy alliances would persist and be developed further in coming decades.

Again the population issue was coupled with environmental concerns. According to Dixie Lee Ray (former head of the US Atomic Energy Commission), David Brower had suggested,

> Childbearing [should be] a punishable crime against society, unless the parents hold a government license. . . . All potential parents [should be] required to use contraceptive chemicals, the government issuing antidotes to citizens chosen for childbearing.[33]

Brower had personally encouraged Paul Ehrlich to write the bestselling book *The Population Bomb* (1968), which popularized Fairfield Osborn's overpopulation alarmism in *Our Plundered Planet* (1948). In his book, Ehrlich predicted that hundreds of millions would starve to death in the 1970s and '80s, even in the West:

> A cancer is an uncontrolled multiplication of cells; the population explosion is an uncontrolled multiplication of people. We must shift our efforts from the treatment of the symptoms to the cutting out of the cancer. The operation will demand many apparently brutal and heartless decisions.

A massive campaign of doomsday alarms of starvation and future disasters was then spread through the media.

Future Shock

The book *Future Shock* (1970), by sociologist, businessman, and futurist Alvin Toffler, described the transition from agrarian to industrial to

a post-industrial society ("the information age") where the pace keeps speeding
up (consumer goods being disposable or
becoming obsolete and where both workers, professions, homes, relationships, body
parts, and even nationalities become more
temporary and replaceable).

All these rapid social and technological
changes would cause a state of "shattering stress
and disorientation" called "future shock."[34]

In 1972, a documentary with the same
title was produced, narrated by Orson
Welles, with scary sound effects and images
to drive home the message of what shocks

Alvin Toffler (1928–2016),
futurist.

the present and near future had in store for humanity, alternating with joy-
ful hippie music and visions of a future with intelligent robots, routine space
travel, designer babies, artificial organs, and happy group marriages.

Toffler, however, was primarily a futurist and not a traditionalist.
The shocking changes were something humanity just had to adjust to,
but *selectively*, and by taking control of the technology for the benefit of
humanity. In the documentary, he states,

> If we can begin to think more imaginatively about the future, then
> we can prevent future shock, and we can use technology itself, and
> build a decent, democratic, and humane society. . . . We have now
> reached the point at which the technology is so powerful and so rapid
> that it could destroy us unless we control it.[35]

These terrifying threats and the promising allures of a high-tech future
would be echoed almost verbatim in 2016 by the World Economic Forum's
Klaus Schwab and his fellow futurists (see chapter 11).

The First Earth Day, 1970

In September 1969, environmental champion US Senator Gaylord Nelson proposed holding teach-ins on college campuses the following spring to create a greater public awareness of the threats to the environment. He then went on a national speaking tour to inspire local activists and set up a national organization to coordinate the teach-ins.[36]

April 22, during most colleges' spring break, was chosen as the first Earth Day. Oil tycoon Robert O. Anderson again opened his wallet to fund it.[37]

Meanwhile, Rockefeller friend Henry Luce's magazines *Time* and *Life* conveyed the image of a planet under threat.

The first Earth Day, which was celebrated in thousands of schools, colleges, and universities across the United States, marked the birth of the modern environmental movement and gathered students, anti-war activists, civil rights activists, hippies, Marxists, and other radicals. They represented the new social revolution, challenging old ways of life.

One of the major Earth Day events, the "ENACT Teach-In on the Environment," was held from March 11 to 14, 1970, at the University of Michigan in Ann Arbor (when this university had its spring break), organized by a university activist group calling themselves Environmental Action for Survival (ENACT). A teach-in is a combination of sitting protest and informal college seminar on a political topic, which became popular in the 1960s following the successful anti-war teach-in held all night at the University of Michigan in 1965. The organizers found it surprisingly easy to raise funds for the project. Alan Glenn of the *Ann Arbor Chronicle* wrote in 2009,

> Money began to pour in, from local sources and others farther afield, including such unlikely benefactors as Dow Chemical, which contributed $5,000. ENACT would eventually raise an astonishing $70,000 to support their teach-in. . . . They raised so much money that they weren't able to spend it all.[38]

Finding venues and attracting media attention was also not a problem.

J. David Allan, a natural science graduate student at the time and now professor emeritus at the university, told the *Chronicle* in that 2009 article, "A lot of what is amazing about the teach-in is that it wasn't that hard. It just kind of happened, in this very organic way. We were able to get venues, we were able to get funding. The media came to us. A lot of things just kind of came together."[39]

The five-day event, held both at campus and all around town, offered 125 seminars, speeches, workshops, debates, forums, rallies, demonstrations, field trips, films, concerts, etc, and attracted around fifty thousand attendees. Over sixty major media outlets, including three TV channels and a film crew from Japan, covered the event.

Make Love Not Babies, 1970 pop poster (photo by Ewa Rudling, ewarudling.se).

April 22.
Earth Day.

A disease has infected our country. It has brought smog to Yosemite, dumped garbage in the Hudson, sprayed DDT in our food, and left our cities in decay. Its carrier is man.

The weak are already dying. Trees by the Pacific. Fish in our streams and lakes. Birds and crops and sheep. And people.

On April 22 we start to reclaim the environment we have wrecked.

April 22 is the Environmental Teach-In, a day of environmental action.

Hundreds of communities and campuses across the country are already committed.

It is a phenomenon that grows as you read this.

Earth Day is a commitment to make life better, not just bigger and faster; To provide real rather than rhetorical solutions.

It is a day to re-examine the ethic of individual progress at mankind's expense.

It is a day to challenge the corporate and governmental leaders who promise change, but who short change the necessary programs.

It is a day for looking beyond tomorrow. April 22 seeks a future worth living.

April 22 seeks a future.

We are working seven days a week to help communities plan for April 22. We have come from Stanford, Harvard, Bucknell, Iowa, Missouri, New Mexico, Michigan and other campuses.

We are a non-profit, tax exempt, educational organization. Our job is to help groups and individuals to organize environmental programs to educate their communities.

Earth Day is being planned and organized at the local level. In each community people are deciding for themselves the issues upon which to focus, and the activities which are most appropriate.

We can help, but the initiative must come from each community. We have heard from hundreds of campuses and local communities in all fifty states. Dozens of conservation groups have offered to help. So have the scores of new-breed environmental organizations that are springing up every day.

A national day of environmental education was first proposed by Senator Gaylord Nelson. Later he and Congressman Paul McCloskey suggested April 22. The coordination has been passed on to us, and the idea now has a momentum of its own.

All this takes money. Money to pay our rent, our phones, our mailings, brochures, staff, advertisements.

No list of famous names accompanies this ad to support our plea, though many offered without our asking. Big names don't save the environment. People do.

Help make April 22 burgeon.

For you. For us. For our children.

The Environmental Teach-In, Inc.
Room 200
2000 P Street, N. W.
Washington, D. C. 20036
I enclose $10, $20, $50____dollars [tax deductible]
How can I help my community?
Name
Address

National Staff: Denis Hayes, Coordinator; Linda Billings, Stephen Cotton, Andrew Garling, Bryce Hamilton, Sam Love, Barbara Reid, Arturo Sandoval, Philip Taubman

Earth Day advertisement in *New York Times* 1970 declaring mankind a pestilence on the planet.

The opening rally, with Senator Gaylord Nelson as speaker, had an audience of seventeen thousand. Besides environmental science, the event planners wanted to make it as inclusive as possible and therefore included topics such as the Vietnam War, women's liberation, racial equality, and social justice under the umbrella of ecology.

Overpopulation and family planning was also debated, under the heading "Sock It to Motherhood: Make Love, Not Babies" (a popular slogan during the hippie era).[40] This was actually one of the core concerns. According to Senator Nelson,

> Central to the theme of the first Earth Day in 1970 was the understanding that U.S. population growth was a joint partner in the degradation of our nation's environmental resources.[41]

The image of mankind as a cancer or a plague on the planet was slowly eating its way into the soul of the people. Full-page Earth Day ads in leading newspapers helped drive the message home, in no uncertain terms.

Despite the misanthropic view of mankind, however, early environmental activism had positive effects, especially on local problems such as air, soil, and water pollution, and on environmental awareness in general.

Nearly five decades later, Earth Day is celebrated in more than 193 countries across the globe, coordinated by the Earth Day Network.[42]

Revolution from Above

Among the young activists of the time, we find a new generation of Rockefellers (including David's daughters Abby, Neva Goodwin, and Eileen) to spread the new message of environmentalism, feminism, New Age, free sex, and global justice, ironically founded in Marxist ideology and opposition to capitalism.

This group also included Dr. José Argüelles, an art teacher, initiator of the Whole Earth Festival, and one of the co-founders of the Earth Day concept who would later whip up global hysteria around the year 2012.

In the anthology *This Cybernetic Age* (1969), the author of one of the chapters, John D. Rockefeller III, praised the new revolutionaries who were clearing the ground for a new global culture.

> Instead of worrying about how to surpress the youth revolution, we of
> the older generation should be worrying about how to sustain it. The
> student activists are in many ways the elite of our young people. They

perform a service in shaking us out of our complacency. We badly need their ability and fervor, in these troubling and difficult times.

A unique opportunity is before us to bring together our age and experience and money and organisation with the energy and idealism and social consciousness of the young. Working together, almost anything is possible. [43]

The '68 revolution and green movement were thus largely orchestrated from above, with active help from the upper classes rather than spontaneously from the grassroots—a disturbing fact that some clearsighted radicals at the time pointed out. Soon, however, these connections would be forgotten and ever new generations of activists shouting slogans at the same forces that had created them. The young environmental and social justice activists thus became little more than useful tools for propagating the destructive Neo-Malthusian ideals of the ultra-rich.

International plutocracy has often pursued policies, used movements, and promoted doctrines that most people would consider to be anti-capitalist. Yet both capitalism and the Left arose during the same period in history, both have the same historical outlook, and both view traditional culture, the family, and nations as obstacles on the path towards a World State. (Kerry Bolton, *Revolution from Above*, 2011)[44]

In a short period of time, numerous new radical organizations with environmental protection on the agenda were initiated. Friends of the Earth and Greenpeace became two of the most vocal and influential.

Greenpeace

Greenpeace was founded in 1971 by Irving Stowe, Jim Bohle, Paul Cote, and others. It developed out of a protest group against nuclear tests called the Don't Make a Wave Committee, and from the Society for the Prevention of Environmental Collapse (SPEC) and quickly grew to become one of the leaders in the environmental activism. Its objective was to expose corporate environmental misconduct using methods such as organising boycotts and campaigns as well as spectacular and sometimes risky environmental actions.

Within Greenpeace there were also visions of creating an ecological spiritual movement (which can be seen in the in New Age symbolism of the Rainbow Warrior, the name of the ship used in Greenpeace's risky protest actions against whale hunting).

As with Friends of the Earth, there was a connection to the Sierra Club (in Greenpeace's case, in the form of financial help with the founding).[45] Unlike other environmental organizations, Greenpeace had an ambition not to accept government funding and were funded mainly by philanthropic foundations, fund-raising and membership fees.[46]

The Stockholm Conference, 1972

From June 5 to 16, 1972, Stockholm hosted the United Nations Conference on the Human Environment. The secretary general of the conference was the Canadian oilman Maurice Strong. Swedish Prime Minister Olof Palme held the welcoming speech.

The conference motto was "Only One Earth" and was introduced with the slogan "Man Builds, Man Destroys" in a UNEP propaganda film about the conference.[47]

Developing countries criticised the conference for being too focused on the needs of industrialised nations and suspected that the latter wanted to limit their economic and population growth. At the preparatory Founex

Maurice Strong in Swedish newspaper, January 25, 1972: "We have ten years to avert the disaster."

Conference, however, which Strong and British economist Barbara Ward (Baroness Jackson of Lodsworth) had organized in Switzerland the previous year, a foundation for cooperation had been laid as the concept of sustainable development was formed. The phrasing of the agreement guaranteed that economic development could go hand in hand with conservation.

Barbara Ward (member of the Fabian Society) was an early advocate of sustainable development and the (ostensible) redistribution of wealth to developing nations (see NIEO, chapter 5).

***

In addition to the official conference, a number of alternative events were arranged in Stockholm. The conference attracted professional activists from other parts of the world and provided an opportunity for the newly formed radical environmental organizations to start applying pressure. During the first days there were some violent demonstrations and two hundred activists tried to interrupt the conference.

At an abandoned air field outside Stockholm, a tent camp was provided by the Swedish government, with the help of hippie collective Hog Farm of Woodstock fame. There were public debates, progressive musical performances, get-togethers with Native Americans, a demonstration for the whales, and a party to celebrate life.

It was a motley crew of peace and environmental activists, communists, hippies, Native Americans, drug users, and local youth. Many were anti-imperialists, protesting against the US war in Vietnam and other controversial issues of the time. Some local activist groups were well organized and arranged alternative bus tours and other activities.[48]

However, there were also sharp dividing lines between groups, especially between revolutionaries and hippies. The well-organised communists were marching around in the city center, while the hippies took the specially provided busses to Skarpnäck to get high with Hog Farm, listen to music concerts, and partake in the various happenings.[49]

***

The Stockholm conference resulted in the Declaration of the United Nations Conference on the Human Environment (the Stockholm

Maurice Strong talking to the activists at the hippie tent camp at Skarpnäck outside Stockholm (photo by Staffan Wennberg).

Declaration)[50] and the United Nations Environment Program (UNEP), with Maurice Strong as chairman.

A system-wide Earthwatch "to monitor major global disturbance in the environment and to give early warning of problems requiring international action" was also initiated, as well as data collection projects through collaboration between the UN organizations UNEP, WMO, and ICSU. The plan had worked out perfectly.

Some radicals, however, protested that the agreement was not as stringent as it could have been and suspected that the environmental crisis was much worse than politicians assumed or revealed. Their enemies were, ironically, the same supercapitalists than had both funded them and crafted the ideals they were now fighting for. They became, in effect, what we today would call controlled opposition and pawns in an illusory dichotomy.

The strategy of inviting environmentalists would become even more professionalized in the decades to come, with activists becoming a staple of every major conference.[51] Not until the end of the 1980s, however, would climate change be added to the list of concerns inspiring radical action.

Despite the many genuine activist groups, the Stockholm Conference was an elite project from the start, with countless connections to the Rockefeller family and affiliated organizations.

- Funding for the conference secretariat and one of its advisory committees came from the Rockefeller Foundation.[52]
- Maurice Strong, who had been an oilman for most of his career and would become president of Petro-Canada, was a friend of David Rockefeller's and had been a board member of Rockefeller Foundation from 1971 to 1976, with later skeptic Frederick Seitz.[53] Strong's conference advisor was Nelson Rockefellers's friend Carroll L. Wilson.
- The chairman of the American delegation, Russell Train (chairman of the Conservation Foundation), was a close friend of and mentor to Laurance Rockefeller.[54]
- Laurance Rockefeller was also part of the US delegation, leading the group Human Settlements.
- Barbara Ward and René Dubos (Rockefeller University) wrote the framework for the conference titled *Only One Earth: The Care and Maintenance of a Small Planet* (through IIEA, funded by the Ford Foundation and the World Bank).[55]
- Ward was also a personal friend of Nelson Rockefeller's. In 1967 she was appointed Albert Schweizer Professor of International Economic Development at Columbia University, established by Nelson when he was governor of New York.[56]

Survival of Spaceship Earth

Just before the conference, the film *Survival of Spaceship Earth* premiered, featuring John D. Rockefeller III and Rockefeller-associated Maurice Strong, Barbara Ward, René Dubos, Dr. Margaret Mead, Walter O. Roberts, and nuclear chemist Dr. Harrison Brown of the Manhattan Project.[3]

---

3   The film was funded by the Charles F. Kettering Foundation, Alfred P. Sloan Foundation, George Gund Foundation, and Tinker Foundation.

Just like *Future Shock*, this film flashed a series of frightening images, including of "dead" dummies in polluted water and children without limbs (portrayed as victims of pollution when in fact they were known victims of the drug thalidomide, offered to pregnant women in the late 1950s and banned from 1961). The primary message of the film was that we were destroying the environment ("we" as in us ordinary people, not the corporations that had created urban sprawl, motor dependency, over-consumption, deforestation, and environmental pollution).

Developing nations were also a problem. Barbara Ward urged them not to repeat the mistakes of the industrialised countries. Instead, she explained, their development would need careful planning and guidance in order to manage their urbanisation and industrialisation in a more con-trolled way.

Lastly, there were too many people on the planet. In the film, John D. Rockefeller III ("Mr Population") was interviewed, reiterating how President Nixon in 1969 had taken the population issue very seriously and initiated the Rockefeller Commission on Population (led by John D. III himself).

***

The Report of the Commission on Population Growth and the American Future contained fifty proactive measures to halt population growth in the US, including:

- Offering family planning services and voluntary sterilisations;
- Providing public and private child care;
- Mandating sex education in schools;
- Legalizing contraceptives and abortion (even for teenagers);
- Facilitating adoptions;
- Affording equal rights for women and for children born out of wedlock;
- Supporting better integration of minorities and the poor; and
- Creating government departments, agencies, committees, and councils for tracking and managing population growth.[57]

On March 27, 1972, the report was presented to President Nixon by the Rockefeller Commission. Nixon—who had earlier been open to such ideas—realized, however, the strong opposition he would face from churches and conservative groups and, being well aware that fertility rate was already declining, he rejected the report. He emphasised that abortion was unacceptable as a means of population control and that contraceptives for teens would weaken the family structure.[58]

***

Immediately after the interview with John D. Rockefeller III in *Survival of Spaceship Earth,* clips of poor and pregnant African or African American women, newborn babies, and starving children of different ethnicities were flashed at the viewer, while some of the Rockefeller Commission's above proposals were being read out by the narrator.

Despite just having cited the commission's recommendation that laws prohibiting abortion be liberalized, that government funds be made available to cover abortion services, and that abortion be covered in health insurance, the narrator specifically points out—seemingly as a *direct address* to President Nixon—that the commission "*never* advocated abortion as a means to control population." The sequence ends with the narrator staging an ominous voice, as if this was the end of the world:

"The Commission's proposal has now been *rejected*."

Immediately thereafter, a gun is fired point blank at the viewer, followed by a clip of newborn brown baby with blood on it. New shot, new clip of the baby. This shocking imagery is repeated six times in rapid succession.[59]

***

On June 17, 1972, the day after the Stockholm Conference, the first Watergate arrests were made, which two years later would lead to President Nixon's resignation.

David Rockefeller

RBF chairman 1980–87

As an intelligence officer during World War 2, I learned that my effectiveness depended on my ability to develop a network of people with reliable information and influence. Some may feel this technique is cynical and manipulative. I disagree.

—David Rockefeller[60]

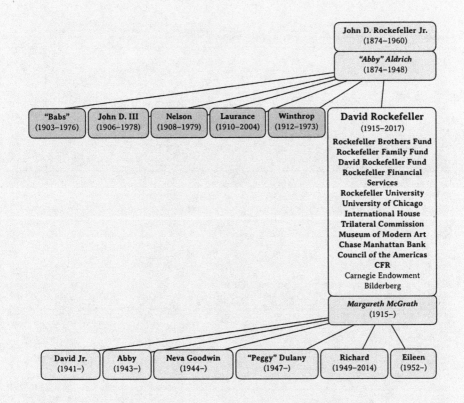

DAVID ROCKEFELLER WAS born in 1915 as the youngest son of John D. Rockefeller Jr. and Abby Aldrich. He married Margaret "Peggy" McGrath. They had six children: David Jr., Abby, Neva, Peggy, Richard, and Elaine. David was the most intellectual of the brothers and made his career in the family bank, Chase Manhattan, where he became CEO in 1960 and chairman in 1969. He founded the Council of the Americas (1963), the Trilateral Commission (1973), and the David Rockefeller Fund (1989). He was also chairman of the Rockefeller University, the Rockefeller Group, Rockefeller Financial Services, MoMA, and the Council on Foreign Relations, as well as being on the board of the University of Chicago, the Carnegie Endowment for International Peace, and the International House of New York. He was very involved in international endeavors and was part of the Bilderberg steering committee from 1964. He co-founded the RBF and became its vice president (1958) and chairman (1980). David died in his sleep on March 20, 2017, at 101 years old.

World Trade Center 1 and 2 ("David and Nelson") 1973–2001.

Chapter Five

Crisis and Opportunity

Population growth, energy shortages, environmental pollution, food scarcities, and the possibility of climate change now raise basic questions about the capacity of the planet to sustain a qualitative life for its inhabitants. How these issues, which are fuelling the competition for natural resources and also are posing new trade and balance of payment problems, are handled will have a decisive influence on the future of world order.
—Rockefeller Foundation, *1974 Annual Report*

POLITICAL BREAKTHROUGH

Soon opportunities would arise to realize the Rockefeller brothers' ambitions for the world. Nelson was getting closer to the ultimate political power, while his youngest brother David worked diligently in the background. During the next few years, world events would take a dramatic turn. This was also true for the family itself, with a power struggle on the horizon.

European Management Forum

In 1971, Professor Klaus Schwab (1938–) founded the European Management Forum (EMF), which in 1987 was renamed the World Economic Forum. Focus areas from the start were "future challenges" and "corporate strategy and structure."

In January 1973, the third annual European Management Forum symposium was held in Davos, Switzerland, sponsored by Prince Bernhard of the Netherlands and the European Commission and gathering participants from European transnational corporations such as Shell Oil, Unilever, and Philips to discuss closer cooperation between leading corporations in Europe and the rest of the world. At this meeting, participants agreed on ethical guidelines for how to solve problems at a global level, resulting in *The Davos Manifesto*.[1] Club of Rome chairman Aurelio Peccei held a speech summarizing the conclusions from *Limits to Growth* (1972). The world was in need of an effective management in order to handle its environment and resources (the crisis described by Peccei would soon arrive).[2]

Within a few decades, the European Management Forum/World Economic Forum would evolve into one of the world's leading global fora, with the power to influence political agendas and reshape the world (see chapter 11).

In 1980, the European Management Forum instituted a prize to mark the Forum's first decade. Henry Kissinger—who had been Schwab's supervisor when studying for his Master of Public Administration at the John F. Kennedy School of Government at Harvard University—was the natural choice for the prize, "for his achievements in strengthening international cooperation."[3]

World Trade Center

Meanwhile, on the other side of the Atlantic Ocean, the efforts to strengthen international cooperation intensified.

In April 1973, the World Trade Center in downtown Manhattan was inaugurated. It was a symbol to the emerging globalization and the growing dominance of the United States in the world arena.

In 1960 David Rockefeller had proposed that a world trade and financial center be built in New York. He managed to persuade the Port Authority of New York and New Jersey to build the iconic twin towers World Trade Center 1 and 2.

The building project was initiated in 1968, after David's brother Nelson, governor of New York, had secured the necessary permits.

The whole complex would take nearly three decades to finish and the twin towers (dubbed "David and Nelson" by the press) became the highest skyscrapers in the world at that time. They included offices for over fifty thousand people and used as much electricity as a city of four hundred thousand inhabitants, with air conditioning alone equating to the energy consumption of all fridges in a city of one million, David bragged in his autobiography.[4]

The Trilateral Commission

A few months later, in July 1973, David Rockefeller and Harvard professor Zbigniew Brzezinski founded the Trilateral Commission (TriCom), with Carroll L. Wilson (MIT and Atomic Energy Commission), and Gianni Agnelli (Fiat and Club of Rome).[5] Henry Kissinger was also a founding member and became a member of the executive committee after leaving his position as secretary of state in early 1977.

TriCom was created from David Rockefeller's vision of expanding the transatlantic cooperation of another private elite club, the Bilderberg Group, to regions beyond Europe and North America. The Pacific-Asian became the third region, represented initially only by Japan but later other Eastern nations were also invited.[6] David Rockefeller became president of the North American division, while the European division was led by Club of Rome member Max Kohnstamm.

Initial funding came from David Rockefeller personally as well as the Ford Foundation, Lilly Endowment, and from the Kettering Foundation.[7] The Trilateral Commission would later be funded by the RBF, the Rockefeller Foundation, and other philanthropies and corporations such as the William & Flora Hewlett Foundation, Andrew W. Mellon Foundation, David and Lucile Packard Foundation, Exxon Corporation, Mobil, IBM, and Cargill.

The goal of the Trilateral Commission was to create consensus around the solutions to the problems identified in its analyses, in areas such as international trade, the environment, law enforcement, population control, and foreign aid. Inspiration came from Henry Kissinger's and Nelson Rockefeller's Special Studies Project.

The philosophy was influenced by Brzezinski's 1970 book, *Between Two Ages*, which described the emergence of a society undergoing a transformation towards a unified global community ruled by an "enlightened" technocratic elite:

> The international equivalents of our domestic needs are similar: the
> gradual shaping of a community of the developed nations would
> be a realistic expression of our emerging global consciousness; con-
> centration on disseminating scientific and technological knowledge
> would reflect a more functional approach to man's problems, empha-
> sizing ecology rather than ideology; both the foregoing would help
> to encourage the spreading of a more personalized rational human-
> ist world outlook that would gradually replace the institutionalized
> religious, ideological, and intensely national perspectives that have
> dominated modern history.[8]

This ideology was clearly linked to the ideals of the Club of Rome, and to those of Brzezinski's inspiration, Pierre Teilhard de Chardin.[9]

According to David Rockefeller, the sharpest minds in the world should be given the authority to manage its future problems. TriCom members were handpicked by David from top political and financial circles of the US, Europe, and Japan.[1] TriCom was offered as a private forum to meet, exchange ideas—and be influenced.[10] Soon, TriCom members would occupy leading positions both in the US presidential administration and in the European Economic Community (EEC), including presidents of the European Commission such as Roy Jenkins and Jacques Delors.[11]

In the coming decades, both the Trilateral Commission and the Club of Rome would play key roles in establishing Rockefeller's transnational climate agenda and TriCom's reports would successfully come to affect the direction of world politics.

---

1   Members included board members of Chase Manhattan Bank and its International Advisory
     Group, including Agnelli, Kissinger, John Loudon (Shell Oil and WWF), William A. Hewitt
     (John Deere), David Packard (Hewlett Packard), Henry Ford II (Ford Foundation), and
     Robert O. Anderson (ARCO and Aspen Institute).

Origins of the Chinese Wonder

Meanwhile, the foundation was laid for China's emergence as an economic superpower. The Trilateral Commission and David Rockefeller played a substantial part in this, too. David had a special gift for making valuable contacts all over the world and, according to friends and collegues, for persuading almost anyone to cooperate with his plans.

> D. R. made you feel important. He would listen to your stories, and what is more important is that he would remember your stories. (John M. Foregach, Yale University)
>
> David can put anyone at ease and work his way into their kind of sphere of confidence and comfort and he uses that ability to sort of then bring people together under kind of his flag or umbrella of safety. (Richard Parsons, Chairman and CEO, Time Warner Inc.)[12]

In 1970, at a press conference in Singapore for Chase Manhattan Bank, David Rockefeller mentioned that it would be a logical next step to establish some form of contact with the People's Republic of China. In his

President Nixon and Chinese Prime Minister Zhou Enlai toasting during Nixon's 1972 visit to China.

autobiography, he wrote that he had a feeling that the Chinese leaders took note of this remark. The following year, Henry Kissinger went to Beijing for secret negotiations with the regime. This opened for a state visit by Nixon to Beijing in 1972.

The meeting resulted in a gradual thawing of the frosty relations between the two countries.[13] In June 1973, David Rockefeller himself, with his wife Peggy and representatives of Chase Manhattan Bank, also visited China.

He got to meet Prime Minister Zhou Enlai and secured an agreement making Chase Manhattan the first American correspondent bank of the Bank of China after the Communist takeover. In a *New York Times* article in August 1973, David described his impressions of China, writing that he was impressed by "the sense of national harmony." Despite the oppression and the bloody cultural revolution (1966–79), he praised Chairman Mao's leadership and social experiment as "one of the most important and successful in human history" and "[w]hatever the price of the Chinese Revolution, it has obviously succeeded not only in producing more efficient and dedicated administration, but also in fostering high morale and community of purpose."[14]

However, thirty years later he wrote that it was a shock to learn about the crimes against humanity that were subsequently revealed.

David would return another five times for meetings with the Chinese leaders. This included meetings where he represented as chairman of the Council on Foreign Relations and which were hosted by the Chinese People's Institute for International Affairs (PIFA). The People's Republic of China was now ready for dramatic changes.

The Bank of China opened an account in the Chase headquarters in New York, while Chase arranged a China Forum to which more than two hundred American corporations were invited. Chinese business-man Rong Yiren, chairman of China International Trust and Investment Corporation (CITIC) became David Rockefeller's most important ally in opening China up to foreign investments.

A meeting was also set up in 1982 between the Trilateral Commission and China's new leadership (Deng Xiaoping) to consolidate the eco-nomic collaboration (later, Chinese members would be included in the organization). David Rockefeller and Chase Manhattan Bank thus became China's gateway into the United States—and vice versa.

In his autobiography, he wrote that "the door to China had swung open, and Chase was waiting on the other side . . . to walk through it.[15]

In 1979, the US established full diplomatic relations with China. That same year, the Rockefeller Foundation also established its presence in China. Soon thereafter, the one-child policy was adopted.

David had opened the door to a China which, during the coming decades, would evolve into a technocratic model state for the rest of the world—a fusion between socialist central planning and capitalist efficiency and a center for the world's manufacturing industry.

Triggering Event: The 1973 Oil Crisis

The Trilateral Commission appeared just as the first oil crisis ground the world to a halt. The concerns raised in previous years now seemed to come true and the world got a taste of the untenability of relying so heavily on oil. The public was alarmed, even if the crisis was political and had nothing to do with a real shortage.

While President Nixon was preoccupied with the Watergate scandal, the diplomat, Trilateral, national security advisor, and newly appointed secretary of state, Henry Kissinger, was deeply involved in the events leading up to the Yom Kippur War (October 6–23, 1973) between Israel and a coalition of Arab countries. This resulted in an oil embargo towards the West from the Arab world, causing an immediate oil shortage at the consumer end and crippling world economy. The price of oil went up by a staggering 400 percent.

Sheik Ahmed Zaki Yamani, oil minister of Saudi Arabia at the time, and officially in charge of coordinating the oil embargo of the Arab nations, revealed in an astonishing interview in *The Guardian*, nearly three decades later, that *the Americans,* with Kissinger as its master architect, was behind the price hike.

> I am 100 per cent sure that the Americans were behind the increase in the price of oil. The oil companies were in real trouble at that time,

they had borrowed a lot of money and they needed a high oil price to save them. . . . King Faisal sent me to the Shah of Iran, who said: 'Why are you against the increase in the price of oil? That is what they want? Ask Henry Kissinger—he is the one who wants a higher price.'[16]

The question of world energy resources had previously been discussed at Aspen's second international environmental workshop The Environment, Energy, and Institutional Structures, organized with IIED in 1972. The idea came from oil magnate Robert O. Anderson, and was funded by the Rockefeller Brothers Fund.[17] Representatives from the Conservation Foundation were also involved. The outcome of the workshop was that an overview of global energy-related problems was developed. It also produced a report, *World Energy, the Environment, and Political Action* (1973).[18]

The consequences of a possible oil crisis had also been discussed at a Bilderberg meeting in Saltsjöbaden, Sweden (May 11–13), five months before the real "crisis." The meeting included political advisor Zbigniew Brzezinski, Robert O. Anderson (ARCO), Gianni Agnelli (Fiat/Club of Rome), Carroll L. Wilson, and representatives of Exxon, BP, and Royal Dutch Shell.

Subjects discussed at this Bilderberg meeting included what would happen to the world economy if oil prices were to increase by 400 percent.[19]

Critical Choices for Americans

In December 1973, Nelson left his position as governor of New York State to chair the Nixon Administration's Commission on Critical Choices for Americans, in response to the identified crises in energy, population, ecology, etc. The project—which had similarities with Nelson's earlier Special Studies Project—was to investigate the rapid changes that were taking place in American society, as well as the role of the US on the world stage.

Just as for the Special Studies Project, Nelson and Kissinger gathered an impressive list of top brains for the project, including future US president Gerald Ford, Carroll L. Wilson, Edward Teller, George Schultz (treasury secretary, 1972–74; secretary of state, 1982–89), Robert O. Anderson, and Laurance Rockefeller. The six panels were:

- Panel I—Energy and Its Relationship to Ecology: Economics and World Stability
- Panel II—Food, Health, World Population, and Quality of Life
- Panel III—Raw Materials, Industrial Development, Capital Formation, Employment, and World Trade
- Panel IV—International Trade and Monetary Systems, Inflation, and the Relationships Among Differing Economics Systems
- Panel V—Change, National Security, and Peace, and
- Panel VI—Quality of Life of Individuals and Communities in the USA.

A critical article in *New York Times*—which also tried to make an estimation of the family's immense wealth and influence—voiced the suspicion that Nelson's new project was little more than another tool for his aspirations to become president.

> Once again, Nelson Rockefeller is apparently a candidate for President. And once again, as he roams across the country trying to convince the Republican faithful that he is worthy of their trust—and as he revs up the engine for what has been called his vehicle for getting the Republican nomination, the $6.5-million Commission on Critical Choices for Americans—the problem central to a candidacy is being ignored.
>
> The problem is this: The Rockefeller family wields enormous power. It controls one of the largest private fortunes in the United States and one of the country's largest banks, Chase Manhattan.[20]

Nelson had originally estimated the project cost to $20 million, which he tried to get from government funds. This was stopped by Democratic congressmen skeptical of his commission. Instead it was funded with $2 million in gifts from Nelson and his brother Laurance, supplemented by funding from the Rockefeller Foundation.[21]

The final reports, *Critical Choices for Americans*, were issued in thirteen volumes in 1976 and 1977.[22] The stage was now set for major shifts in the global economy and trade.

New International Economic Order (NIEO)

The NIEO was a set of proposals from some developing countries through the United Nations Conference on Trade and Development (UNCTAD) in the 1970s, in the wake of the oil crisis. A new economic order was called for; a revision of the international economic system in favour of developing countries, replacing the Bretton Woods system which had benefited mainly the nations that had created it (especially the United States). NIEO was said to be based on equity, sovereign equality, interdependence, and cooperation among all states.

The Declaration on the Establishment of a New International Economic Order and the Programme of Action on the Establishment of a New International Economic Order were adopted by the General Assembly at its sixth special session, on May 1, 1974.[23]

The chosen date was highly symbolic. The aim of NIEO was a redistribution of the wealth of the world and eliminating the widening gap between developed and developing countries. This was an old socialist ambition that had been pushed by the developing countries in G77 Group (formed in 1964 during the first UNCTAD conference) and supported by Social Democrats in the West.[24] The Programme of Action included:

- Transfer of technology
- Regulation and control over the activities of transnational corporations
- Charter of Economic Rights and Duties of States
- Promotion of co-operation among developing countries
- Assistance in the exercise of permanent sovereignty of states over natural resources
- Strengthening the role of the United Nations system in the field of international economic cooperation.

The purpose of NIEO was to strengthen North-South relations. It was widely debated in the seventies but never actually implemented in practice. Instead the market-oriented neoliberal economic order, with free trade, globalization, privatizations, and deregulations, would gain momentum in the coming decades, especially after the fall of the Soviet Union and the end of the Cold War.

But NIEO wasn't completely abandoned. The ambitious action plan would be intertwined with the environmental and population concerns of the time and the aspirations would later evolve into *alter-globalization*—a synthesis or middle ground between neoliberalism and anti-globalization.

The term itself would resurface forty-five years later in a UN resolution as a part of Agenda 2030, and become a stated priority for the G20 Focus Groups (see chapter 11).

Co-opting the South

North-South relations were also seen as crucial by the newly founded Trilateral Commission in their long-term vision for the world. A new economic order, based on interdependence and cooperation between developed and developing nations, was an integral part of their agenda. Several reports were published during the seventies that outlined their idea of NIEO.[25]

Brzezinski, however, was of the opinion that the aspirations from the developing nations was a major threat that could create an unstable and chaotic world. After the liberation of previous European colonies, there were also concerns within TriCom about the continued availability of raw materials, and a fear that the West could be cut off from essential supplies in cases of nationalisation, as the NIEO Declaration (§4.d) included:

> Full permanent sovereignty of every State over its natural resources and all economic activities. In order to safeguard these resources, each State is entitled to exercise effective control over them and their exploitation with means suitable to its own situation, including the right to nationalization or transfer of ownership to its nationals.[26]

The oil embargo after the oil crisis had been a wake-up call for the North. The South had to be co-opted into a cooperative endeavour.[27] Columbia professor of law, Richard Gardner (1927–2019), a US ambassador and founding member of TriCom, was the leading US negotiator at UNCTAD. In a report for the Trilateral Commission (with Gardner as main author) he wrote,

> We categorically reject not only old-fashioned colonialism but also latter-day concepts of neo-colonialism, paternalism and tutelage. All countries should be free to determine their own political, economic and social systems, free of external coercion (*A Turning Point in North-South Relations*, Trilateral Task Force Report, 1974).[28]

Gardner later admitted that the wording was a concession to the spokesmen from the developing countries, in order to "gain a hearing in the rest of the world." The interests he represented in the Trilateral Commission's network (e.g., Exxon Corporation) had no intentions whatsoever of staying out of the domestic affairs of the South.[29] In April 1974, Gardner wrote more candidly about the necessity for global planning and the need for a new order, using deceptive methods.

> In short, the 'house of world order' will have to be built from the bottom up rather than from the top down. It will look like a great 'booming, buzzing confusion,' to use William James' famous description of reality, but an end run around national sovereignty, eroding it piece by piece, will accomplish much more than the old-fashioned frontal assault. . . . The hopeful aspect of the present situation is that even as nations resist appeals for 'world government' and 'the surrender of sovereignty,' technological, economic and political interests are forcing them to establish more and more far-reaching arrangements to manage their mutual interdependence.[30]

The ever increasing global crises would in the end force reluctant nations to accept TriCom's aim of a global management of the global commons. They clearly played both sides of the chessboard.

It was therefore no surprise when US secretary of state Henry Kissinger lent his support for some of the demands for NIEO in a speech made at the Seventh Session of the UN General Assembly on September 1, 1975.[31] He knew what the demands *really* meant: a bait to get the South on the hook. The new economic system become a perfect vehicle for the Trilateral Commission's power ambitions and was merged with the environmental agenda as a way of securing the raw materials for the multinationals. This coincided with some of the aspirations of the closely linked Club of Rome (with TriCom members such as Max Kohnstamm and Carroll L. Wilson).

The ambitions of NIEO would later be mirrored in the Brandt Report, *North-South: A Program for Survival* (1980). The Brandt Commission was initiated by Robert McNamara, president of the World Bank Group and a trilateral in spirit and action.[32] Both McNamara and Fabian Society member Barbara Ward had been consulted in Gardner's *Task Force Report*.

Other leading social democrats, such as Gro Harlem Brundtland and Fabianist Roy Jenkins (president of the European Commission, 1977–81), were also recruited to join the Trilateral cause.

Reforming International Institutions

An underhanded way of securing the TriCom version of the new international economic order was to include the strongest economies of the developing world into the international system of decision making. In 1976, TriCom issued the *Reform of International Institutions* report (with Bergsten, an economist and former assistant to Henry Kissinger, as one of its authors).

The report proposed an extensive program, including recommendations for GATT, IMF, and OECD, in order to make the world "safe for interdependence." The goal was a control system for interdependent states, based on a three-tier system:

1. a small informal inner group;
2. a broader group including the larger countries;
3. formal implementation of decisions through existing or new universal institutions.[33]

C. Fred Bergsten suggested that emerging economies like Saudi Arabia, Iran, Brazil, and Mexico should be brought into the inner circle in order to avoid "outsiders" from "disrupting the system."[34] The remaining developing nations would thereby be excluded. This concept was actualized in 1976 through the G7 Group and its global institutional partners.

The founders of G7 were deeply anchored in the Trilateral Commission (both Brzezinski and David Rockefeller took credit for its inception).[35] These links have remained.[36] In the following decades, the G7/G8 group (and later the G20 group) would evolve into a proto-world government, where many important global decisions and action plans have originated (see chapters 9, 10, and 11).[2]

The "G" Groups

- G6, the Group of Six (formed in 1974 and first convening in 1975) was an unofficial forum for the heads of the richest industrialised countries: France, West Germany, Italy, Japan, United Kingdom, and United States.
- G7, the Group of Seven (first convening in 1976) had grown into an international intergovernmental economic organization and included France, Germany, Italy, Japan, United Kingdom, United States, and Canada.
- G8, the Group of Eight (1998–2014) included G7 and Russia until the invasion of Crimea.
- G20, the Group of Twenty (founded in 1999), is a wider international forum for governments and central bank governors from the world's leading economies: Argentina, Australia, Brazil, Canada, China, the European Union, France, Germany, India, Indonesia, Italy, Japan, Mexico, Russia, Saudi Arabia, South Africa, South Korea, Turkey, United Kingdom, and United States plus invited international organizations and focus groups (see appendix C).

---

2 The Trilateral Commission usually has its annual summit before the G7 summit (see appendix C) and around one-third of the sherpas (the personal representatives of the heads of state who prepare the annual summits) have been members of TriCom.

The Debt Trap

Before the oil crisis, many of the developing nations had high growth rates which increased their spending and imports (especially Argentina, Brazil, and Mexico). They were slowly starting to catching up with the economies of the West. The oil crisis price shock, however, forced them to borrow money in an effort to sustain their economies. They turned to commercial banks and private lenders to obtain funds for their increased payments, while the International Bank for Reconstruction and Development (part of the World Bank Group) and IMF offered loans for infrastructure projects that would "help" their economies. That initiated a debt bubble that would be growing during the coming decade, resulting in a severe debt crisis as higher interest rates made it harder to make payments on the loans.[37] This would in the end make the South more willing to accept the demands for structural reforms from the moneylenders and open them up to the borderless neoliberal world order, such as:

- cutbacks in government expenditures, especially in social spending;
- rollback or containment of wages;
- privatization of state enterprises and deregulation of the economy;
- elimination or reduction of protection for the domestic market and fewer restrictions on the operations of foreign investors;
- successive devaluations of the local currency in the name of achieving export competitiveness.[38]

The reform program was later named the Washington Consensus, consisting of ten neoliberal recommendations from the Washington-based World Bank Group, the International Monetary Fund, and the US Treasury Department, by John Williamson, an economist from the Peterson Institute for International Economics (founded in 1981 by C. Fred Bergsten, who co-wrote the TriCom reform program for international institutions).

World Bank president Robert McNamara introduced the structural adjustment lending before he finished his term in 1981 and became a member of the Trilateral Commission. He was succeeded by original TriCom member Alden W. Clausen, former president of Bank of America, who continued a massive scaling up of the policies through the eighties before fellow Trilateral Barber B. Conable took over as president of the World Bank.

TriCom skilfully used the aims of social democratic internationalism to benefit corporate economic globalization. These two ostensibly contrary political discourses would later be merged to form the basis for the desired technocratic world order (NIEO) where the nation states would be made obsolete.

Power Shift in the White House

After Spiro Agnew had to resign for tax evasion on October 10, 1973, Gerald Ford had been appointed as Nixon's vice president. Nine months later, August 9, 1974, Nixon also had to resign. To fill the vacancy, Ford appointed Nelson Rockefeller as his new VP.

Both appointments were made possible through the 25th Amendment to the US Constitution, adopted in 1967.[39] When President Kennedy was assassinated in 1963, his vice president, Lyndon B. Johnson, was prevented by the constitution to appoint a successor because the VP must be nominated at an official nominating convention and be elected together with the president in the presidential election. In order to avoid such a situation occurring again, Nelson Rockefeller had worked hard to bring about a change in the constitution that would permit the incoming president to appoint a new vice president himself.[40]

This new amendment now came in very timely for Nelson—whose highest ambition was to become president—and the nation found itself in the unique situation of being governed by two non-elected men. Nelson was finally only a heartbeat from the ultimate power in the United States. By his side he had his faithful henchman, Henry Kissinger, as secretary of state. The Rockefeller family practically owned the White House. With Nixon out of the way, the Rockefeller population and environment agenda could now be taken to the highest political level.

Only two days after Nelson became vice president, Zbigniew Brzezinski, executive director of the Trilateral Commission, wrote to Nelson and recommended that Kissinger, as secretary of state, continue to handle diplomatic and power relations while Ford focused on domestic problems.

> I believe you can make the most singular and vital contribution in the area of focusing attention on, and developing required policy responses to, the emerging and increasingly urgent global problems, most of which do not fit traditional bureaucratic patterns or jurisdiction.[41]

These happened to be the same "urgent global problems" which had been highlighted by Nelson's Commission on Critical Choices for Americans.

Gerald Ford, who in the previous year had been a panelist in Nelson's commission, did approve Nelson's (Brzezinski's) suggestion to focus on domestic policy, and to have domestic policymakers report to the president

Henry Kissinger and Nelson Rockefeller at the center of power.

via the vice president. However, even though Nelson initially enjoyed some leeway in both domestic and foreign politics, he would soon find himself blocked, especially by the Chief of Staff Donald Rumsfeld, by the Senate, and by some of the presidential advisors, from getting the amount of influence he had aimed for.

Nelson complained that all he got to do was go to funerals and earthquakes and that the only real decision he was allowed to make was redesigning the vice presidency seal. (He didn't like the old one of an eagle with drooping wings and a single arrow in its claws. He thought it looked like a "wounded quail," and replaced it with a spread-winged eagle clutching multiple arrows.) Alas, Nelson never got to live up to the grandeur of his new seal. He was still only number two—something which he found hard to tolerate as he was "just not built for standby equipment."[42]

As the president of the US Senate, Nelson did, however, manage to get the crucial North American Free Trade Agreement (NAFTA)—which formed the basis of the neoliberal globalization process—"fast-tracked" through the Senate through less-than-democratic means.[43] The democratic model didn't suit the Rockefellers brothers' ambitions for the world. A new way of governing was required in order to realise their utopian vision of the future.

Aspen Workshop and Climate Consultation

In preparation for the third United Nations World Population Conference in August 1974 in Bucharest, Romania, a summer workshop was held at the Aspen Institute in 1973. Topics debated at the workshop included the planet's carrying capacity in relation to climate change, toxic substances, energy, soils and water, and how social and ecological problems could be controlled.[44] The workshop resulted in the report *World Population and a Global Emergency* by Thomas W. Wilson, Jr., with persuasive arguments and action-oriented guidelines for curtailing population growth.[45]

When attending this workshop, Maurice Strong, in a conversation with Joseph E. Slater (CEO of Aspen) and Robert O. Anderson (board member and funder), had expressed a wish to meet leading scientists in a relaxed environment and get more detailed information about the planetary boundaries, human impact, and triggering events that could cause irreversible damage. Inspired by this request, the Aspen Institute, under the supervision of Walter Orr Roberts, organised an experimental "consultative education" in August 1973, where Strong and his recent successor as director of UNEP, Mostafa Tolba, received private tuition from nineteen leading experts in various fields, including Nobel laureate Sir Peter Medawar, René Dubos, Carroll L. Wilson (MIT), B. R. Seshachar from India, Fereydoun Hoveyada (Iran's ambassador to the UN) and the German meteorologist and climate pioneer Hermann Flohn.

Walter Orr Roberts (1915–1990), American astronomer, atmospheric physicist, teacher, and philanthropist.

This unique private consultation resulted in a number of recommendations for UNEP's continued operations.[46] It would also turn out very useful in the 1980s and in the establishment of the IPCC in 1988.

The Aspen Institute Climate Program

In 1974, Aspen also initiated the Project on Food, Climate, and Environment, headed by Walter Orr Roberts (who had been supported by Laurance Rockefeller from early on in his career).

This was part of a larger international project, The Impact on Man of Climate Changes, drafted at the Meteorological Institute in Bonn (where Hermann Flohn worked) and was conducted in collaboration with Club of Rome–related IFIAS in Stockholm, the National Center for Atmospheric Research (NCAR) in Boulder, Colorado, and included sub-projects in Japan and the Soviet Union.[47] The "international" project was, however, primarily an American project sponsored by the Rockefeller Foundation, RBF, UNEP, Lilly Endowment, and John Deere & Co. (led by the Trilateral William A. Hewitt).[48]

The project resulted in a number of studies on the interaction between climate change and food production from different parts of the world.

Droughts had destroyed crops in the USSR and the US Midwest; the United States had suffered an unusually cold winter in 1971–72; the fishery had collapsed in Peru as result of an El Niño–related event; and Sahel in Africa had been threatened by famine due to persistent drought. The price of grain had soared. Taken together, these anomalies gave the impression of a global problem. These studies were then used as indication that climatic anomalies were occurring and increased public awareness that this might be a problem.

In a speech to the UN General Assembly (published in the *New York Times* on April 16, 1974) the US secretary of state, Henry Kissinger, called for better ICSU/WMO research on climatic disasters, and indicated a willingness from the US to lead this research.[49]

After a meeting at the Rockefeller Foundation in August 1974, with experts such as Walter Orr Roberts, Stephen Schneider, and Lester Brown, an international coordina-tion of climate prognoses were rec-

Party Leader about Sweden in 2000: "Climate change is the biggest threat," *Svenska Dagbladet*, November 1974.

ommended in order to manage the threat of climate change.[50]

In November 1974, Olof Palme, prime minister of Sweden, predicted that by the end of the century, climate change would be the biggest threat. As a close friend of Bert Bolin's (later chairman of the IPCC), he was well informed.[51]

The Rockefeller Foundation also predicted that the handling of crises such as population growth, energy shortages, environmental pol-lution, food scarcities, competition for natural resources, and the possi-bility of climate change would have "a decisive influence on the future of world order."[52] RF funded both IFIAS and the Climate Research Unit (CRU) at East Anglia, Norwich, England. The following year, CRU and WMO organized the International Symposium on Long-term Climate

Fluctuations, which lent more support to the theory of carbon dioxide as the main cause of rising temperatures.[53] Avenues were now open for anchoring this theory more firmly within the scientific community.

RBF and Worldwatch Institute

In 1974, the Rockefeller Brothers Fund launched its environmental program. It was a natural progression from the basic focus on population and conservation issues in the previous decades.

Together with the Rockefeller Foundation, RBF also supported the creation of the Worldwatch Institute, founded by Lester Brown (member of the CFR and World Future Society) to develop a warning system for global problems and social trends.[54] It was based on the same futuristic ideology as the Club of Rome. (In 1984, the institute would start publishing their famous annual report, *State of the World*, where crucial issues such as global warming were identified.)

Meanwhile, funds were being channeled to the interest group Union of Concerned Scientists at MIT for spreading public awareness of environmental crises caused by the technological development. This group, too, would help spread climate change awareness in the following decades.

The RBF's *1974 Annual Report* discussed how computer modelling could be used to research problems and propose alternative solutions for policy makers. Co-author of *Limits to Growth*, Donella H. Meadows from Dartmouth College, received a scholarship to further investigate this option.[55]

The RBF, headed by Laurance and David, made careful preparations before the crucial climate decisions in the 1980s. The RBF board now included David's children, David Jr. and Neva, as well as their cousins Abby Milton O'Neill and Hope Aldrich Rockefeller Spencer, who would later carry the family legacy forward.

The 1975 Conference Crisis and Opportunity

In June 1975, the World Future Society (founded in 1966, see chapter 4) hosted the milestone conference, The Next 25 Years: Crisis and Opportunity.

The conference was organized (on commission from Edward Cornish) by Graham T. T. Molitor (1934–2017), lawyer and expert on forecasting who had worked as policy researcher for Nelson Rockefeller during his 1964 and 1968 presidential campaigns and became research director for the White House in 1973. It was thus no surprise that US Vice President Nelson Rockefeller was chosen as opening speaker and author of the foreword to the conference book in which he set the tone of the conference, with references to his Commission on the Critical Choices for Americans:

> What are the prospects for mankind—optimistic or pessimistic? Are the challenges to be met as crises or opportunities? Now is the time for thinking through our most difficult problems.
>
> Throughout the years, the future of America has been a great concern and a constant challenge for me personally. Two years ago, I organized the Commission on the Critical Choices for Americans, composed of a group of distinguished leaders grappling with many of the questions raised at the second general assembly. The need is urgent to focus strenuous efforts on devising actions which will enable us to meet the economic, political, and social challenges in the years ahead.[56]

Other speakers included congressmen such as Edward Kennedy and Hubert Humphrey. The conference drew around two thousand people and was a great success for the society. During the conference a number of renowned futurists discussed how the world could be united under a common project. If the perception of "a world in crisis" was more widely accepted, it would provide opportunities for creating a global civilisation with a unified global consciousness and global governance. (Climate change was later identified as the crisis best suited to motivate the general public to agree to the changes desired, and New Age as a means for rallying the masses.)

The conference and the conference book outlined a direction for how the ideas of the Futurist movement could be promoted. It was a vision that would lead to the Great Transformation—a futuristic utopian vision of creating a perfectly ordered world system, seasoned with equal doses spirituality and environmentalism. Topics discussed had close connections to the concerns summed up in the Club of Rome report *Limits to Growth*. As always, population was a key issue. Drastic means of attenuating population growth were discussed, as well as the methods for altering man and the planet on a fundamental level.

Graham Molitor's theories on how policy issues are developed and implemented had a profound influence on the strategy drawn up for achieving the futuristic goals (see appendix B). He stated that reality lay somewhere between crisis and opportunity, and that, "[e]ven though we may not be doing so bad now, the point is that we can do it better. In short, mankind seeks and strives for PERFECTION."

The Molitor Model
1. LEADING EVENTS, so alarming that rectification by public or private policy is required.
2. LEADING AUTHORITIES/ADVOCATES enter to champion causes.
3. LEADING LITERATURE provides written analysis, rationale and help spread new ideas and concepts.
4. LEADING ORGANIZATIONS enter the fray and provide an institutional base from which the cause can be pursued.
5. LEADING POLITICAL JURISDICTION implements new political solutions.

The fear of the great catastrophe would open the door to fulfilment of the envisioned Utopia. Molitor concluded that the planning of the future would need a new technique of social navigation. He answered to his own call and outlined a forecasting scheme on how things eventually would progress and how to influence the desired road to the future. To Molitor, who would later become vice-president of The World Future Society, it was a choice between chaos or calculation.

Through a better understanding of change, wiser alternatives can be selected.[57]

The other conference attendees, a motley crew of futurists, global planners and spiritual leaders, including several members from the Club of Rome, were already agents in Molitor's model for the decades to come. They had been well prepared after the LEADING EVENT (the 1973 energy crisis) and would now help with LEADING ADVOCACY, writing LEADING LITERATURE, and founding of LEADING ORGANIZATIONS, in order to achieve LEADING POLITICAL JURISDICTION, while Nelson, the Rockefeller family, and their billionaire allies, were standing ready to offer their assistance.

Ervin László

Of significant importance for the coming agenda was Ervin László, a Hungarian pianist, systems philosopher, and project leader of the Club of Rome project Goals for Global Society, which resulted in the book *Goals for Mankind* (1977).[58] The project was presented at the conference with its goal to "raise the problem with inner limits and their paramount role in deciding the future of mankind."[59] László, a special fellow of the United Nations Project on the Future, would in the following years serve as a director of the NIEO project.[60]

So what was László's desired future? He had been editor of the book *Cosmic Humanism and World Unity* (1975), written by professor of philosophy Oliver L. Reiser (1895–1974) and published by the UN think tank World Institute (founded by lumber magnate Julius Stulman, a major financier and backer of World Future Society). Reiser's book can be seen as the blueprint for the goals of the futurist movement. It was strongly influenced by Pierre Teilhard de Chardin's ideas on the Omega Point and the noosphere, evolutionary humanism, and concepts derived from theosophy and freemasonry.[61]

Reiser outlined a grandiose plan for transforming the world and creating a "Cosmic Wisdom Temple" (a world government) with a common religion where mankind would be integrated into the technological system. He would later, together with José Argüelles, become instrumental in advocating these techno-spiritual transhumanist/eugenicist ideas where man was to be upgraded and improved to fit into the new great World Organism.[62]

Barbara Marx Hubbard

Another important attendee was futurist Barbara Marx Hubbard (1929–2019). She was a board member and early financier of the World Future Society and has been called "the mother of the futurist movement" with her ideas of conquering space and promoting conscious evolution.

During the conference she held a mini seminar called SYNCON, using a unique method designed to bring together opposing groups and gradually working towards a "synergistic convergence" in their thinking, exploring ways to build "new worlds on Earth," "new worlds in space," and "new worlds in the human mind." As a part of her engagement as a Soviet American citizen diplomat, such seminars were later held in Washington, DC, and Moscow. The Iron Curtain didn't stop her ideas from spreading across the globe.[63]

Barbara Marx Hubbard, who credited Pierre Teilhard de Chardin with her spiritual awakening, would later describe the emergence of a new species, *Homo universalis*, through the development of new technology and human control of evolution.[64] Perhaps her background as the wealthy heiress to the "Toy King of America," Louis Marx, made her believe in human possibilities beyond what most people would regard as realistic.

In the decades following the conference, László and Hubbard would come to play essential roles in promoting these futuristic ideas to a growing New Age audience, with the threat of catastrophic climate change as a motivating force. Laurance Rockefeller would become a crucial ally in this endeavour.

Laurance and the RBF had been actively promoting the growth of the emerging New Age movement and supported organizations like the Esalen Institute (founded in 1962), the Lindisfarne Association[65] (founded in 1972 by William Irving Thompson), and Planetary Citizens (founded in 1974), and was now ready to support the new spiritual movement, with "lightbringers" such as Barbara Marx Hubbard.

Through virologist Jonas Salk, Barbara also came to meet other "evolutionary souls" who wanted to help humanity to the next level in evolution.

Congressional Clearinghouse on the Future

In 1976, Annie Cheatham, with Senator Charlie Rose and members of the World Future Society, founded the futuristic think tank Congressional Clearinghouse on the Future and became its first president.[66] A young Al Gore joined shortly after he had been elected to the US Congress in 1976.

The clearinghouse was a bipartisan legislative service organisation (LSO), providing members of Congress with foresight into long-term challenges in the legislative process by predicting the future and offering policy solutions to create the desired development.

Cultural anthropologist Margareat Mead with New York mayor John Lindsay (photo by Staffan Wennberg).

These ideas came from the World Future Society (WFS). WFS chairman Ed Cornish and his wife, Sally, served as advisors to the clearinghouse.

Futurists such as Jay Forrester, Dr. Margaret Mead, Barbara Marx Hubbard, and Alvin Toffler were invited for dinner talks.[67] Their goal was a global shift from an industrial to a post-industrial society, with a common global consciousness interlinked via information technology (like the internet, which did not yet exist).

Toffler thought the future would require new political solutions and replacing nation states with large federations such as the European Union and international organizations such as the United Nations. Toffler called this "The Third Wave."[68]

Toffler's ideas had many similarities with those of science fiction writer H. G. Wells (Fabian Society), Pierre Teilhard de Chardin, economist Kenneth Boulding, and philosophy professor Oliver Reiser.

In a letter to the organization's advisory committee, dated March 10, 1983, clearinghouse president Robert Edgar listed nine proposals from a recent meeting with a group of futurists, including:

1. Ask Ted Turner to offer time on his network for futures-oriented subjects and have members of the clearinghouse involved.

2. Identify issues which are important to the present decisions of members of Congress but which have long-range implications and bring these issues to the attention of Congressional staff.

3. Develop strategies for existing futures-oriented legislation and supply witnesses for the hearings. Add a futures component to each hearing.

The group of futurists also identified issues believed to become of growing significance to the nation, covering areas such as education; retraining the work force; national security; biotechnology; demographic shifts; environmental problems; the use of space; interdependence; creating "future thinking" institutions to alleviate crisis management; and economic issues, including the New International Economic Order (NIEO).[69]

Congressional Institute of the Future

In 1978, Al Gore, Newt Gingrich, and John Heinz founded the Congressional Institute of the Future, a research institute under the clearinghouse. Funding came from Siemens, the W. Alton Jones Foundation (Cities Service Company), C. S. Mott Foundation (General Motors), Merck & Co, IBM, and Carnegie Corporation.[70] Institute members later became pivotal in anchoring the climate issue politically in the United States. In 1989 the institute also founded GLOBE for influencing legislation on a global level (see chapter 6).

The Rockefellers' Unfinished Agenda

In 1976, the Rockefeller Brothers Fund launched the Environmental Agenda Project, in cooperation with environmental organizations under Rockefeller patronage and funding (including the National Resource Defense Council, Friends of the Earth, the Environmental Defense Fund, the Conservation Foundation, and the Sierra Club). The project was to result in constructive suggestions on how to solve environmental problems identified by the environmental NGOs as most urgent to tackle during the coming decade. The report, *The Unfinished Agenda* (1977), included seventy-five recommendations covering ten urgent areas (including population, food and agriculture, the energy economy, water and air pollution).

As always, population was a top priority. Suggestions for domestic policy for stabilization or reduction of population included contraceptives, abortions, sterilizations; daycare; tax reliefs for singles and childless couples and higher taxes for those with more than three children. For developing nations, suggestions included paramedical personnel providing contraceptives, abortions, and sterilizations; and foreign-aid measures with "an indirect negative effect on fertility" such as education and employment for women.[71]

These recommendations were very similar to those presented by the Rockefeller Commission to President Nixon in 1972—which got rejected by the president. This may explain the title.

Similar proposals would return a third time in the 1980 *Global 2000* report. The solutions also closely resembled those emerging from the 1975 World Future Society conference. Conference participants Willis Harman, Lester Brown, and Jay Forrester had been involved in this project, as well as the consultants Donella Meadows and Carroll L. Wilson.

Assassination Attempts on President Ford

Nelson Rockefeller, who had become vice president in the Ford administration after Nixon's resignation, failed to reach his life goal of becoming president of the United States when Gerald Ford, persuaded by Donald Rumsfeld and Dick Cheney, decided to run for office without Nelson in the 1976 elections. The family thereby lost a valuable position in the White House and left Nelson resentful and bitter.

Coincidentally, however, Nelson had been *very* close to having his childhood dream realised through two bizarre assassination attempts and a freak traffic accident.[72]

The first incident happened in Sacramento, on September 5, 1975. The perpetrator was twenty-six-year-old Lynnette "Squeaky" Fromme, a member of Charles Manson's cult, the Family. Her motive was to protest the threat of pollution to the redwood trees. Manson (by that time in prison for the 1969 Tate murders) had called her "Red" to signify both her red hair and her passion for the redwood trees. Dressed all in red, she aimed an antique .45 caliber Colt at the president when he came out after speaking at the Convention Center. Due to the quick intervention from Secret Service agents, the attempt failed. She received a life sentence but was released after twenty-four years.

A fascinating coincidence is that Nelson's father, John D. Rockefeller Jr., had been a strong advocate for the protection of the redwood forests.[73] According to David Rockefeller it was the redwood trees which inspired the family's long-lasting interest in conservation.[74] As tribute, part of the Humboldt Redwoods State Park, containing the world's largest remaining contiguous old redwoods, was named the Rockefeller Forest in 1952.

The Rockefeller Forest in Humboldt Redwood State Park.

Only seventeen days later, a left-wing radical and FBI informant, Sara Jane Moore, made another attempt. She was a forty-five-year-old book-keeper fascinated by Patty Hearst (the young heiress who was kidnapped in 1974 by the Symbionese Liberation Army and joined their militant activities). The SLA wanted to coordinate the radical left's activism for feminism, anti-capitalism, and anti-racism, and gather all races, genders, and ages to fight against the fascist capitalist class.[75] Sara's assignment was to keep an eye on SLA activities. Despite being an FBI informant, one day she was suddenly overcome by a strong impulse to start a revolution. She had somehow gotten hold of a weapon and fired one shot at the president, missed, and was promptly disarmed by an ex-marine during her second attempt. The bullet hit a bystander (who luckily survived). Sara also received a life sentence, but was released after thirty-two years.

And then, on October 14, a nineteen-year-old male accidentally smashed into President Ford's limousine when passing in a motorcade through Hartford, Connecticut. The limousine was, however, well armored and Ford was shaken but unscathed.

Another curious coincidence is that one of the revelations from Bobo Sears Rockefeller during her divorce from Winthrop in 1954 was that the brothers had used to brainstorm on how to make Nelson president without being elected.[76]

David Rockefeller wrote in his autobiography, "He [Nelson] knew what he wanted. He wanted to be President of the United States. Knew it early. And I think he felt that the family was there to help him achieve that.[77]

Family Discord

After Nelson's failure to reach his goals in politics he was forced to return to the family office at Rockefeller Center, where he found his position as the leader of the family challenged. When trying to resume control over the RBF and the family office, a conflict with his eldest brother, John D. III, and his brother's children ensued. Nelson wanted to consolidate all power to himself and considered only Laurance, David, and John's son Jay qualified to lead a reformed family office. The rest were not cut out for the task.

Several of the cousins had also tarnished the family reputation by criticising the brothers in the 1976 bestseller *The Rockefellers: An American Dynasty*.[78] On top of this they had given the authors, David Horowitz and Peter Collier, access to the family's archives without control of the end result. The book, describing the emergence of the Rockefeller family's empire, was largely based on interviews with the cousins and came to be genuinely hated by the brothers. The authors predicted a family empire that would crumble; the brothers were ambitious but had flaws, while the cousins were all too mediocre to live up to the legacy of their parents.[79] Some of the cousins had also spoken in negative terms about their "reactionary and unsympathetic" parents.[80] David saw it as "Marxist propaganda" and would work hard for the rest of his life to prove them wrong. The bitterness was still tangible in his 2002 autobiography.

In the ensuing family power struggle, Laurance felt that he had only been temping for Nelson over the last two decades and had no problem handing the reins back to his brother. David also supported Nelson but John thought Nelson had failed and that it was time for David to assume the position as family patriarch. After a few months of bitter conflict, Nelson had to admit defeat and David took over as the leader of the family.[81]

Before withdrawing from the RBF, Nelson made sure to get his and David's friend, Henry Kissinger, elected to the board of directors. The Nobel Peace Prize laureate had been one of the family's most valuable assets and Kissinger's

Nelson Rockefeller and Jimmy Carter in the White House, October 27, 1977.

loyalty paid off both financially (including a personal gift of $50,000 from Nelson in 1969)[82] and in terms of influence (in 1977 David made him chairman of Chase Manhattan Bank's International Advisory Council and board member of the Council on Foreign Relations). A decade later, when David handed the reins of the RBF over to his son and the younger generation, he would praise his old friend: "Henry's guidance, wisdom, and friendship as a member of the board were for me without equal. I am very grateful."[83]

Family conflicts subsided when the eldest brother John D. III died in a car accident in July 1978. Nelson himself expired only six months later of a heart attack in the arms of his mistress Megan Marshack.[84] A few years earlier, Winthrop (1973) and Babs (1976) had passed away from cancer.

After Nelson's death, Henry Kissinger wrote the following of his friend and mentor:

> That Nelson Rockefeller is dead is both shattering and nearly inconceivable. One thought him indestructible, so overpowering was he in his energy, warmth and his deep faith in man's inherent goodness. For twenty-five years, he had been my friend, my older brother, my inspiration and my teacher.[85]

Despite the personal conflicts within the family, the family's control over the White House remained, now primarily through David.[86] The Carter administration was dominated by members from the Trilateral Commission, and Zbigniew Brzezinski became national security advisor.

Cyrus Vance, Trilateral and earlier chairman of the Rockefeller Foundation, replaced Kissinger as secretary of state.

In 1980, David Rockefeller became chairman of the RBF, replacing Laurance. John D. Rockefeller's favorite grandchild was now in charge. The following year, David resigned as president of Chase Manhattan Bank to focus his full attention on conservation, population, and the climate. His extensive network of connections with world leaders and global corporations would come in handy. He also continued as chairman of the Trilateral Commission's North American branch, and as chairman of the International Advisory Committee at Chase Manhattan Bank.

The Global 2000 Report

In May 1980, *The Global 2000 Report to the President* was presented. The report included population trends and resource estimations for the coming two decades. It concluded that global prospects were bleak and in need of firm guidance.[87] This conclusion—which happened to be right in line with the Rockefeller family's view on global development—is hardly surprising as the head of the commission behind the report was Gerald O. Barney, study director of the RBF's *The Unfinished Agenda* (1977).[88]

Both the RBF and Rockefeller Foundation had long-reaching plans for curtailing population growth. Large sums had already been invested in proactive measures in the developing nations and in biomedical research.[89]

First Global Conference on the Future

On July 24, 1980, in Toronto, Canada, the World Future Society organized another groundbreaking futurist conference, the First Global Conference on the Future, sponsored by leading tech companies.[3] Again, Graham Molitor was on the planning committee.[90] This futurist conference was international, with around five thousand delegates from North America, Europe (including the Soviet Union), and developing countries (including China). During the opening session, two groups of futurists were set against each other:

- A neo-Malthusian group, with Aurelio Peccei from the Club of Rome and Lester Brown from Worldwatch Institute, held the view that the planet's natural resources would not be able to sustain an ever-increasing population and that a tight regulation therefore was necessary.
- A technophile group, which included Herman Kahn from Hudson Institute (Kahn had been a frequent guest of Nelson Rockefeller's, who had appreciated his ideas), was more optimistic and thought we had barely started exploiting human ingenuity and capacity for technological innovation.

---

3   Official sponsors included Bell Canada, Control Data Corporation, General Motors of Canada, IBM Canada, Imperial Oil, Kodak Canada, Petro-Canada, Royal Bank of Canada, Shell Canada, Sun Oil Company of Canada, Trizec Corporation, Xerox of Canada, and the United Nations.

After the session, however, the conference participants unanimously adopted the view of the first group. Thereafter the conference was divided into three themes (similar to the panels in Nelson Rockefeller's Commission on Critical Choices for Americans):

1. **Human** concerns (including food, human environments, health, and medicine)
2. **Global** concerns (including population, resources, environmental quality, science, and technology)
3. **Management** concerns (including "new economy," social and institutional change, and global governance).

Discussions included deforestation, pollution, extinction of animal species, depletion of natural resources, and climate change.

The world was facing massive changes and effective solutions were called for to handle the crisis of complex problems—*world problematique*—as the Club of Rome called it. There was a need for rational management and a new international order where the environment, resources, and social justice would be managed within the framework of a system built around interdependence and a network between the nations.[91]

The conference motto was "Think Globally, Act Locally" which became a popular slogan during the coming decade.

In the conference book *Through the '80s: Thinking Globally, Acting Locally*, W. Warren Wagar wrote that technocracy was the highest form of capitalism and predicted its impending implementation.

> The most likely scenario for the future involves, therefore, the welding of the governments and business communities of the major industrial powers into a single, more or less monolithic, more or less coordinated system of control that will manage the capitalist world-economy in the twenty-first century.
>
> The executives of the chief multinationals, the department heads of government ministries, and their counterparts in the nominally socialist countries will work together easily and pleasantly, speaking

the same language and pursuing the same goals. International councils and commissions, informal networks of technocrats of all kinds
will gradually erode national and even corporate authority in their
common dedication to a higher cause: The empowerment of the new
class itself.[92]

Wagar (1932–2004) was a historian, futurologist, and the vice president of the H. G. Wells Society, which was initially created to champion the political ideas proposed by Wells in *The Open Conspiracy* and
The World Brain. Nearly two decades earlier he had written, "There is
no better time to implement radical changes than after a world-wide
catastrophe."[93]

Through the '80s also included other concepts from the 1930s technocracy movement, such as "circular economy"—a concept which would
reemerge thirty years later (see chapter 11). More far-reaching technological
solutions were also presented at the conference. For the first time, transhumanist ideas were introduced, presenting the vision of a technologically
upgraded human with a brain connected to a computer (*Symbionic Minds*).[94]
The development of a technologically improved human and artificial intelligence was also discussed at the conference by later New Age guru Barbara
Marx Hubbard from the International Committee for the Future. These
ideas were thereafter presented to the general public via science fiction films
and popular science magazines. For example, in the January 1982 Swedish
edition of *Popular Science*, Lewis M. Branscomb, head of research and development at IBM stated, "The ultimate computer will be grown in a petri
dish, implanted inside the skull, and interfaced with the brain."

Chairman of the closing session was Maurice Strong from the Club of
Rome, who would later include many of the conference ideas in the UN
Sustainable Development Goals (and after his chairmanship of UNEP
become executive director of Petro-Canada oil company).[95]

Soon, the Rockefeller family's longtime goals of reducing population
in developing nations would be transformed into climate change action
and reducing fossil fuel consumption.

It was time to create "a common future" for humanity.

David Rockefeller Jr.
RBF chairman 1987–1992

*Unfortunately a minority in the country discovered how to
manipulate public opinion by sowing doubt about science*

—David Rockefeller Jr.

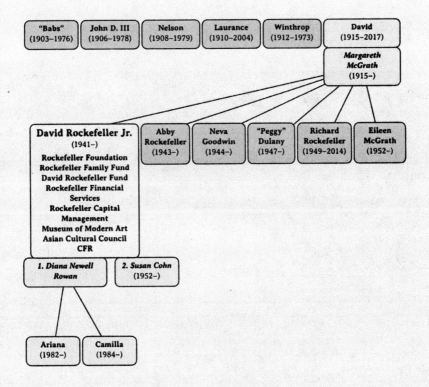

DAVID JR. WAS born in New York in 1941 as eldest child of David Rockefeller and Margaret "Peggy" McGrath. Between 1980 and 2005 he was married to Diana Newell-Rowan with whom he had the daughters Ariana and Camilla. In 2008, he married documentary filmmaker Susan Cohn. Initially, David Jr. had no interest in the traditional family businesses and devoted himself to art and music, and was active in the Boston Symphony Orchestra and was on the board of the National Endowment for the Arts. In December 1974, one year after his sister Neva, he joined the board of the RBF and was in 1982 put in charge of revising the foundation's programs. In 1988 he became chairman of the RBF. Three years later he succeeded his father as chairman of Rockefeller Financial Services. In 2004, he became board member of the Rockefeller Foundation and between 2010 and 2016 he was the third family member to serve as its chairman. He is also a member of the CFR and the Asian Cultural Foundation and a board member of MoMA. After his father's death in March 2017, he became the new family patriarch.

The planet in safe hands.

Chapter Six

One World

When the Fund began making grants to address climate change issues in 1985 it was difficult to persuade citizens and policy makers of the significance of this threat to life on the planet. Today, the knowledge base as well as perceptions of the problems that could result from global warming are dramatically different from what they were just five years ago.
—Rockefeller Brothers Fund, *1989 Annual Review*[1]

ONE FAMILY

After the family turbulence at the end of the 1970s, the Rockefellers now agreed to strive for unity by sharing a common vision about creating a more just world without racism and prejudice; eradication of poverty; improving education; and finally figuring out how humanity could survive without degrading the environment.[2]

The overarching goal was global interdependence. The family would work in unison to realise the utopian vision of sustainable development. Few, however, realised what this vision would mean when implemented in the real world.

The One World Program

From the mid-1980s, family efforts were primarily focused on getting climate change awareness onto the international political agenda. They had already laid the foundation decades earlier by founding scientific institutions and funding climate research. RBF now openly started funding organizations and scientists working to make climate change a political issue.

The RBF began grantmaking on climate change in 1984 and has consistently maintained an interest in climate change through this entire period. The RBF's work on climate change can be thought of in four phases, which we will briefly describe here.

The first phase, stretching from 1984 to 1992, focused on basic research on science and policy. Two strategies underpinned this phase of grantmaking: 1) distilling consensus on climate science and, 2) moving the discussion of climate change from the scientific community into the policy arena. (Rockefeller Brothers Fund, *Sustainable Development Program Review 2005–10*)[3]

The ambitions for what later came to be known as "sustainable development" were implemented within the framework of RBF's new One World Program, which had been developed by the leader of the next generation, David Rockefeller Jr., at request of chairman David Rockefeller.

The purpose of the project was to realize the Rockefeller family's common vision. It was initiated in 1982, and implementation started in 1983.

The planning committee included cousins Larry Rockefeller and Abby Milton O'Neill; Lucy Rockefeller's husband, Jeremy Waletzky; and board members Gerald Edelman (Nobel laureate in biology from Rockefeller University) and Peter J. Goldmark. Goldmark would later implement the ideas of the One World Program in his position as president of Rockefeller Foundation (1988–97).

Committee members were carefully chosen. David wanted the most suitable to take the lead in passing down the family legacy. There was no room for compromise or chance.

The RBF Board of Directors

In 1981, Thornton Bradshaw (ARCO oil company and Aspen Institute) was elected to the board of directors of the RBF, joined in 1984 by former chairman of the American delegation at the Stockholm Conference and friend of Laurance Rockefeller, Russell Train.

Laurance himself resigned from the RBF in 1982 and stayed on as advisor only, while delving into fringe interests such as New Age, Egyptology, and UFO research (later including Steven Greer's Disclosure Project).[4]

Some of the older cousins had already held seats on the RBF board since the end of the 1950s. The younger ones had been trained in charity work in the smaller foundation Rockefeller Family Fund (founded in 1967 by John, Laurance, Nelson, David, and Martha Rockefeller). Now they were considered experienced enough to take over the RBF and able to add their own priorities.

In addition to David Jr., who succeeded his father as president of the RBF in November 1987, the board included Nelson's son Steven, Babs's daughter Abby Milton O'Neill, Laurance's son Larry, and David's daughter Neva.

During the 1970s, John's son Jay (John D. IV) had been active in the Rockefeller Foundation, and was also invited to the Trilateral Commission in 1977, served as governor (1977–85) and senator for West Virginia (1985–2015). With the exception of Jay and Steven, the cousins kept a much lower profile than their famous fathers.

RBF's Green Profile

The One World program consolidated the environmentally oriented profile that the RBF had developed in previous years. It stated that the nations had become more economically and ecologically dependent on each other and that the problems of resource consumption, environmental degradation, and international security could only be solved through cooperation at regional or global level.[5]

These goals were shared by the Trilateral Commission (which included David Rockefeller, Henry Kissinger, and Gianni Agnelli on the TriCom Steering Committee); the world had to be united to deal with these threats from several fronts.[6] David had undertaken becoming one of the central driving forces in implementing them. Chase Manhattan's private airplane and David's well-established contacts with world leaders would now come in handy.

In 1984, Larry Rockefeller (member of the RBF board of directors and lawyer at the Natural Resource Defence Council) had wanted to strengthen organizations working with climate, acid rain, the greenhouse effect, biodiversity, population, toxins, and water—now with climate as top priority.[7]

Sir Crispin Tickell

The climate concerns had previously been raised in the European community and at the G7 meeting in 1979 by British diplomat Sir Crispin Tickell (related to Julian Huxley of UNESCO). Tickell was a sherpa for G7 and *Chef de Cabinet* from 1977 to 1981 to British Roy Jenkins, chairman of the European Union (and also a member of the Trilateral Commission).[8]

Here, too, Henry Kissinger played a part. During his time as a fellow at the Harvard Center for International Affairs in 1975–76,[1] Tickell had been commissioned to analyze how climate change could affect world politics.[9] This resulted in the book *Climatic Change and World Affairs*.[10] As advisor to Margret Thatcher, Tickell also brought it up the climate issue at the 1984 G7 Summit.[11]

Just like the Rockefeller family, Tickell was deeply engaged in the population issue and later became involved with the British organization Population Matters which advocates a drastic population reduction in order to preserve the planet and its resources.[12]

The Geneva World Climate Conferences

In 1979, 1990, and 2009, the World Meteorological Organization (WMO) organized a series of World Climate Conferences in Geneva, Switzerland. At the first of these conferences, prepared by William W. Kellogg[2] and

---

1   The Center for International Affairs was founded in 1958 by Kissinger and Robert R. Bowie with the purpose of analysing various world problems identified by its founders. Like Kissinger, Bowie was part of the inner circle of the Trilateral Commission and the Council on Foreign Relations.

2   William W. Kellogg (Aspen Institute and National Center for Atmospheric Research) was a veteran and had been part of both the SMIC report and the Rockefeller Foundation/National Academy of Sciences 1956 study, "The Biological Effects on Atomic Radiation."

Jesse Ausubel (IIASA and the Sloan Foundation), global cooperation for predicting and preventing any potential impact of human activity on the climate had been called for. The scientific foundation prepared for the conference was based to a large extent on Carroll L. Wilson's 1971 SMIC climate report.

The Charney Report

Later in 1979, together with veterans George Woodwell (from the Conservation Foundation), David Keeling, and Roger Revelle, geophysicist Gordon J. F. MacDonald[3] wrote a memorandum about the effect of carbon dioxide on the atmosphere for the Nixon administration's Council on Environmental Quality (of which MacDonald was a member (1970–72), when Russell Train from the Conservation Foundation was chairman).[13]

The Carter administration and the chairman of the Council on Environmental Quality, James Gustave Speth (from the National Resources Defense Council), responded by assigning the NAS Climate Research Board, under Jule G. Charney from MIT, to investigate it further. The NAS panel included Bert Bolin from University of Stockholm.

The resulting *Charney Report* gave additional support to MacDonald's and Woodwell's memorandum.[14] Bolin would thereafter be given one of the leading roles in anchoring the theory.

World Resources Institute

In 1982, by government mandate, James Gustave Speth (who would later become a member of the board of RBF) founded the World Resources Institute, with funding from the MacArthur Foundation. Speth gave Woodwell and later Bert Bolin and Maurice Strong places on the board.

There were close ties between the carbon dioxide theory proponents, and all were more or less linked to the Rockefeller network. The pieces were slowly but surely falling into place, with the RBF as a coordinating force.

---

3   McDonald had earlier been involved in the 1965 report *Restoring the Quality of the Environment*, written with Roger Revelle and later climate skeptic Frederick Seitz.

The 1980 Villach Climate Conference

In 1980, the year after the first World Climate Conference, a series of intergovernmental climate conferences were initiated.

The first was held in the little Austrian town of Villach, sponsored by UNEP, ICSU, and (WMO) with Bert Bolin as chairman. The goal was, according to UNEP director Mostafa Tolba, to provide nations with guidance on the climate issue. This laid the foundation for the creation of the Intergovernmental Panel on Climate Change (IPCC).

As there was still at this point much uncertainty about the scientific foundation of the CO_2 theory, it was decided at the first meeting that a thorough analysis of causes and consequences would be prepared by Bert Bolin at the International Meteorological Institute (IMI) in Stockholm.[15] The results were to be presented at the next Villach meeting in 1985.

The RBF Climate Program

In June 1985, RBF vice president Russell Philips assigned an official at the RBF, Thomas Wahman, to start looking into the research on climate change. Part of Wahman's assignment was to award one or two $100,000 donations for policy-driven research which would alert policy makers, private leaders, and the general public to the problem, as well as attracting funding from other foundations. Wahman was expected to present a proposal that would make Larry Rockefeller "enthusiastic." The climate issue was to be the flagship of Rockefeller's One World Program.[16]

In September 1985, after consulting Gus Speth and other experts, Wahman submitted the report on climate change to RBF executive officials. He stated that he had never faced a research field with such large uncertainties.

He pointed out that it would be very difficult to secure an international treaty on the regulation of carbon dioxide.

Strategies for moving the climate issue to the political forefront were suggested. This was something the RBF was very good at. Wahman suggested that they support biologists and ecologists focused on climate change, and to use the Brundtland World Commission on Environment and Development to draw up a worldwide action plan.

The idea was accepted by the Brundtland Commission's Chief Secretary Jim MacNeill and by British professor and ecologist Gordon Goodman from the Beijer Institute.[17]

The Beijer Institute was thereafter chosen by the RBF to execute the plan, together with the World Resources Institute, Woods Hole Research Center, and Environmental Defense Fund.[18] The four institutes soon became the RBF's main grantees in the areas of environment and energy. It was a very strategic plan.

The 1985 Villach Climate Conference

In October 1985, the second Villach Climate Conference was held, organized by IMI and Beijer Institute and sponsored by the ICSU, UNEP, and WMO.

At this conference, Bert Bolin's study, *The Greenhouse Effect, Climatic Change, and Ecosystems*, initiated after the first Villach Conference, was presented.[19]

Rockefeller henchman Gordon Goodman, chairman of Bolin's panel on CO_2 emissions in the atmosphere, now declared that scientists must play a more politically active role in eliciting an international response to climate change.

It was clear to Goodman that science "was at a new dawn" due to the climate issue. Despite uncertainties, he felt that the global climate was changing due to human activities and that the debate needed to be focused on how an intervention could best be handled. This meant "channeling available resources in such a way that we can understand, predict, and possibly make direct changes in the global climate for the benefit of mankind."[20]

UNEP chairman Mostafa Tolba (recipient of the private climate tuition at the Aspen Institute) declared in his speech that time was ripe for a serious discussion between politicians and industry leaders on how to lower carbon dioxide emissions. He called for an international committee that could encourage research, evaluate data, and issue action plans for governments, international organizations, and the general public.[21]

This second Villach meeting resulted in, among other things, the creation of the Advisory Group of Greenhouse Gases (AGGG), with Bert
Bolin, Kenneth Hare, and Gordon Goodman as members, and funding
from the RBF, the Rockefeller Foundation and the German Marshall
Fund.[4]

While Tolba was to approach the US presidential administration, time
had now come to issue specific policy recommendations, with the assistance of the RBF.

Wahman had also been invited to Villach but was unable to attend and
instead sent Michael Oppenheimer, chief scientist of the Environmental
Defense Fund, to report to the RBF and, together with Gordon Goodman,
develop strategies, using the Brundtland Commission as a mechanism for
drawing up a ten-year action plan for battling climate change.[22]

Goodman, Oppenheimer, and Woodwell were then assigned by
Wahman to organize an international scientific symposium, led by
Woodwell, with a policy workshop led by Goodman. The steering com-
mittee included Bolin and Hare from the AGGG, Jill Jäger from the
Beijer Institute, Carl Christian Wallén from UNEP, and W. C. Clark
from IIASA.

The planned activities were intended to result in a report to be included
in the final Brundtland Report. In December 1985, the RBF granted
$100,000 to the Brundtland Commission, but the main responsibility for
the sub-report was later transferred to the Beijer Institute.[23]

Trilateral Assistance
Heeding the advice from RBF and Goodman an increased activity to
put the climate on the political agenda now ensued, with assistance from
David Rockefeller's TriCom network.

---

4   The German Marshall Fund was founded in 1972 through a donation from West Germany
     to commemorate the twenty-fifth anniversary of the Marshall Plan. It was headed by Henry
     Kissinger's earlier doctoral student and close friend Guido Goldman from the Harvard Center
     for International Affairs; Nicholas Siegel, "The German Marshall Fund of United States: A
     Brief History," The German Marshall Fund, March 9, 2012.

In June 1986, Woodwell and Oppenheimer, with GISS/NASA scientist James Hansen, had warned about climate change at a US Senate hearing, initiated by Republican TriCom member John H. Chafee.[24]

In October 1986, Woodwell's symposium was held in collaboration with the World Resources Institute.

A month later, Karl-Heinz Narjes, vice president of the European Commission (member of the Trilateral Commission's executive committee and Council on Foreign Relations), held a meeting at the European Parliament with sixty leading climate scientists from Europe and the US to evaluate the consequences of carbon dioxide and other greenhouse gas emissions.[25]

The RBF Climate Workshops in Villach and Bellagio

The RBF went on to organize two climate workshops in Europe. The first, Management Issues Workshop, was held on September 28 to October 2, 1987, in Villach, Austria. The other, Policy Development Workshop, was held at Rockefeller Foundation's conference center in Bellagio, Italy, on November 9–13, 1987. The latter workshop included representatives from the European Commission, the Swedish government, the German Bundestag and the British Commonwealth Secretariat. It was funded by the RBF, the Rockefeller Foundation,[26] the German Marshall Fund, the W. Alton Jones Foundation (founded by oil magnate W. Alton Jones) and the governments of Sweden and Austria.[27]

The final draft, written by Jill Jäger from the Beijer Institute (with the help of Goodman, Oppenheimer, and Woodwell), included recommendations for governments to immediately rethink their energy strategies for meeting emission requirements. In order to tackle the climate challenges, an institutional change in world governance was also called for.

This proposal happened to coincide with the wishes of their powerful sponsors—and with the RBF's stated goals in *Prospect for America: the Rockefeller Panel Reports* (1961).[28] The initiative and the agenda were clearly intertwined with Rockefeller family interests. The operation had been planned and orchestrated from RBF's Room 5600 in Rockefeller Center.

The workshop was described by several participants as a way of transforming scientific facts into political truths.[29] Said Thomas Wahlman, "A little money, some perseverance, some strategic thinking and planning coupled with a perception of the probable" could "get some things done on the world stage."[30]

The Bellagio conclusions would then serve as basis for the upcoming climate conference in Toronto the following year.

The Brundtland Report

The World Commission on Environment and Development project was initiated by UNEP, following a UN decision in 1983. The mission was to formulate a global agenda for change. As always, a more effective international cooperation was called for in order to manage the ecological and economic interdependence.[31] The project was launched just as the RBF was planning its new One World strategy with a very similar goal.

The Brundtland Commission was headed by former Norwegian Prime Minister Gro Harlem Brundtland (member of the Trilateral Commission) and included Maurice Strong (UNEP, TriCom), Susanna Agnelli (sister of Gianni Agnelli) and the American representative, William D. Ruckelshaus (Monsanto, TriCom, CFR, World Resources Institute).

> A sustainable development can be defined as a development that satisfies the needs of today without compromising the possibility of future generations to fulfil their needs. (*Our Common Future: The World Commission on Environment and Development* 1988:7)

The report's main author, Jim MacNeill, from the Institute for Research on Public Policy in Canada, had close contacts with the RBF and later became co-author of the Trilateral Commission's *Beyond Interdependence* (1991). Gordon Goodman was special advisor and wrote the chapter on energy, in which he inserted concerns over climate change, referring to the Villach meeting in 1985, where it had been established that "no nation has either the political mandate or the economic power to combat climatic change alone."

Willy Brandt (1913–1992), Social Democrat, Chancellor of West Germany 1969–1974.

Olof Palme (1927–1986), Social Democrat, Prime Minister of Sweden 1969–1976; 1982–1986.

Gro Harlem Brundtland (1939–), Social Democrat, Prime Minister of Norway 1990–1996.

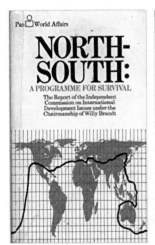  

North-South (1980). Common Security (1982). Our Common Future (1987).

The Brundtland Report was the last in a series of three reports (the titles had a "common" theme which would be reused by Pope Francis in 2015).

1. The first was the Brandt Commission's *North–South: A Programme for Survival* (1980). This commission was initiated by Robert McNamara in 1977 and led by Willy Brandt. The *Brandt Report* described international development and proposed the need for a

large-scale transfer of resources from industrialised to developing countries (in other words, related to NIEO). Three years later, the Brandt Commission felt compelled by the financial situation to publish a follow-up report, *Common Crisis North–South: Cooperation for World Recovery.*

2. The second report was the Palme Commission's *Common Security: A Blueprint for Survival* (1982), about nuclear weapons and peace. It, too, concluded that individual nations could no longer seek security at each others' expense and proposed that lasting peace could only be achieved through cooperation.

3. The Brundtland Report, *Our Common Future: World Commission on Environment and Development*, was published in June 1987. It was a continuation of the work initiated at the Stockholm Conference and included guidelines on how future politics needed to change in order to achieve the visions of "sustainable development."[32]

Within a few decades, sustainable development would grow into a global doctrine for humanity, based on fear of climate change—all to realise the vision of One World.

The 1988 Toronto Climate Conference

In the beginning of 1988, Oppenheimer reported to Wahman that the meetings in Villach and Bellagio had been successful. A policy response to the final document from Bellagio was needed. The report was published on June 7, and presented at coordinated press conferences in Washington, Stockholm, and Toronto.[33] A few weeks later, the 1988 Conference of the Atmosphere took place in Toronto, led by Ontario Premier Howard Ferguson (who had also attended the Bellagio and Villach conferences) with Gro Harlem Brundtland, and with Jill Jäger, Gordon Goodman, and Michael Oppenheimer on the steering committee. It gathered a number of political leaders and heads of state and had the desired effect on participants.

The climate issue had now grown into a million-dollar venture.

The climate disasters threatening the planet can only be compared to a worldwide nuclear war (Swedish Liberal Party leader Bengt Westerberg paraphrasing the conclusion of the Toronto Conference).[34]

The 1988 Senate Climate Hearing

One of the Toronto Conference attendees was US Senator and TriCom member Timothy Wirth.

Only a week before the conference, Wirth had organised the famous US Senate hearing in which GISS/NASA scientist James Hansen declared that he was 99 percent sure that the high summer temperatures that had been registered in the US in 1988 were due to the greenhouse effect.[35] In a candid interview for the investigative TV show *Frontline* two decades later, Wirth openly admitted to brazenly manipulating the hearing by picking a strategic date and sabotaging the air conditioning (!) to create unbearable heat in the hearing room:

> Believe it or not, we called the weather bureau and found out what historically was the hottest day of the summer. Well, it was June 6 or June 9 or whatever it was, so we scheduled the hearing that day, and bingo: It was the hottest day on record in Washington, or close to it. It was stiflingly hot that summer. . . . What we did was [we] went in the night before and opened all the windows, I will admit, right? So that the air conditioning wasn't working inside the room. . . . So Hansen's giving this testimony, you've got these television cameras back there heating up the room, and the air conditioning in the room didn't appear to work. So it was sort of a perfect collection of events that happened that day, with the wonderful Jim Hansen, who was wiping his brow at the witness table and giving this remarkable testimony.[36]

Just as during Kissinger's address to the UN General Assembly fourteen years earlier, local weather became a powerful ally in getting the message across. Few agreed with Hansen on his conclusions at the time.[37] The hearing, however, still gave echoes around the world and the climate issue got its political breakthrough.[38]

In a speech to the Royal Society in London in September 1988, British Prime Minister Margaret Thatcher warned of three changes in the atmospheric chemistry: the increase of greenhouse gasses, the thinning of the ozone layer, and the effects of acid rain on soil, lakes and forests. She also pointed out the success of previous efforts to reduce city smog and clean up the Thames.[39] The speech had been written by her scientific advisor, Sir Crispin Tickell.

In Sweden, the Toronto Climate Conference resulted in a parliamentary debate on October 19, 1988, where Conservative Party leader Carl Bildt (later TriCom member) pointed out the seriousness of the climate threat:

> We must be able to address the new threats to our environment, besides the previous ones. I think Swedish environmental debate largely seems stuck in the rather outdated views of the 1970s nuclear power debate. This risks leading politics down the wrong path. Now, on a global scale, there are often other and far more serious problems requiring our attention. Around the world the threat of climate change, as a result of fossil fuel use, is being discussed.
>
> What both Gro Harlem Brundtland, from her political viewpoint, and Margaret Thatcher, from a completely different political viewpoint, are talking about is the possibly greatest global threat to

Margaret Thatcher (1925–2013), Conservative Party, UK Prime Minister 1979–1990.

Ronald Reagan (1911–2004), Republican, US President 1981–1989.

George H. W. Bush (1924–2018), Republican, US President 1989–1993.

our environment for the rest of this century. . . . This spring we on the Conservative side—against Social Democrat vote—actually pushed through the resolution that carbon dioxide emissions will not be permitted to increase.[40]

It may be worth noting that in the 1980s, the political interest in climate change came as much from the Right as from the Left of the political spectrum—a fact later forgotten.

The Founding of the IPCC

After the 1985 Villach meeting, Mostafa Tolba had advised the US secretary of state, George Shultz, to adopt policies for mitigating climate change.[41]

The Reagan administration suggested an intergovernmental panel for studying climate science. The Department of State, the Department of the Interior, and the Environmental Protection Agency (EPA) were involved in the process of drafting a proposal. The proposal was then sent to the World Meteorological Organization (WMO) where it was adopted after a few amendments.[42]

On March 25, 1988, an invitation was sent to member states to join the panel.[43] The coordinated efforts led to the creation of the Intergovernmental Panel of Climate Change (IPCC) by WMO and UNEP.

On December 6, 1988, the UN General Assembly adopted a resolution where climate change was declared as a common concern for all of humanity and advised all organizations and programs within the UN system to support the IPCC. Now there was an organization which could provide analyses with greater authority and issue policy recommendations to governments and NGOs.[44] Bert Bolin became its first chairman, at the recommendation of Mostafa Tolba.

The IPCC was not created to carry out original research but only to assess published literature, from both peer-reviewed and non-peer-reviewed sources, and make assessments and recommendations. The mission intended to analyse—in a comprehensive, objective, open, and transparent manner—the scientific, technical and socio-economic information relevant to understanding the scientific basis of risk of human-induced climate change, its potential impacts and options for adaptation and mitigation.

The founding of the IPCC coincided with the US presidential election. The new president-elect, George H. W. Bush (vice president to Reagan) took the climate issue very seriously, which was hardly surprising considering his membership in the Trilateral Commission and friendship with his advisor David Rockefeller.[45]

According to the State Department, the US, which had the capacity to create a consensus around crucial issues, should take the lead in addressing global warming. The United States could, however, not act unilaterally since climate change was a global issue which could only be solved by an effective global response. The key was to bring all nations together and create both a scientific understanding and a united effort.[46]

The Trilateral Commission and the Soviet Union

The day after the UN had declared the climate as a common threat to humanity, Soviet general secretary Mikhail Gorbachev made a speech to the UN General Assembly where he expressed that

further world progress is now possible only through the search for a consensus of all mankind, in movement toward a new world order. The world community must learn to shape and direct the process in such a way as to preserve civilization, to make it safe for all and more pleasant for normal life. It is a question of cooperation that could be more accurately called 'co-creation' and 'co-development.'

On January 18, 1989, at the Central Committee of the Soviet Union Communist Party, Gorbachev met with Trilateral Commission leaders Georges Berthoin (European chairman of the Trilateral Commission), Valéry Giscard d'Estaing (president of France, 1974–81), Yasuhiro

Nakasone (prime minister of Japan, 1982–87), Henry Kissinger, and David Rockefeller to discuss a merging of the capitalist and the socialist system.[47] Internationalists from both the West and the East would now be joining forces.

After the collapse of the Soviet Union, Gorbachev founded Green Cross International and the Gorbachev Foundation and began to actively work for the implementation of the transformation to a new world system. Several parallel projects were initiated, all with the mission to create a better political structures for addressing global environmental problems.

GLOBE

In 1989, GLOBE (Global Legislators for a Balanced Environment) was founded under Congressional Clearinghouse on the Future,[48] by a group of cross-party legislators from the US Congress, the European Parliament, and the Japanese Diet. Founding members included US Democratic senators James H. Scheuer (founding president of GLOBE), Al Gore (international president of GLOBE, 1990–92), and John Kerry; US Republican John Heinz III (president of GLOBE USA); Dutch Socialist Democrat Hemmo Muntingh; and Liberal Democrat Takashi Kosugi from Japan. (The fact that GLOBE was founded with representatives from these three regions can be viewed in relation to some of its founders' involvement in TriCom.) In 1991, legislators from the Soviet Union/Russia were also invited.[49]

GLOBE International is registered as an environmental NGO in and operated from Brussels.[50] Its purpose was to function as an international hub for information exchange between legislators, enabling them to "respond to urgent environmental challenges through the development and advancement of legislation," working with international institutions, national parliaments, and the media.[51]

Just like the Congressional Clearinghouse on the Future in the US and the Fabian Society in England, but with environment and climate as its main focus and an international reach, the aim of GLOBE was—and still is—to secure international environmental legislation on problems identified by the Club of Rome, the Brundtland Commission, and

The discreet red entrance (the door in the middle) to Fabian Society's former office in London, shared with GLOBE and Left Foot Forward (photo taken 2013). In 2015, GLOBE moved to Brussels and the Fabian Society office is now located at another London address.

The Shaw Window (AKA the Fabian Window) at the London School of Economics, designed by author George Bernard Shaw and created by Caroline Townsend in 1910, depicts the Earth ready to be forged into a new and better world by Fabian Society founders Sidney Webb and E. R. Pease. The inscription reads: REMOULD IT NEARER TO THE HEARTS DESIRE.

others as requiring global solutions. Initially, these were climate change, ozone depletion, acid rain, waste disposal, deforestation, overfishing, and overpopulation.[52]

Part of GLOBE's mission was "to seek free market solutions" to environmental problems.[53] Transnational corporations were invited and GLOBE EU funders included Unilever, Dow Chemical, Proctor & Gamble, and Toyota.

During its first years, GLOBE received funding from the IFAW, the German Marshall Fund, and the W. Alton Jones Foundation (the two latter closely related to the Rockefeller sphere). GLOBE would later be funded by the European Commission; the governments of Norway, Denmark, and Germany; and several UN agencies such as UNEP, the Global Environmental Facility, and the World Bank.

Initial organising assistance also came from Britain through Edward Seymour-Rouse from the International Fund for Animal Welfare (IFAW), who became GLOBE's first executive director.

GLOBE was developed as basically an Anglo-American project representing political and economic interests primarily in Great Britain, USA, and the Netherlands. In the background we find the British Fabian Society (with which GLOBE shared headquarters for a time) and the London School of Economics. It is also related to the Club of Rome via shared members. After the G8 meeting in Gleneagles in 2005, GLOBE has also been invited to its annual global summits.

GLOBE still gathers MPs from across the political spectrum under one umbrella, with a common agenda, using the Molitor model (see appendix B) to its own ends:

1. First, a goal is set;
2. Thereafter, the voter base is influenced by NGO activist campaigns and the media (e.g., via COM+ Alliance of Communicators for Sustainable Development);[5]

---

5   COM+ Alliance of Communicators for Sustainable Development's founding partners include BBC, CGIAR, GEF, ICUN, IPS, UNEP, the World Bank, and the Reuter Foundation.

3. Politicians then "respond to the public opinion" and implement the very goals set out from the start.

This circumvents regular democratic procedures and shapes public opinion to align with the interest of the ruling elite without the real players being visible in the process.

The London School of Economics and Fabian Society

The London School of Economics and Political Science was founded in 1895 by Beatrice and Sidney Webb, both members of the Socialist think-tank Fabian Society, working for the gradual introduction of socialism through legislation rather than by violent revolution.[54]

The Rockefellers have had close ties to the school—which was described in the 1930s as "Rockefeller's baby."[55] These ties have remained. Besides funding from the Rockefeller Foundation, Republican David Rockefeller got a part of his education at this socialist school in 1937–38, before moving on to get his PhD at the Rockefeller-funded University of Chicago.

CLIMATE ACTIVISM

Getting the Green NGOs on Board

It was now time for the next step in the policy cycle (see appendix B).[56]

The Rockefeller foundations increased their funding of both research supporting the hypothesis of CO_2 impact on climate and policy responses to mitigate its effects. The RBF would later describe their efforts during Phase One (1984–92) in their Sustainable Development Program, and take credit for both the Rio Summit and the creation of the IPCC.

> A review of correspondence between then-RBF president Bill Dietel and program staff clearly indicates that the Rio negotiation and treaty, and the creation of the Intergovernmental Panel on Climate Change, were specific aspirations of the RBF program at the time. Total RBF funding committed during this eight-year period was under $1,000,000.[57]

Within the green movement there had been some skepticism towards the CO_2 theory, due to its connection with scientists and politicians proposing nuclear power as the solution.

This view would, however, soon change when funding from large foundations and governments started flowing into their organizations and campaigns. The Brundtland Report had given clear guidelines. In February 1989 Gordon Goodman (Beijer Institute) noted that international attitudes towards greenhouse gas were dramatically different from those in 1986,

> . . . largely due to the very high level of exposure given by the news media to a series of unusual climatic anomalies.

Goodman suggested to RBF:

> To ensure that public concern would be translated 'into positive action,' there would be 'an important, behind-the-scenes role to be played by thoughtful and well-placed nongovernmental organizations that are free from the political considerations that often constrain government initiative.'[58]

Foundations such as RBF, Rockefeller Foundation, MacArthur Foundation, Charles Stewart Mott Foundation, and Ford Foundation followed the advice and increased their financial support for green NGOs.[59]

In 1987, Adam Markham, pollution campaigner of Friends of the Earth, collected the available climate science in the report *Heat Trap: Threat Posed by Rising Levels of Greenhouse Gases*, published in 1988—the same year as the founding of the IPCC.[60] It became the starting point for the Friends of the Earth involvement in the climate issue.

Soon after, Friends of the Earth launched the first campaign in Great Britain on possible consequences of climate change.[61]

Greenpeace started their climate lobbyism in 1989 through Australian diplomat Paul Hohnen and oil geologist Jeremy Legget. Hohnen later became involved with the energy and environment program at Chatham House (the British equivalent of Council on Foreign Relations).[62]

Climate Action Network

In 1989, Michael Oppenheimer from the Environmental Defense Fund founded the umbrella organization Climate Action Network which gathers green NGOs to exert pressure on policy makers on the climate issue. Today it includes more than nine hundred organizations.[63] In the background we find RBF.

CLIMATE SKEPTICISM

Global Climate Coalition

That same year, the skeptical lobby organization Global Climate Coalition (1989–2002) was also created by a number of oil producers, petrochemical companies, and car manufacturers. Chairman was William O'Keefe from the conservative think tank George C. Marshall Institute. Rockefeller's crown jewel, Exxon, held a leading position in GCC, as well as the Ford Motor Company, BP, GM, DuPont, and Royal Dutch Shell.[6]

The GCC opposed regulations for limiting climate change and challenged the theory of global warming.[64] After the Kyoto Protocol 1997, however, oil companies started leaving GCC to instead embrace the carbon dioxide theory of global warming.

George C. Marshall Institute

In the 1990s, the George C. Marshall Institute was described as the leading climate skeptic think tank in the US, founded in 1984 by Frederick Seitz (1911–2008), Robert Jastrow (1925–2008), and William Nierenberg (1919–2000).

---

6   Exxon chairmen during GCC's existence, Lawrence Rawl and Lee Raymond, were both members of the Rockefeller-dominated Council on Foreign Relations. Raymond was also closely connected to Rockefeller family by his membership in the Trilateral Commission and his position as board member in Rockefeller's Chase Manhattan Bank.

There were, however, many connections with the CO_2 proponent camp. Jastrow founded and from 1961 headed the Goddard Institute for Space Studies (GISS) at NASA (at his retirement 1981 succeeded by James Hansen).[65] Nierenberg succeeded Roger Revelle in 1965 as head of the Scripps Institution of Oceanography.

Seitz, who assumed the role as the greatest climate skeptic in the US together with Fred Singer, had for decades collaborated closely with the Rockefeller family. In 1968, Seitz succeeded Detlev Bronk as president of the Rockefeller University and had been on the executive board for the Rockefeller Foundation from 1967 to 1977 (the board at this time also included Maurice Strong).[66] Seitz was also a member of Council on Foreign Relations. Despite his skepticism he was given the "David Rockefeller Award for Extraordinary Service to The Rockefeller University" in 2000.[67]

Thus, both the leading skeptics *against* and advocates *for* the carbon dioxide theory at the time were part of the Rockefellers' vast network.

CLIMATE POLICY

Stockholm Environment Institute

In 1989 Stockholm Environmental Institute (SEI) was founded, with Gordon Goodman as its first executive director. Officially initiated by the Swedish government through the Minister for Energy, Birgitta Dahl, with ideological kin Bert Bolin as co-founder, it had actually evolved out of (and received staff from) the privately owned Beijer Institute, founded by Swedish businessman Anders Wall, former CEO of Kol & Koks AB (Coal & Coke Inc.) which became Beijerinvest AB.

The stated purpose of SEI was to bring politics and science together to implement the defined goals for a sustainable development. Its goal was thereby directly political and the science was mainly used as leverage for realising the desired political ambitions.

SEI has primarily been funded by the Swedish International Development Cooperation Agency (SIDA), multilateral and bilateral organizations, NGOs, and—right from the start—the Rockefeller Foundation and Rockefeller Brothers Fund. The first year RBF donated $215,000 to SEI for the organizing of conferences with climate scientists and policy experts, in collaboration with the Environmental Defense Fund and Harvard's John F. Kennedy School of Government.[68]

During the coming decade (with the support from the Rockefellers' foundations and Nelson's son Steven Rockefeller), SEI, Gordon Goodman, and the American Tellus Institute would produce scenarios for how a future ideal society could be designed and function. The goal was to create a planetary civilization with a global consciousness—One World!

Abby Milton O'Neill
RBF chairman 1992–1997

In my view, it is important for foundations at this time both to make their voices heard more effectively, and to make clear to the larger society the critical power of philanthropy, and this is precisely what the RBF is attempting to do.

—Abby Milton O'Neill

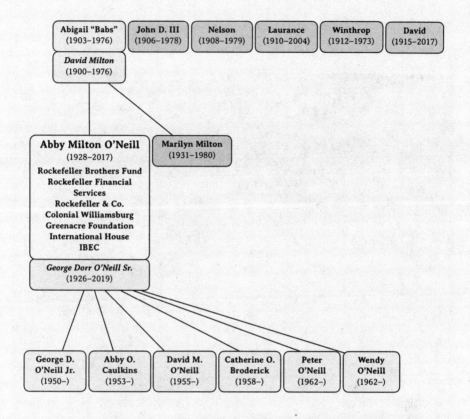

| Abigail "Babs" (1903–1976) | John D. III (1906–1978) | Nelson (1908–1979) | Laurance (1910–2004) | Winthrop (1912–1973) | David (1915–2017) |

David Milton (1900–1976)

Abby Milton O'Neill (1928–2017)
Rockefeller Brothers Fund
Rockefeller Financial Services
Rockefeller & Co.
Colonial Williamsburg
Greenacre Foundation
International House
IBEC

Marilyn Milton (1931–1980)

George Dorr O'Neill Sr. (1926–2019)

| George D. O'Neill Jr. (1950–) | Abby O. Caulkins (1953–) | David M. O'Neill (1955–) | Catherine O. Broderick (1958–) | Peter O'Neill (1962–) | Wendy O'Neill (1962–) |

ABBY MILTON O'NEILL was born in 1928 as the oldest daughter of Abigail "Babs" Rockefeller Mauzé and David Milton. She married George Dorr O'Neill, chairman of Meriwether Capital, and had six children: George Jr., Abby, David, Catharine, Peter, and Wendy. Abby started her career at Bradford College and, in 1958, became the first and oldest of the cousins to join the board of the RBF. Thirty-four years later she became chairman, after her cousin David Rockefeller Jr. In 1977, Abby became a board member of Rockefeller Financial Services and Rockefeller & Company and served as chairman from 1998 to 2004. She was also CEO of the Greenacre Foundation, vice president of Colonial Williamsburg, and board member of the International House of New York. Abby died in her sleep on May 3, 2017, at eighty-nine years old.

The attack on World Trade Center, September 11, 2001—the beginning of a new era.

Chapter Seven

The Great Transition

Global climate change is one of the emerging transnational threats to peace, justice, and sustainable development that have moved closer to the top of the international security agenda since the end of the Cold War. Like other cross-border challenges—crime and terrorism, infectious disease and the strains of gross economic inequity—its management will require collaboration and social innovation at all levels of human activity.
— Rockefeller Brothers Fund, *1998 Annual Review*

TOWARDS A UNIFIED WORLD

At the dawn of the new decade, the Cold War had just ended. The fear of a nuclear holocaust had subsided.

The Rockefeller Brothers Fund, with David Rockefeller Jr. as its president, now started funding programs and activities in Eastern Europe. After the fall of the Berlin Wall, the East should now be reintegrated with the West. Effective management of the earth's resources was on the agenda and climate change was reaffirmed as a top priority.[1]

The efforts were coordinated with the Rockefeller Foundation, whose president, Peter C. Goldmark Jr., had been involved in RBF's One World Program. On the RF board (1989–98) we also find Peggy Dulany, founder of Synergos and sister of David Jr. It was time to launch the Great Transition towards a planetary civilisation and create new commandments for a new world order.

Beyond Interdependence

In 1991, the Trilateral Commission published its report, *Beyond Interdependence: The Meshing of the World's Economy and the Earth's Ecology*. The title indicated its message. The report was a direct sequel to the Brundtland Report and written by its main author, Jim MacNeill, together with Dutch management professor and Minister of the Environment Pieter Winsemius and Japanese political scientist Taizo Yakushiji.

Beyond Interdependence presented a global action plan for realising the objectives drawn up in *Our Common Future*. The report, with a foreword by David Rockefeller (founder and chairman of the Trilateral Commission) and Maurice Strong, stressed the importance of fully implementing a number of goals before 2012. These goals included a series of global conventions on animal species, forests, chemicals, the atmosphere, the oceans, and fresh water. This would result in a final global agreement.

As the environment was a global concern, the report recommended international laws and institutions for dealing with environmental problems, initially in the form of a World Environment and Development Forum (WEDF) which would function as a forum for world leaders and provide leadership, guidance and support to the UN system. This council would have an annual meeting to assess progress and propose appropriate further action. It was intended to be the highest platform for policy development and coordination of climate action.[2]

The IPCC would provide additional scientific assessments and policy analysis. As consensus around the climate was building, this council could be developed into an Earth council, with a much stronger mandate to make decisions and enforce regulations.

In 1989, the Hague Declaration (with twenty-four signatories) had called for a new international institutional authority that could preserve the Earth's atmosphere and fight global warming.[3]

Beyond Interdependence also predicted that national sovereignty would become limited due to the common environmental problems and the ongoing economic integration. Climate change was said to challenge old forms of governance, and global cooperation was presented as essential for addressing the problem.[4]

The action plan recommended that a number of organizations be used to reach the targets, besides the UN organizations, including the World Bank, the IMF, the International Union for Conservation of Nature (IUCN), and various philanthropic foundations. If the existing organizations failed to deliver results, new super-coordinated ones would be formed.

The policy process could be controlled through coordination of research and data collection, drafting of policy proposals, distribution of resources to organizations, and by pressing "free riders" to join the club. These recommendations were based largely on Pieter Winsemius's management theories from McKinsey & Company.[5] The plan was included in the Trilateral Commission's project for reshaping the future political architecture in Europe, the US, and the rest of the world. This was right in line with the Rockefeller family's vision of global interdependence.

The 1992 Rio Conference

From June 3 to 14, 1992, the United Nations Conference on Environment and Development (UNCED, also known as the Rio Conference or the Earth Summit) was held in Rio de Janeiro, almost exactly twenty years after the Stockholm Conference. Again Maurice Strong served as secretary-general and also led the conference committee, with Jim MacNeill as advisor.

The Rio Conference had been well prepared. In accordance with the Trilateral Commission's suggested action plan, Rockefeller Brothers Fund in 1991 began funding a number of organizations in order to create an international response and to ensure that all nations represented at the Rio Conference would have a clear understanding of the need to limit emissions of greenhouse gases.

Through a series of carefully planned meetings and publications, this informal network worked to increase public awareness about climate change. Among beneficiaries were the Stockholm Environment Institute (SEI) which received funding to develop Climate Network Europe (now Climate Action Network, CAN). This project coordinated activities for thirty associations involved in the climate issue in the European

Community. Environmental Defense Fund (EDF), Woods Hole Research Center, and World Resources Institute also received the funding to provide leadership, disseminate information and arrange seminars in the United States, Europe, Japan, Southeast Asia, and developing nations.[6]

The foundation for this second environmental conference had been laid by the Brundtland Report (1987), the Trilateral Commission's *Beyond Interdependence*, and the new Club of Rome report *The First Global Revolution*, with its famous quote:

> The common enemy of humanity is man. In searching for a new enemy to unite us, we came up with the idea that pollution, the threat of global warming, water shortages, famine and the like would fit the bill. All these dangers are caused by human intervention, and it is only through changed attitudes and behaviour that they can be overcome. The real enemy then, is humanity itself.[7]

In the foreword to *Beyond Interdependence*, David Rockefeller stated that sustainable development was the only viable alternative to the doomsday scenarios painted by the Club of Rome.[8] According to David and Maurice Strong, mankind had no other choice but to submit to the dictates of a global stewardship. In his opening remarks at the Rio Conference in 1992, Maurice stated,

> Our essential unity as peoples of the Earth must transcend the differences and difficulties which still divide us. You are called upon to rise to your historic responsibility as custodians of the planet in taking the decisions here that will unite rich and poor, North, South, East and West, in a new global partnership to ensure our common future.[9]

The Rio summit resulted in the *Rio Declaration*, based on the *Stockholm Declaration*, and the adoption of a global action plan, Agenda 21, to achieve a sustainable development.[10] This would later evolve into the highly important Agenda 2030.

The summit also resulted in the unilateral treaty United Nations Framework Convention on Climate Change (UNFCCC) aiming at preventing harmful anthropogenic impact on the climate system, through CO_2 reduction measures.[11] With the signing of this framework, one of the Trilateral Commission's goals—which the RBF had worked for behind the scenes—was reached.

The Rio Conference motto "In Our Hands" was symbolized by a dove-like hand clutching the globe (our common future in the hands of the Rockefeller family and their allies?). The same symbol would two years later reappear in the logo of the Earth Charter, this time with the dove feature emphasized.

Bill Clinton and Jay Rockefeller

After winning the presidential election in 1992, Democrat Bill Clinton moved into the White House, with Al Gore as vice president and Warren Christopher as secretary of state (all TriCom members). These three would play important roles in continuing the agenda laid out by George H. W. Bush. Both Bill and Al were good friends with Senator John D. "Jay" Rockefeller IV. Early in his career Clinton had received financial support from Winthrop Rockefeller's stepdaughter, Ann Bartley.

Clinton grew up in Little Rock, Arkansas, where Winthrop was governor from 1967 to 1971 and would himself hold that position twice.[12]

Bill and Jay had become friends when the both served as governors during the 1970s and Jay became one of the first to support Bill's presidential candidacy.[13] Right at the time of the US intervention in Bosnia, Clinton's family spent their vacation with Jay and Sharon Rockefeller at their farm in Jackson Hole and met other family members such as Laurance.[14]

As senator, Jay, who had funded his own political campaigns, also campaigned for the position as vice president but found himself beaten by his old friend Al Gore. He still thought Al was "rock-solid," the smartest man in the Senate, and gave him his full support.

Pocantico Conferences

Through the continued good relations with the White House, the Rockefeller family, via the RBF, now led by Jay's cousin Abby Milton O'Neill, continued its endeavors to create the world they desired. The challenge for the RBF's

John D. "Jay" Rockefeller IV in West Virginia where he started his political career (photo by Staffan Wennberg, 1968).

board of directors was to "start modelling the practical options" which would form the basis for protocols on climate and biological diversity. This meant developing strategies for speeding up the process and enhancing the capacity of NGOs for the implementation.[15]

RBF support for SEI/Climate Action Network continued, now including the coordination of environmental NGOs (ENGOs) to support the creation of an international climate protocol for limiting greenhouse gas emissions.[16] The RBF also sponsored Media Natura Trust Limited to issue newsletters for Climate Action Network, coordinated and directed by Environmental Defense Fund and Michael Oppenheimer.[17] This gave the RBF direct control over the message issued to all ENGOs in the network.

The support for climate change action grew ever stronger from the environmental organizations and the theory was not questioned by them. The grassroots in these ENGOs were most likely unaware who was pulling the strings.

In 1994, another conference on climate change was held at the Rockefeller Brothers Fund's Pocantico Center outside New York City.

It gathered representatives from academia, conservation organizations, and governments, as well as multilateral corporations and organizations. Participants came from the US, Europe, Asia, and Africa. The meeting focused on developing strategies for mitigating climate change and on promoting international cooperation and resulted in the pamphlet *Turning Up the Heat* written by the German environmental activist Konrad von Moltke (1941–2005).[18]

The RBF also offered their Pocantico Center to the Ford Foundation for its Independent Working Group on the Future of the United Nations, working with the challenges faced by the United Nations, including recommendations on how the organization could be reinforced.[19]

The foundation also funded SEI for the education of the public in preparation for the upcoming first Climate Convention (UNFCCC) meeting, COP1, in Berlin during the spring of 1995. The climate issue was now getting more firmly anchored with decision makers, academia, the business community, and among bureaucrats.

The Club of Budapest

In 1993, the Club of Budapest was founded by Ervin László, as an international spiritual and cultural sister organization to the Club of Rome. The visions of a global consciousness had been a focus for László since 1978 when, during a discussion with Club of Rome founder Aurelio Peccei, he presented the idea of initiating an informal association of creative people.

Both László and Peccei thought a "cultural and cosmopolitan consciousness" was needed in order to tackle the "enormous challenges" of mankind.[20] The Club of Budapest would soon be involved in an effort to bring this goal to fruition with the help of a former communist leader.

The Earth Charter

Two years after the Rio Conference, the project of drafting an Earth Charter, based on the Rio Declaration, for the "emerging global civil society on values and principles for a sustainable future" was initiated.[21]

The vision of an Earth Charter had existed since the founding of the International Union for Conservation of Nature in 1948.[22] It had also been proposed by the Brundtland Commission in 1987 but did not receive enough support at the Rio Conference.[23] In *Beyond Interdependence* (1991), the Trilateral Commission pointed out that common values and commitments had to be gradually developed to protect the global community and future generations.[24]

In 1994, Queen Beatrix of the Netherlands invited Dutch Prime Minister Ruud Lubbers, Jim MacNeill, Maurice Strong, and former Soviet leader Mikhail Gorbachev to a meeting for the development of an Earth Charter.

Nelson Rockefeller's son Steven (vice-president of the RBF), was then appointed to lead a small team drafting the document—all under the patronage and funding of the Dutch government.[25] The drafting project was initiated at the Rockefeller family estate Pocantico in Sleepy Hollow, New York, which the RBF had just started using as a conference center.

Like his uncle Laurance, Steven Rockefeller (professor emeritus of religion) was especially interested in matters of religion and its relationship to nature. In 2001 he was editor of the book *Spirit and Nature: Why the Environment Is a Religious Issue—An Interfaith Dialogue*. With the help of religion the world could be remoulded according to their visions. During the draft for the Earth Charter, Steven Rockefeller was clearly inspired by Teilhard de Chardin's ideas about the development of mankind and the world.[26]

The Earth Charter consists of sixteen "commandments" for the New Age (see appendix E).[27] Consensus around the phrasing of the paragraphs was reached at a meeting at the UNESCO headquarters in Paris in March 2000, before it was made public in a ceremony at the Peace Palace in Hague, Netherlands.

> We stand at a critical moment in Earth's history, a time when humanity must choose its future. As the world becomes increasingly interdependent and fragile, the future at once holds great peril and great promise. To move forward we must recognize that in the midst of a magnificent diversity of cultures and life forms we are one

human family and one Earth community with a common destiny. (The Earth Charter, June 2000)[28]

The document was then placed in a richly painted wooden chest called the Ark of Hope, complete with "unicorn horns" as carrying poles in order to "render evil ineffective." It was presented at the Earth Charter celebration "For the Love of Earth" at Shelburne Farms, Vermont, on September 9, 2001, featuring global peace walker Satish Kumar, musician Paul Winter, Dr. Steven Rockefeller (as member of the Earth Charter Commission), with conservationist Jane Goodall as keynote speaker.

The Earth Charter, handwritten on papyrus, in the Ark of Hope 2001, complete with "unicorn horns" as carrying poles in order to "render evil ineffective."

Two days later, ark designer Sally Linder began a two-month pilgrimage where the ark was carried in procession from Vermont to New York. It was described as a spontaneous reaction after the terrorist attack on the World Trade Center when the two towers, "David and Nelson," were destroyed.

As the United Nations was seen as central to solving global problems threatening humanity, the ark was displayed at the UN headquarters during the preparations for the UN environmental conference in Johannesburg in 2002.[29] Attempts to have it formally recognized during the same conference did, however, not fully succeed.[30] The religious symbolism, however, was more than obvious. The original Ark of the Covenant, with its Ten Commandments, was now to be replaced by the new environmental religion.

From the 1990s, religion also came to play an increasingly important role in the area of climate and conservation, and became another tool for the Rockefeller family in their efforts to realise the vision of a unified global community.

In 2014, Steven wrote that all the religions of the world must change and be influenced by the new global spiritual awareness if the transformation to a just, peaceful and sustainable planetary civilisation was to be achieved.

> Finding ways to foster and encourage the further evolution of the world's religions is a necessary task if humanity is to find its way to a just, sustainable, and peaceful future. The formation of a planetary civilization is itself the major force driving the further evolution of religion today.[31]

The War on Terror

Soon after the terror attack, the new Bush administration launched the "war on terror," led by President George W. Bush and Vice President Dick Cheney. In a speech at the Council of the Americas conference at the Department of Foreign Affairs on May 6, 2002, David Rockefeller commended Dick Cheney's and the Bush Administration's prompt action against terrorism.[32]

After a few years, the threat of terrorism was merged with the threat of global warming in the international political arena and became one of the main points at the Trilateral Commission's annual meeting in Washington, DC, in April 2002, attended by Cheney, Colin Powell (secretary of state), Donald Rumsfeld (secretary of defense), Prince El Hassan bin Talal (chairman of the Club of Rome), and the ever-present Henry Kissinger.[33]

Vice President Cheney had previously been a board member of the Council on Foreign Relations and was a member of the Project for a New American Century, in which the US's aggressive military strategy for the early 2000s was laid out. The project was influenced by Chicago professor Leo Strauss's neoconservative ideology. The policy document stated that, "this process of transformation is likely to be a long one, absent some catastrophic and catalyzing event—like a new Pearl Harbor." [34]

The Great Transition

Meanwhile, the future global and sustainable civilization which the Earth Charter would serve as commandments for, was prepared elsewhere.

> The global transition has begun—a planetary society will take shape over the coming decades.[35]

In 1991, Gordon Goodman from SEI and Paul Raskin from the Tellus Institute had initiated the PoleStar Project to explore scenarios of how a transformation to a planetary civilization could be achieved.[36] This project, too, was sanctioned and supported by the Rockefeller family and funded by the Rockefeller Foundation.[37]

In 1995, the project was developed further in the Global Scenario Group, which was funded by the UNEP, the Rockefeller Foundation, the Nippon Foundation, and SEI. (The scenarios developed later came to be used by, among others, IPCC, OECD, and UNEP.) Steven Rockefeller had contributed both financially and with ideas during the early stages of writing the report. The referee group included Gordon Goodman, Gus Speth, and Bert Bolin from SEI and IPCC.

The project resulted in the report *The Great Transition—The Promise and Lure of the Times Ahead* (2002).[38] The ideas from the report were then further developed in the network The Great Transition Initiative, coordinated by the Tellus Institute, resulting in strategies for bringing the message of necessary changes to the world. It was all based on the visions of a sustainable Utopia—a sort of global socialism under the control of the United Nations. To implement the goals, a coordinated global citizens movement was required.

The dreams of a sustainable Utopia were very similar to Oliver Reiser's vision about the development of a World Sensorium and Teilhard de Chardin's theory about the Omega Point: a technologically interlinked humanity, subject under a world government, where a new religion based on evolutionary humanism (also called transhumanism) would replace traditional faiths.[39] These Utopian ideals, which had been spread in futurist and New Age circles since the 1970s, were now being revitalized.

Implementing this agenda required a coordinated global citizens movement.[40] For this purpose, The Widening Circle—Campaign for Advancing a Global Citizens Movement (TWC) was initiated for working towards implementing the transition to a planetary civilization and fostering the idea of global citizenship. TWC was formally launched in September 2010 in California by leaders of twelve organizations (including Earth Charter Initiative, *Kosmos Journal*, and the Club of Budapest) that were welcomed with "an open invitation for others to join this process of widening and proliferating circles linked in common purpose."[41] TWC believed, "The global transformation requires the awakening of a vast movement of global citizens expressing a supranational identity and building new institutions for a planetary age."[42]

A few years earlier, a large number of organizations had already been launched to implement this new agenda. In 2005, the Gorbachev Foundation, the Japanese Goi Peace Foundation, the Club of Rome, the Club of Budapest, and others launched the Creating a New World Civilization Initiative.[43]

The Kyoto Protocol

In 1997, the UNFCCC arranged its third climate conference, COP3, in Kyoto, Japan. The Rockefeller Brothers Fund had worked with a clear focus to build support for the adoption of the Kyoto Protocol through education and the media. The internet was also used as a forum for climate propaganda.[44] The RBF saw the legally binding national targets as a success for the fund's efforts (including the financial support to a number of organizations). According to the RBF, it was all carefully orchestrated by themselves.

> Carefully orchestrated media and communications strategies supported by the Fund at the Kyoto meeting itself played a helpful role encouraging negotiating progress that resulted in Al Gore making an unplanned trip to Kyoto during the penultimate day of the two-week negotiation to announce U.S. support for a reductions target.[45]

Their network presented evidence for anthropogenic climate change being a real phenomenon. Climate change was also connected to human behavior and thereby to the need for changing mankind at a fundamental level.

Al Gore signed the Kyoto Protocol in 1998, but it was not ratified by the Senate. In March 2001, President George W. Bush announced that climate change was a global problem that had to be addressed jointly by the nations of the world. However, the Kyoto Protocol was found disadvantageous to the United States, which meant that they would now withdraw from the agreement. Still, efforts were being made. Bush suggested technological solutions to the climate problem, such as carbon capture and storage (CCS) instead of emissions reductions.[46]

RBF Climate Funding

During Steven Rockefeller's presidency of the RBF, it now started funding organizations and creating networks across national borders, with the ambition of educating corporations, conservationists, researchers, the health sector, and religious groups on climate change. [47] After the turn of the millennium, an impressive network was created, ready to inform the masses about the climate threat. During the period 1998 to 2004 the RBF invested $10 million on the climate issue.[48] In 2001 they funded ten environmental groups working on climate-related projects. The Greenpeace Fund received $75,000 for their Global Warming Campaign.[49] The purpose of this campaign was to lobby the one hundred largest companies and encouraging them to work together with the rest of the world in the battle against climate change.[50]

It was launched before the UN World Summit on Sustainable Development (WSSD) in Johannesburg held from August 26 to September 4, 2002. Despite this summit failing to address the climate threat (Friends of the Earth accused ExxonMobil for having pressured President Bush to undermine the process)[51] the support from leading corporations and oil companies for the climate issue would increase during the following years.

Climate Business

Getting major corporations motivated to join in was an essential step in RBF's climate strategy, just as it was for GLOBE.

> The business community is a critical voice for countering the oft-heard argument that policy regulating carbon dioxide will harm the U.S. economy. Forward-thinking business leaders have been quite vocal about the opportunities associated with the new energy economy and are positioning their companies—both internally and externally—to take advantage of climate change policy. Further, many of these companies are recognizing that 'going green' is good for their bottom lines. Grantees: Ceres, Clean Economy Network, American Council on Renewable Energy, The Climate Group (Rockefeller Brothers Fund, "Building Constituency Support for Policy Action," *Sustainable Development Program Review 2005–10*)[52]

In 2004 the Rockefeller Brothers Fund established the organization known as the Climate Group in London.[53] It started to engage "newly converted" large corporations and local authorities to implement climate measures that favored continued economic growth. There was money to be made on saving the world from the great climate disaster.

> The Climate Group is an international coalition of some of the world's most powerful leaders. It is globally recognized for its exceptional impact on the climate debate, and respected as one of the

world's most influential non-profits. Its membership is made up of over 100 major brands, sub-national governments and international institutions. The combined revenue of its corporate members is estimated to be in excess of US $1 trillion, while its regional government partners represent almost half a billion people.[54]

UK Prime Minister Tony Blair also got involved and took part in the organization's events. The Climate Group started advocating the use of smart technology for reducing CO_2 emissions, including LED lamps, smart grids, carbon capture and storage (CCS), and electric vehicles.

This was the path of ecological modernization and differed from the environmental movement's focus on a simpler way of life and zero growth. Both perspectives were, however, flagships in the Rockefeller family's armada for creating a world shift.

The fear of climate change had not yet taken hold with the general public. However, this would soon change when a self-confident former vice president would make a grand entry on the climate arena, just as media got obsessed with reporting on freak weather events.

Steven Rockefeller
RBF chairman 1998–2006

The Earth Charter is one expression of what Teilhard [de Chardin] calls the spirit of the Earth that is struggling to come to full consciousness in all of us.

— Steven Rockefeller

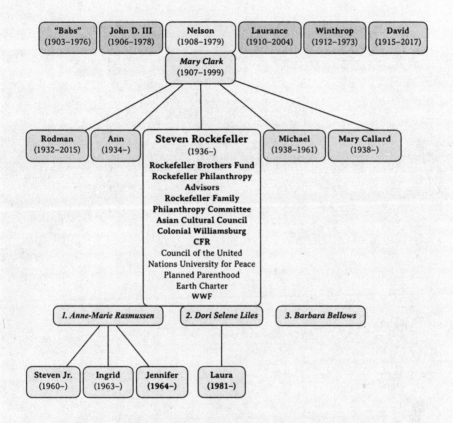

STEVEN WAS BORN in 1936 as the second eldest son of Nelson Rockefeller and Mary Clark. In 1959, he married Norwegian Ann-Marie Rasmussen, the family housekeeper, and had three children: Steven Jr., Ingrid, and Jennifer. In 1969, Steven divorced Mary and married Dori Selene Liles and had a daughter, Laura. In 1991, he married again, this time with history professor Barbara Billows. Like his uncle Laurance, Steven had an interest in spiritual matters and became professor of religion at Middlebury College, Vermont. In 1977, he and Henry Kissinger became board members of RBF and, in 1998, he became its chairman. During the 1990s, Steven was coordinator of the Earth Charter and has served as chairman of the Earth Charter International Council since 2006. He has also been a board member of Colonial Williamsburg and the Asian Cultural Foundation and, since 2009, a member of the Council on Foreign Relations.

In the eye of the storm—Hurricane Katrina.

Chapter Eight

Earth Is Running a Fever

Earth is running a fever. We have measured it. We know the cause: the carbon dioxide and other heat-trapping gases that we are pumping into the atmosphere. We also know if nothing changes, Earth's fever will continue to rise and things will get much worse. And yet there is a cure; in fact, there is an array of real and executable remedies, and there are many physicians poised to tackle this most consequential challenge of our time.
—The Rockefeller Brothers Fund Annual Review, 2005

CLIMATE POLITICS

From 2005, everything was about climate change. That February, just as the Kyoto Protocol was implemented, the European Commission issued a press release headlined "Winning the Battle Against Global Climate Change." If nothing was done to curb CO_2 emissions, it warned, the result could be rising sea levels, extreme weather events, and eventually even "catastrophic events" such as an interrupted Gulf Stream.[1] The doomsday rhetoric was now reaching biblical proportions.

RBF's Sustainable Development Program

This was the year when the Rockefeller Brothers Fund decided to use the lion's share of their Sustainable Development Program funding on fighting climate change.[2] David Rockefeller also decided to bequeath some of his personal funds to RBF after his death, which would increase the fund's assets by 30 percent. The RBF and its next generation chairman, Steven

Rockefeller, were now convinced that there was no longer any doubt that global warming was real and that the debate was over.

> From our vantage point as a philanthropy that has been supporting work on climate change for more than 20 years, it is clear to us that the scientific certainty of global warming is no longer worth debating. The naysayers have been revealed to be few, well paid, and partisan—self-serving ideologues on a premeditated mission to distract us from properly tending to the burning issue of our time. From now on let's just supply them with a toga and a fiddle and pack them off to Rome. We have no time to waste in shouldering the burden of responsibility that falls on our shoulders. (Rockefeller Brothers Fund, Annual Review, 2005)[3]

The RBF's report also quoted James Hansen's claim that the earth would be a completely different planet if carbon dioxide levels were not stabilised within the next ten years. The climate was seen as a global issue, affecting all aspects of human existence.

> The warming of the climate is no longer merely, or primarily, an environmental issue. It is an energy issue; a business issue; an investor issue; a moral issue; a security issue; an agricultural issue; a coastal issue; a religious issue; an urban issue; in short, a global issue that touches every conceivable facet of human existence.[4]

The Rockefeller family's foundations were very well prepared and had assembled an army of agents of change.

> The RBF has supported "allied voices for climate action" that include businesses, investors, evangelicals, farmers, sportsmen, labor, military leaders, national security hawks, veterans, youth, and governors and mayors. (Rockefeller Brothers Fund, *Sustainable Development Program Review 2005–10*)[5]

A total overhaul of the global economy, with a phasing out of fossil energy, was suggested by RBF. CO_2 emissions needed to be lowered by 60 to 80 percent before 2050. The Rockefellers now decided to make global warming their top priority during the critical decade to come.

The climate threat was presented as a global problem requiring multilateral collaboration, under American leadership. However, the initiative on the global political arena initially came from Britain, just like in the 1980s (through Margret Thatcher and her advisor Crispin Tickell).

The G8 Gleneagles Summit

The Fourth Ministerial Meeting of G8 Gleneagles Dialogue on Climate Change, Clean Energy and Sustainable Development was held in Gleneagles, Scotland, from July 6 to 8, 2005. At this Summit, climate change was a top priority on the global political agenda, along with the threat of terrorism (the 2005 London bombings occurred on July 7).

Tony Blair and world leaders at the G8 summit in Gleneagles 2005.

GLOBE was the only NGO invited.[6] The chairman of the G8, British Prime Minister Tony Blair, assigned GLOBE with the mission to gather legislators from leading political parties in the G8, the European Parliament, Brazil, China, India, Mexico, and South Africa. This was a first step towards including more nations and was initially called G8+5.

After the Gleneagles G8 Summit, GLOBE got a more important role in anchoring the climate agenda in national parliaments around the world, and also got direct access to the important guidelines drawn up by this increasingly influential forum.

In the European Parlament this can be illustrated by how easily Anders Wijkman, chairman of GLOBE EU and later chairman of the Club of Rome, in November 2005 got his report (based on the European Commission memo), *Winning the Battle Against Global Climate Change,* adopted by the Parliament.[7]

Through the 2007 G8+5 Climate Change Dialogue, initiated in 2006 by Tony Blair, legislators were provided with advice and information by business executives to ensure that legislation was not only politically viable but also practicable. GLOBE thereafter came to function as a direct link to the European Parliament and other arenas for political policymaking around the world, and become a major player in the global political machinery.

In 2012, three hundred MPs from eight-six nations were gathered at GLOBE's World Summit for Legislators. As of 2018 there were twenty-seven GLOBE subdivisions over the world, coordinated by the head office in Brussels.[1]

Global Warning

Before the Gleneagles Summit, psychoanalyst David Wasdell had prepared a document called *Global Warning*. It was first presented during the World Environment Day on June 5, 2005, at a symposium led by Crispin Tickell[8] before it was handed to the delegates at the G8 in order to push the climate issue higher up on the agenda, alarm the general public, and

---

1   GLOBE funding and support included the governments of Great Britain, Germany, Norway, Denmark, the EU-Commission; the UN, through the World Bank, GEF, UNEP, UNDP, and UN-REDD; and foundations such as Zennström Philanthropies (Skype), Global Challenges Foundation, MAVA, Com+ Alliance of Communicators for Sustainable Development, Zoological Society of London, and London School of Economics Grantham Research Institute on Climate Change and Environment (founded in 2008 by Jeremy Grantham, investment guru and former economist from Shell Oil).

get a planetary emergency declared. *Global Warning* stated that earth's resources were inadequate for sustaining a growing population. It reiterated the recommendations of the Brundtland Report, that humans should not consume more than 88 percent of sustainable planetary resources per annum (leaving 12 percent for other species), and warned that this number was now 120 percent and increasing.

> We have a narrow remaining window to engage global strategic planning and mobilisation, followed by a maximum of fifty years to achieve the transition, to scale down resource usage, to terminate inequitable capital accumulation and begin the long time reduction of global population.[9]

Clearly, the size of the world's population was viewed as the greatest problem, echoing the Neo-Malthusian message of RBF and the Rockefeller Commission's population report from 1972.

The solution advocated was—again—global governance of the environment, reduction of ecological footprints, and addressing the psychodynamics of human behavior in order to reduce the risk of a "global social psychosis" and "pre-traumatic stress syndrome" (denial, paralysis, paranoia, aggression, or spiritual refuge into "the passivity of a meditative trance state awaiting rescue by forces from the beyond") in response to the threats and stress experienced during the transformation to a sustainable Utopia.[10]

The document had the desired effect. At the meeting, world leaders declared their unified support for these ideas.

Tony Blair (who had been recruited as advisor to the Chase Manhattan Bank after his political career) seized the opportunity to step forward as a champion for this cause:

> What I wanted to do therefore at this summit was establish the following, and I believe we have done this. I wanted an agreement that this was indeed a problem, that climate change is a problem, that human activity is contributing to it, and that we have to tackle it;

secondly, that we have to tackle it with urgency; thirdly, that in order
to do that we have to slow down, stop and then in time reverse the
rising greenhouse gas emissions; and finally, we have to put in place a
pathway to a new dialogue when Kyoto expires in 2012.[11]

The *Global Warning* report marked the starting point of an intensified
campaign to give the climate issue greater impact politically and in
the media. The United Nations Panel on Climate Change (IPCC), the
UN Climate Convention (UNFCCC), and the UN Environmental
Programme (UNEP) had, according to Wasdell, not generated the desired
results. The national interests of several countries and the fossil industry
had blocked efficacy, and there was also political opposition.[12]

A few months after the G8 meeting, Wasdell was invited to the Club
of Rome's annual conference by its chairman, Prince El Hassan bin Talal
of Jordania (also a member of the Trilateral Commission and Council on
Foreign Relations). In Wasdell's view the Club of Rome had the perfect
competence to bring the message to the world. It had a sizable network
and *Global Warning* had been issued as an ebook to all members.

Wasdell's advice to the Club of Rome included
a. declaring "a global emergency";
b. branding an excess of CO_2 as an "ecological toxin" that could
    have catastrophic effect on the global biosphere;
c. presenting a strategy that could take the world to "zero CO_2"
    emissions as quickly as possible; and
d. developing institutional instruments for handling the transi-
    tion. The survival of the planet required a "psychodynamic
    renaissance."

He closed by saying, "Now is the time for all people to come to the aid of
the planet. Its future is in our hands."

Wasdell was offered to head the British branch of Club of Rome
but declined so as to not cause an imbalance in relation to other institutions. However, a close collaboration still remained.[13] He also worked on

a conceptual background analysis called *The Feedback Crisis in Climate Change* (2005) which concluded,

> The analysis indicates that there is a critical threshold beyond which the process becomes self-sustaining and can no longer be brought back under control by any reduction in GHG emissions. Should that threshold be crossed, the resultant 'extreme event' in the climate system could lead to the extinction of life as we know it within the global biosphere.[14]

His thoughts were very similar to those of Potsdam Institute Director Hans Joachim Schellnhuber's theory of "tipping points." Wasdell and Schellnhuber would later combine their ideas in the Apollo-Gaia project, proposed by the father of the Gaia hypothesis, James Lovelock, and Martin Lees (British Royal Society and Club of Rome). This resulted in *Beyond the Tipping Point* (2006).[15] Wasdell presented the report at the Climate Institute in Washington and was—surprisingly—met with skepticism from its chief scientific officer who said that both temperatures and CO2 levels had been significantly higher historically and that there wasn't really a problem.

This objection, however, didn't have much impact. The report was welcomed by the political elite and UN diplomats such as Sir Crispin Tickell (Club of Rome member and founding chairman of the Climate Institute).

Now the doors swung open to the European Environment Agency and the European Commission. Schellnhuber had been appointed scientific advisor to the president of the European Commission, José Manuel Barroso, and to the government of Germany. He could thereby develop the climate agenda of both the EU and the G8 group.

The idea of "tipping points" rapidly spread to important players such as the newly appointed executive director of Stockholm Environmental Institute, Johan Rockström, as well as to Al Gore (to whom Wasdell had become advisor on recommendation of the president of the European Environment Agency, Dr. Jacquie McGlade).[16]

The Meridian Program

Wasdell had earlier been involved in a project called the Meridian Program (originally called the Manhattan Project of Behavioral Science) which had been initiated in 1997, during the last phase of the Cold War.

It originated from discussions on board a ship in Moscow in 1985 where twenty-four behaviorists had gathered to develop technologies that could help create world peace. During the meeting, "A second Manhattan Project" was called for.[17] In their technocratic plan the threat of climate change would later come to play a very important role.

In preparation for a coming peace conference, attended by American and Soviet behaviorists, Wasdell had reviewed the Brundtland Report. The question was how the environmental agenda most effectively could be implemented on a global scale. He thought the report failed to answer the question how the desired changes would be implemented and saw a need for a psychodynamic analysis on how large social systems function under stress, as well as a practical ability to get these insights spread to and adopted at all levels of all institutions working with the future well-being of humanity. This would require the creation of a transnational network of social scientists, analysts, researchers and agents of change, including his own educational research trust, Unit for Research into Changing Institutions (URCHIN), established in 1981.

Hurricane Katrina

In August 2005 hurricane Katrina struck with full force and caused devastation in New Orleans, while Central Europe was flooded and Portugal ravaged by wildfires. In political debates and the media these events were linked to climate change. The concept of climate change was starting to penetrate public awareness.

Around the same time, the Rockefeller foundations intensified their funding of green NGOs campaigning for the threat of climate change and related issues. The campaigns grew ever more intense and aggressive until, only a few years later, it started to resemble a revivalist movement. The world needed saving. For some, climate change became a cash cow. The

The flooding of New Orleans after Hurricane Katrina in August 2005.

super-rich philanthrocapitalists now joined in the fight against climate change and unholy alliances were formed.

Al Gore—the Climate Messiah

In 2006 the threat of climate change became truly mainstream due to the documentary *An Inconvenient Truth*, written by and starring former US Vice President Al Gore.

Gore entered the political arena in the late 1970s and was part of the inner circle through membership both in the Trilateral Commission and the Council on Foreign Relations. Back in the late 1960s, he had received tuition and been shown diagrams over rising CO_2 levels by Roger Revelle at the Harvard Center for Population and Development Studies, and had been giving lectures on the human impact on climate since 1989. He was already experienced.

In 1981, Gore had arranged a congressional hearing on global warming, with fellow Democrat James H. Scheuer (who had been part of the

Rockefeller Commission on Population and
the American Future, 1969–72).[18] Among
expert witnesses were Roger Revelle and
Stephen Schneider (one of his last public
performances before passing away in July
1991).

Gore was deeply anchored in the
same Neo-Malthusian worldview as the
Rockefellers and the Population Council,
and had been influenced by the fear of
overpopulation in Fairfield Osborn's *Our
Plundered Planet* and William Vogt's *Road
to Survival.*

In the early 1970s Gore, who was a
Baptist, just like John D. Rockefeller and John
D. Jr., had been admitted to the Vanderbilt

Albert "Al" Gore (1948),
Democrat, US Vice President
1993–2000.

Divinity School in Nashville through a Rockefeller Foundation curriculum designed to attract promising young students to theology studies.
Here, in a course combining theology and natural science, he was indoc-
trinated with the Club of Rome's bleak outlook on the future.[19]

In the mid-1980s, Al and his wife Tipper (who had co-founded the
Parents Music Resource Center, PMRC) led the famous crusade against
"immoral" rock music. In 1985, Al Gore arranged a Senate hearing with
rock musicians such as Frank Zappa, John Denver, and Dee Snider. The
eloquent musicians won the debate but Tipper Gore still got her Parental
Advisory labeling.[20]

Religion and environmental concerns also merged with Gore's futurist
world view, inspired by the World Future Society and Pierre Teilhard de
Chardin.[21] Gore was a given prophet in this techno-spiritual movement
and preached how environmental crises could lead to a global spiritual
awakening and a transhumanist Utopia.[22]

Gore collaborated with futurists and spiritual leaders such as Barbara
Marx Hubbard, Ervin László, and Steven Rockefeller in projects like the
World Commission on Global Consciousness and Spirituality, which was

a fertile ground for such ideas.[23] In his book *Earth in Balance* (1992) Gore had presented the idea of a "Global Marshall Plan" to save the world.

> The new plan will require the wealthy nations to allocate money for trans-
> ferring environmentally helpful technologies to the Third World and to
> help impoverished nations achieve a stable population and a new pattern
> of sustainable economic progress. To work, however, any such effort will
> also require wealthy nations to make a transition themselves that will be
> in some ways more wrenching than that of the Third World.[24]

The transformation to a planet in balance would require harsh measures.

A decade later in 2003, the Club of Rome and Ervin László's Club of Budapest gathered in Frankfurt (with, among others, ATTAC and Eco-Social Forum Europe) to actually launch Gore's Global Marshall Plan Initiative.

After having lost the 2000 presidential election, Al Gore was perfect for the part as the new Climate Messiah.

An Inconvenient Truth

The film *An Inconvenient Truth,* premiering on May 24, 2006, meant a major breakthrough for Al Gore and for public climate awareness. It was shown on TV and in schools all over the world. However, after a high court ruling in May 2007, it may not be shown in UK schools without the teacher pointing out the film's nine scientific errors and one-sided presentation.[25]

The idea for the film came from the film producer and environmental activist Laurie David (1958–) after seeing a presentation which Gore had held in connection with the climate catastrophe film *The Day after Tomorrow* (2004). The following year, she started working on getting the climate issue into popular culture.[2] The production company and co-fi-

---

2   Laurie David had roots in the entertainment business and was now, together with Laurance
    Rockefeller Jr., George Woodwell, and Leonardo DiCaprio, a board member of Natural
    Resources Defence Council (NRDC), which was generously funded by the Rockefeller foundations and had been founded in 1967 by James Gustave Speth (who had also founded World
    Resources Institute).

nancier was Participant Media, founded in 2004 by Jeffrey Skoll (earlier CEO of eBay).

After having made a fortune making him the seventh richest man in Canada, he became a philanthrocapitalist and founded the Skoll Foundation, Skoll Global Threats Fund (2009), and Participant Media in order to "save the world." During a trip to India with the chairman of IPCC, Rajendra Pachauri, Skoll became convinced that the climate threat was much greater than he had thought. He saw it as a problem that the two billion at the bottom of the pyramid would start using fossil fuels, which would lead to the world losing the battle against climate change.[26]

In 2011, Skoll started funding Al Gore's grassroots Climate Reality Project which provided climate education for leaders and activists.[27] It had been initiated by Rockefeller Philanthropy Advisors and was funded by the RBF and the Skoll Foundation.[28] The board of the Climate Reality Project included veteran James Gustave Speth (board member of RBF from 2007).

Merchants of Doubt

Skoll's partner in eBay, philanthrocapitalist Pierre Omidyar, would later become executive producer of the film *Merchants of Doubt* (2014), based on the 2010 book by Naomi Oreskes and Erik M. Conway. In the book and the film, climate skepticism is linked to the tobacco lobby and to conservative and libertarian think tanks such as the Heartland Institute (founded 1984) and fossil industry representatives such as coal magnates Charles and David Koch. The Koch brothers were later identified by Greenpeace and Desmogblog as the major funders of the "climate denial lobby."

Like the Rockefeller family, the Koch brothers were philanthropists who had made their fortune on petroleum products in the multinational corporation Koch Industries and were thought to have good reason to create doubt about climate science. Their foundations funded think tanks, cancer research, and political lobby groups.

While supporting climate skepticism, however, David Koch was also a board member of the Aspen Institute, Rockefeller University (from 1996), and Memorial Sloan Kettering Cancer Center. Buildings named after him can be found at the Aspen Institute, the Rockefeller University campus, and Lincoln Center, New York. In 2010, he joined David Rockefeller and Henry Kissinger in the celebration of Rockefeller University's medical success. Here we find a striking parallell to the leading climate skeptic of the 1990s, Frederick Seitz, and his close ties to the Rockefeller family.

Philanthrocapitalism

There was now a growing network of philanthrocapitalists eager to save the planet. For would-be philanthropists in need of guidance, the Rockefeller Philanthropy Advisors was founded in 2002, offering a series of guides, the Philanthropy Roadmap, as a "multi-faceted international campaign to engage and educate donors in planning, implementing and maintaining an effective philanthropy programme with a long-term goal of creating a new culture of giving."

It was a continuation of the efforts orchestrated through the Rockefeller Family Office since 1891.[29] It rapidly grew to be the largest philanthropy advisory business with 235 full-time employees and 150 clients in 50 countries by 2008.[30] The family was also deeply involved in the Environmental Grants Association (EGA) which was founded in 1985 by the Rockefeller Family Fund, with a host of large corporations and 250 foundations as members.[31] The EGA shared office space and objectives with both RBF and RFF.

According to an investigation by the US Senate, the EGA acted as an important hub for environment-related charity where the donors' activities were coordinated according to the billionaire club's goals. This included the creation of a faux environmental movement aimed at implementing "the low carbon economy."[32]

This made the Rockefeller family, despite the fact that their foundations had moved down on the list as the nation's most wealthy, able to benefit from other major foundations' program activities—especially foundations built up with capital from the IT sector and companies such as Google, Hewlett Packard (HP), and Intel. These corporations were pivotal to the construction of the coming Brave New World. The coordination took place at John D. Rockefeller Jr.'s Interchurch Center ("the God Box").

The family's largest foundation, the Rockefeller Foundation, was also co-founder and financier of the Council on Foundations, which had coordinated philanthropy in the United States since 1949, was also a strategic partner to the globally oriented Global Philanthropy Forum.[33] In 1995, the RF launched the Philanthropy Workshop. Through these channels, the family could influence the direction of philanthropy both in the US and across the world.[34]

In a 1999 report, included in the RBF Project for World Security, Amir Pasic noted that foundations could not change the world on their own.

> By devising well-orchestrated grantmaking endeavors, however, they can serve as catalysts in forging new policy directions, furnishing incubators for innovative ideas, and establishing and sustaining networks of scholars, activists, and public officials.[35] (Foundations in Security: An Overview of Foundation Visions, Programs, and Grantees, Rockefeller Brothers Fund.)

The Doomsday Clock in 2007

In early 2007, the *Bulletin of the Atomic Scientists* moved the hands of the Doomsday Clock to five minutes before midnight and announced that the climate threat, together with the threat of atomic weapons, was the biggest threat to humanity—fifty years after the journal's board member and hydrogen bomb inventor Edward Teller had warned of melting glaciers and ensuing floods.

> Climate change also presents a dire challenge to humanity. Damage to ecosystems is already taking place; flooding, destructive storms, increased drought, and polar ice melt are causing loss of life and property.[36]

Design to Win

In 2007, the report *Design to Win: Philanthropy's Role in the Fight against Global Warming* was published by California Environmental Associates.[3] The mission was to save the world from the climate threat:

> As we prioritized the initiatives, we were guided by philanthropy's comparative advantages. Politicians are fixated on the next election; CEOs are focused on next quarter's numbers. Philanthropists, by contrast, have longer time horizons and can tolerate more risk. Besides being more patient investors, philanthropists have a strong tradition of filling gaps, spurring step-changes in technology and pursuing programming that transcends both national boundaries and economic sectors. Such capacities are exactly what are needed to tackle global warming.[37]

One result of the report was the founding of two foundations entirely focused on the fighting climate change: Climateworks in the United States and the European Climate Foundation (ECF). These rapidly grew to become leading funders of climate-related philanthropy in the United States and Europe. Beneficiaries included GLOBE and the Club of Rome. Soon, chapters were also opened in China and India. Both the RBF and Eileen Rockefeller Growald's Growald Family Fund Foundation became financial supporters of ECF's operations.[38] Strategic partnerships were also formed between the Rockefeller Foundation and the world's wealthiest foundation, the Bill & Melinda Gates Foundation (which had been founded at the turn of the millennium). This partnership also meant keeping the population matter on the agenda—now increasingly clearly linked with the threat of climate change.

---

3 Funding came from US foundations included in the Environmental Grants Association, including the William and Flora Hewlett Foundation and the David and Lucile Packard Foundation.

The 2007 G8 Summit

The climate threat also became a focus during the G8 meeting in 2007, headed by German Chancellor Angela Merkel in Heiligendamm. Behind Merkel, giving directions, was Hans Joachim Schellnhuber from the Potsdam Institute. At this meeting the Heiligendamm process was initiated, which strengthened cooperation between the G8 group and developing economies such as China, South Africa, Mexico, and India. One of the four areas of cooperation was the climate.[39]

Smart Globalization Program

That same year, the Rockefeller Foundation launched the Smart Globalization Program for transforming the world, concluding that climate change and pollution were the biggest threats to developing countries.[40] Luckily, the family's foundations had created an impressive armada of organizations now ready to act.[41] A number of philanthropist-funded green NGOs, such as the Alliance for Climate Protection (now the Climate Reality Project), 1Sky, Energy Action Coalition (now Power Shift Network), Friends of the Earth, and Greenpeace, were created or jumped on board.

The Nobel Peace Prize

In December 2007, Al Gore and IPCC (represented by Rajendra Pajauri) jointly received the Nobel Peace Prize in Oslo, "for their efforts to build up and disseminate greater knowledge about man-made climate change, and to lay the foundations for the measures that are needed to counteract such change." The crisis, however, opened up for "hopeful" solutions. In his acceptance speech, Gore warned,

> We, the human species, are confronting a planetary emergency—a threat to the survival of our civilization that is gathering ominous and destructive potential even as we gather here. But there is hopeful news as well: we have the ability to solve this crisis and avoid the worst—though not all—of its consequences, if we act boldly, decisively and quickly.

Richard Rockefeller
RBF chairman 2007–13

I realize, in thinking back over my years with the RBF, that compassion is woven into the Fund's very fabric: its traditions, its leadership (RBF's current president, Stephen Heintz, epitomizes the trait), and its staff.

—Richard Rockefeller

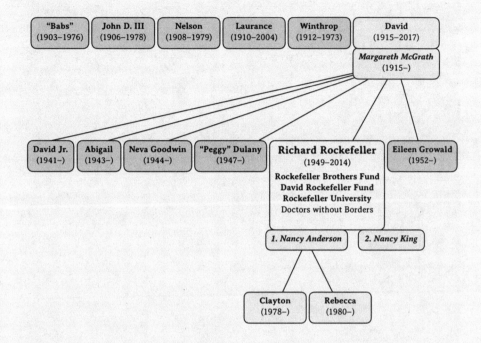

RICHARD ROCKEFELLER WAS born in 1949 as the second eldest son of David Rockefeller and Margaret "Peggy" McGrath. He married Nancy Anderson and had two children with her, Clayton and Rebecca. After their divorce he married Nancy King. Richard became a medical doctor at Harvard University and from 1989 to 2010 was involved in Doctors without Borders. He founded and was chairman of Hour Exchange Portland and was chairman and board member of Rockefeller University until 2006. In 1989 he joined the board of the RBF and became its chairman in 1998. Richard died in a plane crash on Friday, June 13, 2014, on his way home from celebrating his father's ninety-ninth birthday. In 2016, the RBF and the Chinese foundation Lao Niu Brother & Sister Foundation started the Richard Rockefeller Fellowship program to honor his memory and involvement in China.

The new One World Trade Center, 1,776 feet tall, inaugurated in November 2015.

Chapter Nine

The Future We Want

The ultimate challenge is to shape the common concern of most countries and all major ones regarding the economic crisis, together with a common fear of jihadist terrorism, into a common strategy reinforced by the realization that the new issues like proliferation, energy and climate change permit no national or regional solution.

—Henry Kissinger, "The Chance for a New World Order,"
New York Times, January 12, 2009[1]

PREPARING THE GREAT TRANSFORMATION

Now the time was ripe for action, creating the future that the family and their allies in the philanthropic billionaire's club wanted. Politicians, environmentalists, youth, futurists, and the New Age movement were enlisted to help them achieve their goal. Among the agents of change were the newly elected president, Barack Obama, the Soviet Union's last president, Mikhail Gorbachev, the ever-present Club of Rome and its sister think tank, the Club of Budapest, and José Argüelles.

The time had come to move ahead into a new era—the Big Global Shift in 2012. The dream of sustainable utopia with a united world civilisation was now about to be made into reality.

The Rockefeller Family Attacks ExxonMobil

In 2008, the Rockefeller family started a revolt at the Exxon Mobil Corporation's annual meeting and declared that it was time to leave the oil era. They submitted resolutions that the company should take the global warming issue more seriously, reduce emissions, and invest more in the development of renewable alternatives. In addition, they wanted to divide the posts as chairman of the board and managing director (both of which were currently held by Rex Tillerson). The revolt was led by John D. Rockefeller's descendents, Neva Goodwin Rockefeller and Peter O'Neill, and was supported by Neva's father and family patriarch, David Rockefeller: "I support my family's efforts to sharpen ExxonMobil's focus on the environmental crisis facing all of us."[2]

Both Peter O'Neill and Neva were RBF board members. As of 2006, the RBF was headed by Neva's brother Richard, an MD who had worked for Doctors without Borders and been a board member at Rockefeller University. Peter (grandson of David Rockefeller's sister Babs Rockefeller Mauzé) was chairman of Rockefeller Family Office and Rockefeller Family Council, and board member of Rockefeller Financial Services. Both Richard and Peter had been chairmen and presidents of the Rockefeller Family Fund.

The gambit was coordinated and supported by seventy-three of John D. Rockefeller's seventy-eight adult heirs. The family did, however, support Rex Tillerson, whom they saw as a phenomenal oil and gas manager. But the company needed to reinvent itself into an agent of change instead of being an obstruction.[3]

The Rockefeller family's direct personal shares in ExxonMobil were at the time just under 1 percent, but they pointed out that they were the oldest continuous shareholder. The family, however, had a much greater influence by their holdings in the various funds and banks under their control.[4] (The actual value of the family's assets is a closely guarded secret. In 1974, the family was willing to sacrifice Nelson Rockefeller's vice presidency in order to keep it from getting out). The campaign did not achieve much attention beyond media acclaim for the family's "new" priorities but they would soon be back to exert more pressure on their old flagship company.

PROFESSIONAL ACTIVISM

350.org

The climate project was still under way, and climate activism intensified. Again, young activists were to be used to forward RBF's agenda:

> Youth is a growing constituency with mobilize-able members around the country. Because they will inherit the planet, their voice brings a moral element to the debate. In recent years, this constituency also has become more organized politically.
>
> Grantees: Energy Action Coalition, Focus the Nation, 350. org. (RBF, "Building Constituency Support for Policy Action," *Sustainable Development Program Review 2005–10*)[5]

In 2008, the NGO 350.org was founded (with funding from the RBF and the Rockefeller Family Fund).[6] The name came from "350 parts per million—the safe concentration of carbon dioxide in the atmosphere."[7] It was founded by author Bill McKibben, together with some of his university friends. McKibben, who became its chairman, had written "the first book on global warming for a general audience," *The End of Nature* (1989).[8]

In 2006, McKibben had led the Rockefeller-funded campaign Step It Up (organizing a protest walk against coal-fired power plants), followed in 2007 by 1Sky (for clean energy economy). The 350.org was built on the earlier projects.[9]

> 350.org has the look and feel of an amateur, grassroots operation, but in reality, it is a multi-million dollar campaign run by staff earning six-digit salaries. . . . Back in 2007, the 1Sky Education Fund had starting revenues of US$1.6-million. Of that, US$1.3-million was from the Rockefeller Family Fund. In 2008, 1Sky received a further US$920,000 from the Rockefeller Brothers Fund as well as US$900,000 from the Schumann Center, tax returns show. What this means is that from the get-go, McKibben's campaign was bankrolled by the Rockefellers and the Schuman Center.[10]

It was founded as a project under the Sustainable Markets Foundation (SMF), which in turn was founded with the aid of the Rockefeller family and led by "fractivist" attorney Jay Halfon. Through the SMF, funds were channelled to various projects with capital from Environmental Grantmakers Association members such as the Park Foundation, the RBF, and the RFF. Between 2008 and 2017 the RBF granted the 350.org $1,825,000.

Soon, 350.org would start gathering young environmental activists in street protests, demanding changes in both governance and lifestyle.

TechRocks

Jay Halfon had close ties to the Rockefeller family and had a past in Rockefeller Family Fund's TechRocks project from 1999 to 2003 which helped foundations and activist organizations use modern communication technologies in the promotion of their causes. Chief executive of TechRocks was the former head of Rockefeller Family Fund, Donald Ross, founder of the political consulting and PR agency M&R Strategic Services with a host of clients from the Rockefeller network, including Greenpeace USA, where Ross served as chairman of the board. While TechRocks never become a major success the Rockefellers, through Ross, gained a significant influence on how environmental and climate issues were communicated.[11]

TOWARDS GLOBAL GOVERNANCE

Obama, the Climate President

Two thousand and eight was also the year of the presidential campaigns of Barack Obama and John McCain. Obama won and promised to take climate change seriously. His campaign was all about changing the world and he was well prepared for his mission, declaring, "My presidency will mark a new chapter in America's leadership on climate change that will strengthen our security and create millions of new jobs in the process."[12]

The Obama administration included eleven members of the Trilateral Commission. Neo-Malthusian John Holdren, who had worked with Paul

Ehrlich, became scientific advisor (he had also been advisor to Bill Clinton).

Obama had also been advised on which climate policies to pursue, through the Presidential Climate Action Project (financed by the RBF and headed by William S. Becker). The project had been initiated in 2007 by the RBF and the University of Colorado Foundation with the objective to identify which policies the president could implement without involving Congress.[13] It was supported by Holdren and John Podesta

Barack Obama (1961–), Democrat, US President 2009–2017.

(TriCom member) from President Obama's Transition Coordinating Council.[14] Everything was in place. It was now time to act.

The week before the inauguration of President Obama on January 20, 2009, Henry Kissinger, in an article in the *New York Times,* as always took the liberty of outlining the strategies for the new administration. The world was at that time facing one of the worst economic crises of the postwar era. In this dire situation Kissinger called for a new international order. The economic crisis was coupled with other risks, such as climate change and the fear of terrorism. According to Kissinger, these crises could not be solved on a national or regional level but required international coordination.[15]

The conclusions from the RBFs Special Studies Project, which Kissinger had led in the late 1950s, were still as applicable. The solution was always the same. Only the problems and motivations shifted.

The Good Club

Luckily, some of the world's leading philanthrocapitalists were ready to come to the rescue in these difficult times.

On May 5, 2009, David Rockefeller Jr. called Bill Gates and Warren Buffet for a meeting at the Rockefeller University campus on the Upper East Side of Manhattan, New York, in the President's House from 1958 (formerly used as office by Detlev Bronk and Frederick Seitz). It now became the discreet meeting place for the "The Good Club" (as they called themselves) and included venture capitalists such as Ted Turner and George Soros, TV show host Oprah Winfrey, and New York mayor Michael Bloomberg.[16] Together, the members were worth $125 billion. Their foundations, including Soros's Open Society Foundation (founded in 1993), Ted Turner's United Nations Foundation (founded in 1998), Bloomberg Philanthropies (founded in 2006), and the Bill & Melinda Gates Foundation (founded in 1994), were tasked with "saving the world" from the deep global crises identified by Kissinger as the key to the construction of a unified new world.

The 2009 G8 Summit

These crises also became central topics during the G8 Summit in Italy, July 2009, hosted by Italian Prime Minister Silvio Berlusconi. There was much hope of being able to reach the desired results before the upcoming climate summit in Copenhagen in December. At the end of the summit, G8 leaders were in agreement. Besides the G8 group, China, India, Mexico, South Africa, and Brazil were also represented, although among these leaders there was less agreement on binding climate measures.

In order to manage the crises the world was facing, there was now a call for an expansion of the collaboration initiated with the G8+5 to include more nations. The prospect of a strengthened international political architecture had been explored through initiatives such as the UN High-level Panel on Threats, Challenges and Change (2004),[17] the Swedish-French initiative International Task Force on Global Public Goods (2006),[18] and Managing Global Insecurity (Brookings Institution, 2008).[19]

A common link was UN advisor Bruce Jones (vice president and director of the Project on International Order and Strategy at the Brookings Institution and Center on International Cooperation, New York

University) who was involved in all three reports, as well as participation from trilaterals such as C. Fred Bergsten, Brent Snowcraft, John Podesta, Enrique Iglesias, and Madeleine K. Albright.

The RBF funded both the UN report and the Brookings reports, along with foundations and EGA members such as the William and Flora Hewlett Foundation, the John and Catherine MacArthur Foundation, and the UN Foundation, as well as a number of governments. The *Global Public Goods* report was mainly funded by the foreign ministries of Sweden and France. Both reports highlighted problems and provided a foundation for change.

G20—the Global Steering Committee

This created an opportunity for the G20 group (made up of twenty finance ministers and central bank governors) which since 1999 had assembled finance ministers of the twenty largest economies, to develop into a Council for Global Governance—an informal world government. In the fall of 2008, at the initiative of British Prime Minister Gordon Brown and French President Nicolas Sarkozy, George W. Bush hosted a G20 summit in Washington, DC, to deal with the financial crisis in a coordinated way.

After two subsequent summits in London and Pittsburgh, the upgraded status of the G20 became permanent and more areas of cooperation, such as the climate, were added after 2010. The OECD in Paris eventually got the role as the "quasi-secretariat"[20] and G8 lost its exclusive position.

In June 2009, Brazil, Russia, India, China, and South Africa founded BRICS. The G8 summoned the Western nations, while BRICS summoned the emerging economies. A new international political architecture was beginning to take shape, with G20 at the hub as leading forum (see chapter 11). That same year, the Trilateral and president-elect of the European Council, Herman von Rompuy, said, "2009 is also the first year of global governance with the establishment of the G20 in the middle of a financial crisis; the climate conference in Copenhagen is another step towards the global management of our planet."[21] Kissinger's and the Rockefeller family's goal now finally seemed close to being realized.

Countdown to the 2009 Climate Summit

In preparation for coming climate summit in December there had been an intense propaganda campaign and there were great expectations among leading politicians of reaching a binding agreement—at least outwardly.

In a speech to the UN General Assembly, Gordon Brown said that we could not hope for a second chance if we missed this opportunity to protect our planet.[22] This, however, required funding. The richest part of the world would have to fund the developing nations' path towards a low-carbon economy. Otherwise, developing nations would not cooperate. He concluded his speech on a hopeful note:

> And as we learn from the experience of turning common purpose into common action in this our shared global society, so we must forge a progressive multilateralism that depends on us finding within ourselves and together the qualities of moral courage and leadership that for our time and generation can make the world new again— and for the first time in human history, create a truly global society.[23]

On October 24, 2009, 350.org organized the International Day of Climate Action with 5,200 synchronized demonstrations in 181 countries to influence the delegates at the upcoming COP15.[24] It was a joint manifestation between a large number of green NGOs, including Greenpeace and Friends of the Earth. Media also helped whip public sentiments into a frenzy.

The Copenhagen Climate Summit

On December 10, 2009, President Obama received the Nobel Peace Prize, "for his extraordinary efforts to strengthen international diplomacy and cooperation between peoples."

The following week, he landed in an exceptionally cold Copenhagen (the so-called Al Gore effect) to join the other world leaders at the United Nations Climate Change Conference (COP15). Prospects were bleak. Several nations stubbornly refused so sign the agreement. The far-reaching demands for emission reductions had to be removed from the text. The Grand Finale, with President Obama as the world's climate savior, was not to be.

The summit finally agreed on what came to be called the Copenhagen Accord, with the goal of limiting global mean temperature rise to two degrees, and a proposal to create a green climate fund.[25]

The green NGOs found this compromise a huge disappointment and a death warrant for millions of people.[26] Their reaction was an echo of the environmental activists after the Stockholm Conference in 1972. The "failure" was, however, not that surprising. Climate negotiations followed the Fabian strategy of gradualism, where smaller subgoals are reached step by step. The new climate regime would not be built in a day. The efforts to reach a binding agreement, complete with binding implementation measures, had to be given enough time for final success.

China, among others, was viewed as an obstacle due to its unwillingness to reduce emissions as radically as the West wished. The RBF and the Blue Moon Fund (previously W. Alton Jones Foundation) sponsored a dialogue between China and the United States (Track II) to bridge earlier misunderstandings and strengthen the climate collaboration. Such collaboration was a priority for the RBF. Their *Annual Review 2009* was solely dedicated to this subject and stressed the good relations with China nurtured by the Rockefeller family since the early 1900s.

Richard Chandler, IPCC author from the Carnegie Endowment for International Peace, leading the dialogue, expressed RBFs gradualist strategy:

> A global climate policy, if it is to work, must eventually be truly binding and include enforcement mechanisms for all countries. Incremental steps, though short of perfection, will help persuade governments and their citizens that the costs of climate action are manageable and worthwhile. Several steps can be taken to increase confidence and trust both officially and outside official channels. (Richard Chandler, Rockefeller Brothers Fund, *Annual Review 2009*)[27]

Climategate—Hide the Decline

Just as the Copenhagen Summit climate conference started, the "Climategate" scandal occurred: leaked email conversations from a number of leading climate scientists at the Climate Research Unit, University of East Anglia in Norwich, England (the same institute that had organized the 1975 WMO conference, which paved the way for the carbon dioxide theory as the major force driving climate change).

According to critics, the emails appeared to show irregularities such as manipulation of climate data in order to create the so-called "hockey stick" by combining data sets from different sources and trying to "hide the decline" of one of the proxies, and preventing skeptics from being published.[28]

The incident, which got some media coverage and was highlighted by skeptical science blogs, did not have any major impact on the negotiation process at the Copenhagen summit. It did, however, raise some eyebrows and concerns both within and outside the scientific community.

Several committees were appointed to investigate the case, which either toned down or dismissed the seeming irregularities as misinterpreted quotes taken out of context. The chairman of one of the inquiries was Baron Oxburgh,[1] earlier director of GLOBE International, who was accused by critics of having a vested interest in alternative energy.[29]

NEW AGE TRANSHUMANISM

The Noosphere Congress

In July 22, 2009, a peculiar conference took place in Bali, Indonesia, called the Noosphere Congress, organized by the Foundation for the Law of Time, founded by Dr. José Argüelles, a University of Chicago–educated art teacher and New Age prophet and one of the originators of the Earth Day concept.

---

1   Baron Oxburgh had been chairman of the UK branch of Shell Oil (where he had warned of climate change), honorary president of the Carbon Capture and Storage Association, advisor to Climate Change Capital, and chairman of Falck Renewables, and biodiesel D1 Oils, plc.

The conference declaration of the Noosphere Congress spelled out the challenges the world was currently facing:

> Today many voices are speaking about the state of affairs of our planet. The consensus is that we are in a crisis: Climate change, environmental degradation, economic collapse, social disorder, war and the potential for mass destruction. It is a crisis because no one really seems to have a solution, much less does it seem that anyone understands the whole situation and can communicate clearly how and why it got to be this way.[30]

But this, according to Argüelles, was only part of the plan: "One important factor rarely discussed is that the mega crisis is actually the function of an evolutionary shift—we are about to enter a new geological era called the noosphere. Yes, the noosphere, the planetary sphere of mind."[31]

Barbara Marx Hubbard, who saw Argüelles as a genius,[32] explained with her typical blissful smile that the world was in a state of global emergency, and that this was good news. It would bring about the birth of a new era with "a global awakening of conciousness."

Hubbard said, "We celebrate the "planetary birthing experience" as we come together as one planetary body embracing the next epoch of human evolution as a Universal Humanity."[33]

The Global Brain

The noosphere concept, a membrane with humanity's collective thoughts and experiences enveloping the world, originated with Pierre Teilhard de Chardin, French philosopher Edouard Le Roy, and Russian scientist Vladimir Vernadsky, in turn inspired by esoteric teachings on spiritual evolution. These ideas were then further developed by philosophy professor Oliver Reiser, Buckminster Fuller, and Peter Russell. Russell's book *The Global Brain* (1982) was also made into a film (titled *Awakening Earth* in the UK).[34] Again, mankind was portrayed as a cancer on the planet, while at the time on the threshold to the next stage in evolution, connected by electronic networks into a unified global consciousness (the internet did not yet exist).

Argüelles felt that he had formed a synthesis of the work of these pioneers.[35] According to their theory, the world would soon be entering a new geological era. It had been preceded by a period in which man, over the past two hundred years, had affected the planet in a negative way, resulting in chaos and destruction. The creation of the technosphere had escalated into a global mega-crisis. But there was now a chance for redemption. According to Argüelles, the cybersphere with its electronic communication network—the internet—was the road to the noosphere and the future paradise:

> This is the prelude to the manifestation of the noosphere, which is dependent on these two factors: technospheric breakdown and a worldwide electronic communications network. How we respond to the global crisis and at the same time utilize and learn from the cybersphere is of the greatest importance. To meet the challenge requires an effort of the human mind and will that is virtually super mental in nature.[36]

The Mayan Calendar

Argüelles had spent much of his life interpreting the secrets of the Mayan calendar. He was obsessed with time-keeping and had created a new Mayan-inspired calendar for "the new galactic era," which he unsuccessfully tried to get adopted globally by sending an ultimatum to the UN and the Vatican in 1995.[37] The Gregorian calendar, in his opinion, gave an erroneous and artificial concept of time. Instead he proposed a thirteen-month year with twenty-eight-day months and the uniting of all of humanity into a "shared galactic consciousness."

José Argüelles (1939–2011).

Argüelles is probably best known for initiating the Harmonic Convergence, a synchronized global meditation event on August 16–17, 1987, to mark the countdown to the end of the Mayan Calendar in 2012. According to Argüelles, this was the beginning of an energy shift (the biosphere–noosphere transition) which would bring universal peace. All evils such as war, violence, injustice, and oppression would magically disappear when "the fifth earth" and "the sixth sun" were born on December 21, 2012.[38]

These ideas had been spreading in New Age circles during the 1990s. Now the magic date was fast approaching and a number of influential spiritually minded people, such as Ervin László, Ashok Gangadean (initiator of the World Commission on Global Conciousness and Spirituality), Roger Nelson (Global Conciousness Project), Boris Petrovic (Tesla Academy), Barbara Marx Hubbard (World Future Society), and Russian scientists from the Institute for Scientific Research in Cosmic Anthropoecology, had joined his mission.[39]

Argüelles considered himself a reincarnation of the Mayan king Pacal Votan and assumed the name Volum Votan. On March 3, 2002, he was honored by nine Indigenous elders atop the Pyramid of the Sun in Teotihuacán, Mexico, as Valum Votan, Closer of the Cycle and bringer of new knowledge. He was awarded a ceremonial staff to be used to wake humanity up to the meaning of 2012, the conclusion of the 5,125-year Great Cycle of the Mayan Calendar.[40]

Technospirituality
Behind the New Age façade—with a José Argüelles seemingly stuck in a drug-related psychosis from experimenting with LSD in the 1960s and becoming a visionary—one finds a futurist high-tech plan for building a new "Temple of Solomon" comprised of a global satellite system for interconnecting all of mankind, described in Oliver Reiser's book *Cosmic Humanism* (1975) as Project Prometheus and Krishna.[41]

Argüelles would spread Reiser's ideas of a World Sensorium (a World Brain) to a New Age audience, with illustrations of a future man, upgraded with brain implants and connected to a central database (the Global Mind). The internet was only the first phase.

Argüelles's version of history also included the story of Arcturus and the Arcturian involvement with the development of our solar system, their creation of life on Mars, and the battle between the two kingdoms Atlantis and Elysium.[42] He claimed that the Mayans had been in contact with the "Galactic Brotherhood" on Sirius (also referred to by Theosophists H. P. Blavatsky and Alice A. Bailey as "the Masters from Sirius").

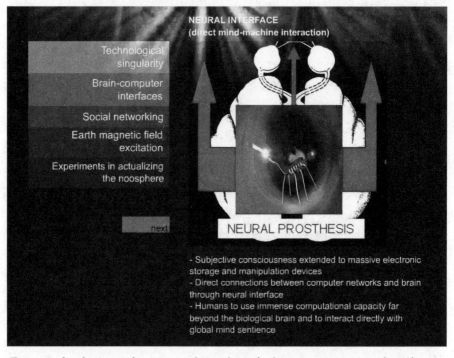

Future technologies such as a neural prosthetis for brain-computer interface (from a series of infographics on Argüelles's noosphereforum.org website, 2009).

In his *Intergalactic Bulletin*, Argüelles describes how the Sirius Star Council, which "oversees the development of mankind" and had chosen Earth for their evolutionary experimentation in biochemical engineering, would send a beam to Earth to activate the noosphere and telepathically impress certain codes upon man's new mental body, the Holomind Perceiver.[43]

> The Holomind Perceiver (HMP) is to be telepathically imprinted as an act of self-evolution. Its origin is in the beam codes of the Sirius Star Council in their activation of Earth's noosphere.[44]

Surreal as these visions might seem, they seem to have been embraced at the highest echelons of global society. On June 26, 2009, the president of the United Nations General Assembly, Miguel d'Escoto Brockmann, at a UN conference on the economic crisis in New York, officially declared the advent of the noosphere and praised Pierre Teilhard de Chardin.[45]

Co-Creating the New Human Species

Barbara Marx Hubbard's *The Book of Co-Creation*, based on channeled messages from a spirit guide identifying itself as "Christ," also described how mankind would develop a "Christ consciousness" and gain "Christ powers" through transcendent cybernetic technology and space colonisation.

> To work with me to save the world, you must develop your own Christ consciousness (love) and your own Christ capacities (transcendent technologies—astronautics, genetics, robotics, cybernetics, microtechnology, psychic powers). Accelerate the co-creative revolution. Begin to build new worlds on earth and new worlds in space. Make your life a conscious act.[46]

Hubbard's voice of "Christ" gave specific instructions on how humanity would be united and collectively transformed from *Homo sapiens* to *Homo universalis*. A new species was to be created, using cybernetic technology (similar to the transhumanist ambitions of Humanity+, see chapter 11). In this way, a violent Armageddon could be avoided. Those electing not to join her and the other enlightened souls in stepping into the new era would, however, be eliminated:

> This act is as horrible as killing a cancer cell. It must be done for the sake of the future of the whole. So be it: be prepared for the selection process which is now beginning. We, the elders, have been patiently waiting until the very last moment before the quantum transformation, to take action to cut out this corrupted and corrupting element in the body of humanity. It is like watching a cancer grow; something

must be done before the whole body is destroyed. . . . The destructive one-fourth must be eliminated from the social body.[47]

Laurance Rockefeller, who funded the publication of *The Book of Co-Creation*, loved it and saw it as a catalyst to a Christ experience for all of mankind.[48] Humanity was now ready to resume control over the evolutionary process and the creation of a new, technologically enhanced, human. Barbara Marx Hubbard called this "co-creation"—a concept inspired by Pierre Teilhard de Chardin and developed at the Foundation for Conscious Evolution, founded in the early 1990s through a donation from her "beloved patron" Laurance Rockefeller to realize the spiritual, social and scientific, and technological potential of man.[49]

Barbara Marx Hubbard's concept the Wheel of Co-Creation (a wheel with twelve interdependent sections which together would build the new global Utopia) was founded on the ideas of fellow futurist Buckminster Fuller.

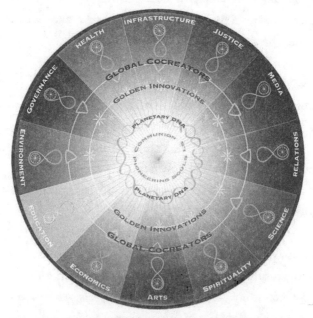

The Fuller-inspired Wheel of Co-Creation (from Barbara
Marx Hubbard's website barbara-marxhubbard.com, 2016).

Fuller, with whom she had been working during his final two years, praised Barbara's contributions for the futurist agenda.

> There is no doubt in my mind that Barbara Marx Hubbard—who helped introduce the concept of futurism to society—is the best informed human now alive regarding futurism and the foresights it has produced.[50]

Before his collaboration with Marx Hubbard, Fuller had also worked closely with Laurance's niece Neva Rockefeller Goodwin to establish and build the Design Science Institute (now Buckminster Fuller Institute).[51]

Hermeticism and Occultism

The background of the transhumanist world view can be traced to ancient hermetism, passed down from its rediscovery during the Renaissance through alchemy, Rosicrucianism, freemasonry, and theosophy.

During the twentieth century these ideas have been spread through the neo-theosophy of Alice A. Bailey (originator of the term *New Age*) and "the great Plan," allegedly dictated telepathically to her by her spiritual master, Djwal Khul, and described in a number of books, including *Externalization of the Hierarchy* (1958). The vision was one of a united humanity with a common religion and a world government (through the United Nations). Bailey's publishing company, Lucis Trust, is a United Nations–affiliated NGO.

Similar visions were also expressed by Danish spiritual philosopher Martinus (1890–1981) who taught "intellectual Christianity" and prophesied that a world government and a common culture were part of humanity's future in which mankind had evolved from the state of animal-man to becoming a "real human."

Neither of these teachings, however, included technological upgrades, only spiritual practices to develop a higher state of consciousness with enhanced intuition, compassion, clairvoyance, and a spiritual state of oneness.

Cosmic Humanism

The Theosophy-inspired but more materialistic "evolutionary humanism" of American physicist Oliver Reiser, outlined in his in *Cosmic Humanism and World Unity* (1975), also includes a world population merged into a single world organism, with a common religion, under a world government. These visions in turn echo those described in *The Open Conspiracy*[52] and *World Brain* by Fabian Society member H. G. Wells (included in Adler's Great Books).[53]

The plans, drawn up during the 1970s, now came to the surface. These ideas were connected to the visions from the World Future Society's 1975 conference, The Next 25 Years: Crisis & Opportunity, about using crises as means to create a "friendly Utopia" and the document *Changing Images of Man* (1982) from the Stanford Research Institute, which looked into the possibilities of creating a new and better human—post-industrial man.[54]

Worldshift 2012

In 2009, Argüelles's and Marx Hubbard's partner, Ervin László (founder of the Club of Budapest), together with futurist David Woolfson, created the network Worldshift 2012 to spread the notion of an impending shift to a global consciousness. Both the Club of Budapest and Worldshift 2012 were part of a partnership with the Noosphere Forum and the Foundation for the Law of Time.

The thoughts of a global consciousness had been a focus of László since he, in 1978, during a discussion with the Rome Club founder Aurelio Peccei, developed the idea of starting the sister think tank, the Club of Budapest. Both were convinced that a new "cultural and cosmopolitan consciousness" needed to be developed to deal with humanity's "enormous challenges."[55]

László, who had been the editor of Reiser's book *Cosmic Humanism*, described the plan in a more accessible way in his books *Macroshift: Navigating the Transformation to a Sustainable World* and *Worldshift 2012: Making Green Business, New Politics, and Higher Consciousness Work Together*. In dramatic terms, mankind was described as facing a critical "tipping point" if nothing was done. Despite being described as a predetermined evolutionary process, people still had to make a choice, or be doomed.[56]

The Worldshift 2012 initiative was part of a plan that had been discussed four years earlier at a conference in Tokyo called Creating a New World Civilization. The conference had gathered four thousand delegates, including ambassadors and other diplomats, with Ervin László, Ashok Gangadean, and Mikhail Gorbachev from the World Wisdom Council as speakers.[57]

Fourteen organizations (including Japanese Goi Peace Foundation, the Club of Rome, the Kabbalist Bnei Baruch, and the Club of Budapest) were represented. The goal was to combine the four S's (Sustainability, Systems theory, Science, and Spirituality). There was also a close collaboration with the UN system and UNESCO.

Among supporters of the Worldshift 2012 initiative besides José Argüelles and Barbara Marx Hubbard were Mikhail Gorbachev (Club of Rome, Club of Budapest, Earth Charter, Green Cross International), yoga philosopher Deepak Chopra, and primatologist Jane Goodall.

The dates 9/9 2009, 10/10 2010, 11/11 2011, and 12/21 2012 were chosen as Worldshift days. The countdown to 2012 had begun.

2012

The ending of the Mayan Calendar on December 21, 2012 represented what Teilhard de Chardin called the Omega Point when humanity would reach "Christ Consciousness" (the Second Coming of Christ) and develop the Mystical Body of Christ (the World Brain). For this "end date" the self-styled prophet Argüelles had wanted to gather a critical mass of 144,000 New Age believers for a final synchronised peace meditation event, Harmonic Convergence 2012, at sacred places such as the pyramids at Giza, the Dead Sea, Mount Olympus, and the San Francisco Bay Area.[58] Then in 2013, Valum Votan and his Timeship Earth, with a calendar that would liberate mankind, would be ready for takeoff.

Alas, as with many manifestations of mass hysteria, this Grand Finale was not to be. José Argüelles, the Closer of the Cycle, had just finished his book, *Manifesto of the Noosphere: The Next Stage in the Evolution of Human*

Consciousness, when, during a trip to the Australian outback in March 2011, he suddenly passed away from peritonitis and never got to partake in the Great Shift for which he had dedicated much of his life and "Valum Votan" did not get to close the circle with his ceremonial staff.

However, his disciple, Stephanie "Red Queen" South, director of the Foundation for the Law of Time, still continued working diligently to spread the visions of Votan.

CLIMATE POLITICS

Others also had great expectations for 2012. While the New Age movement meditated for oneness in eager anticipation of the Age of Aquarius, the building of a New Earth, and the coming shift in consciousness, others were preparing for the upcoming UN environmental conference in Rio de Janeiro—the same city as twenty years earlier.

Before the first Rio Conference, Maurice Strong and David Rockefeller had discussed the possibilities of establishing an Earth Council for 2012. This had been proposed in those New Age circles that had supported the Noosphere Forum (via organizations such as Earth Council 2012) but concrete proposals were also developed within more "serious" political and academic circles. However, there were some common denominators.

The Future We (the Rockefellers) Want

> *The best way to predict the future is to design it.*
>
> —Buckminster Fuller

From April 13–15, 2009, Rockefeller Brothers Fund and Michael Northrop gathered thirty sustainability and communication experts and philanthropists to Rockefeller's estate and conference center Pocantico Hills outside New York City to find out why the public was not engaging in the struggle for a more sustainable world.

They concluded that the problems seemed overwhelming and thus made people see mitigating actions as pointless. Too much doom and gloom. The experts decided to create a more positive starting point for building a new world together. The approach was clearly influenced by

Barbara Marx Hubbard and her futuristic approach. This resulted in the The Future We Want project, financed by the RBF, Surna Foundation, and the Capstone Turbine Production company. It was headed by William S. Becker from the Presidential Climate Action Project, one of the attendees at the conference. Visionaries, artists, technologists, planners, and designers were invited from around the world to describe their vision of how society should look and act twenty years into the future if we took on the challenge and built a world based on people's hopes for the future.

In 2011 the project was presented to the United Nations, which was in the process of planning the new environmental summit in Rio 2012. They liked the idea and Secretary General Ban Ki Moon chose "The Future We Want" as the official motto for the conference.[59]

Becker was also part of the Climate Change Task Force, together with its founder, Mikhail Gorbachev. The organization had been formed just before the climate conference in Copenhagen and was supported by the Club of Rome, the Club of Madrid, and Gorbachev's Green Cross International. The Climate Change Task Force included Sir David King, climate advisor to the British prime minister. Sights were now set on 2012.[60]

Planet under Pressure

In January 2009, the Earth System Governance Project was launched, under the direction of German political scientist Dr. Frank Biermann. It was based on a UN project running for over a decade, the International Human Dimensions Program, and was sponsored by the International Council for Science (ICSU), International Social Science Council (ISSC), and United Nations University. The science network gathered nearly 1,700 scientists from around the world.

The results of the project would be discussed March 26–29, 2012, at the Planet Under Pressure conference preceding the Rio +20 conference. As an echo of Teilhard de Chardin, the conference declaration asserted that there was a growing consensus that the planet had now entered a new era caused by human activity—the Anthropocene. A number of scientists had worked to identify thresholds and limits that could generate unacceptable environmental and social changes.[61]

Planetary Boundaries
after Johan Rockström, Stockholm Resilience Centre et al. 2009

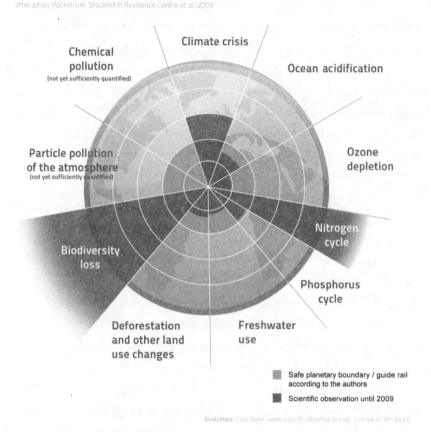

Chemical pollution
(not yet sufficiently quantified)

Climate crisis

Ocean acidification

Particle pollution of the atmosphere
(not yet sufficiently quantified)

Ozone depletion

Nitrogen cycle

Biodiversity loss

Phosphorus cycle

Deforestation and other land use changes

Freshwater use

Safe planetary boundary / guide rail according to the authors

Scientific observation until 2009

Illustration: Felix Müller (www.zukunft-selbermachen.de). Licence CC-BY-SA 4.0

Source: Steffen et al. (2015), **Planetary Boundaries**, Jan. 2015 (design by Felix Mueller).

This project was led by Johan Rockström from the Stockholm Resilience Center in collaboration with, among others, Hans Joachim Schellnhuber, Paul Crutzen, and James Hansen. The result had been presented at the Club of Rome Conference in Amsterdam on October 26, 2009.[2] Rockström and the recently appointed chairman of the Club of Rome, Anders Wijkman, would later publish the 33rd Club of Rome

2   Among participants at the Amsterdam conference were Mikhail Gorbachev, Ruud Lubbers, Jorma Ollila (chairman of Royal Dutch Shell), Paul Hohnen (the architect behind the Greenpeace climate agenda), Anders Wijkman, and Crispin Tickell. The Club of Rome's royal patron, Queen Beatrix of the Netherlands, attended. Sponsors included Philips, Royal Dutch Shell, and KLM Airlines.

Report, *Bankrupting Nature: Denying Our Planetary Boundaries* (2012) which became widely read.[62]

The Conference Declaration of the Planet Under Pressure Conference concluded that the earth system was a complex interconnected system which included the global economy and society. Just like in Barbara Marx Hubbard's Wheel of Co-creation, each part was interconnected and interdependent with all others.

The analysis also concluded that current political structures were not equipped to deal effectively with global challenges, such as climate change and the threats to biodiversity. A Sustainable Development Council within the United Nations system was called for, mandated to integrate the three pillars of sustainable development (ecological, economic, and social sustainability). This was a carbon copy of the conclusions made in the Trilateral Commission's *Beyond Interdependence* (1991) and the proposal of a World Environment and Development Forum. These ideas in turn went all the way back to the RBF Special Studies Project from 1958 (under the name International Development Authority).

An article in *Science* in March 2012, signed by thirty-two scientists headed by Frank Biermann, summarised the recommendations on institutional changes:

> Human societies must now change course and steer away from critical tipping points in the Earth system that might lead to rapid and irreversible change. This requires fundamental reorientation and restructuring of national and international institutions toward more effective Earth system governance.[63]

They suggested that the major economies of the G20 should be given greater power in relation to other nations in the new Policy Council:

> The most promising route is creating a high-level UN Sustainable Development Council directly under the UN General Assembly. To be more effective, such a council should rely not on traditional UN modes of geographical representation, but give special predominance

to the largest economies—the Group of 20—as primary members that hold at least 50% of the votes in the council. Only such a strong novel role for the Group of 20 will allow the UN Sustainable Development Council to have a meaningful influence in areas such as economic and trade governance.[64]

The 2012 UN Summit Rio+20

The United Nations Conference on Sustainable Development (UNCSD) was held from June 20 to 22, 2012, but did not attract the same attention as its predecessor twenty years earlier, despite its forty-five thousand participants. Several of the most prominent world leaders, who had just met at the G20 summit in Mexico a few days earlier, declined participation and instead sent lower-ranking representatives.[65]

In connection with to the Rio Summit, GLOBE International held its first World Summit of Legislators, with three hundred parliamentarians just a few days prior to the opening of the conference. It was arranged in close collaboration with the UN and the World Bank. Ban Ki Moon praised their efforts to implement the sustainability agenda in their respective national Parliaments: "And in today's increasingly interconnected world, you are also a link between the global and local—bringing local concerns into the global arena, and translating global standards into national action. I very much welcome your engagement in the process."[66]

No binding agreements were signed during the Rio+20 Summit, and the conclusions of the conference consisted mostly of vaguely formulated ambitions to "continue efforts to create sustainable development." The only action decided upon was the establishment of an intergovernmental policy forum, the High Level Political Forum on Sustainable Development (HLPF) for overseeing the implementation of sustainable development and its three pillars.

The green NGOs were dissatisfied with the result, as always. They claimed that the leaders had not lived up to their responsibility but had let themselves be guided by corporate interests. Greenpeace thought that much more needed to be done in terms of global agreements and governance.

> We still need a global deal, we need global governance to support and
> foster a great transition where equity, economy and ecology are not
> in competition but in harmony to deliver sustainable development.[67]

On September 10, 2012, Ban Ki Moon made a speech where he praised
the Rockefeller family for their generosity both to the UN and its predecessor, the League of Nations.[68]

There was no doubt who the architects of the future were. The Trilaterals
and the Rockefeller family now seemed to succeed in swaying the global
agenda toward the future they wanted. Both the G20 and the Sustainability
Forum were now firmly established—although these forums had not been
given the muscles which the Trilaterals had in mind for enforcing the Great
Transition. The world's four corners needed to be united.

That same year, the finance families Rockefellers and Rothschilds
had pulled their weight and entered a partnership through Lord Jacob
Rothschild being included on the board of Rockefeller Financial Services—
the company that handled all Rockefeller's business. At the same time the
Rothschild family consolidated their British and French operations under
a single umbrella.[69] In the press release Lord Rothschild was welcomed by
the family patriarch at the time, David Rockefeller:

> Lord Rothschild and I have known each other for five decades. The
> connection between our two families remains very strong. I am
> delighted to welcome Jacob and RIT as shareholders and partners in
> the ongoing development of our investment management and wealth
> advisory businesses.[70]

Now the march towards the global agreement planned to be signed in
Paris 2015 would begin. . . .

> All the problems that we face, from climate change, to financial con-
> tagion, to nuclear proliferation, are too complex and cross-cutting for
> ANY one government or indeed governments to solve alone. (Hillary
> Clinton, Clinton Global Initiative, 2013)

Valerie Rockefeller
RBF chair 2013–22

Because the source of the family wealth is fossil fuels, we feel an enormous moral responsibility for our children, for everyone—to move forward.

—Valerie Rockefeller

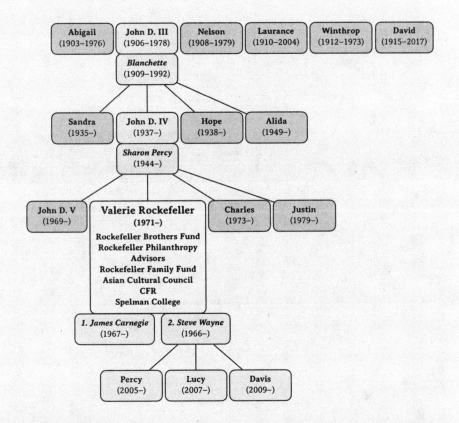

VALERIE WAS BORN in 1971 as the daughter of Senator John "Jay" Rockefeller IV and Sharon P. Rockefeller. Valerie was named after her mother Sharon's twin sister, who was murdered in the family home by an unknown intruder in 1966 (the case was never solved). In 2000 Valerie married Australian investment banker James Carnegie but soon divorced him and in 2004 married Steve Wayne, CEO of Russian real estate company Jensen Group. They had three children—Percy, Lucy, and Davis—before divorcing in 2017. Valerie earned a degree in international relations at Stanford University and has worked as a special needs teacher for adults with learning disabilities. In 2003, Valerie joined the RBF as a board member and became its chairman ten years later. She is also a board member of the Asian Cultural Council and Rockefeller Philanthropy Advisors and a member of the Council on Foreign Relations. In 2022, Valerie was succeeded by Joseph Pierson as chair of the RBF.

The wishing tree in the public section of the 2015 Paris Climate Summit COP21.

Chapter Ten

The Road to Paris

We have been involved in the entire United Nations Framework on Climate Change (UNFCCC) process going back to Kyoto. We have supported grantees in different areas including in the think-tank community who have been working on the substance of the global climate agreement. We've also focused on supporting advocates working globally to increase the ambition and push for agreement. And, along with some other foundations, we also provided support to the process itself.
—Stephen Heintz, The Rockefeller Brothers Fund[1]

A NEW GENERATION AT THE HELM

For the Rockefeller family, 2013 was a special year as their oldest charity, the Rockefeller Foundation, celebrated its one hundredth anniversary.

As of 2010, the family once again had direct control over Rockefeller Foundation when David Rockefeller Jr. (who had been a board member since 2006) was appointed chairman—a position no family member had had after John D. III was chairman from 1952 to 1971.

In 1913, Valerie Rockefeller Wayne (daughter of John D. "Jay" Rockefeller IV) became chairman of the RBF, after Richard Rockefeller (whose private aircraft crashed on June 13, 2014, on his way home from David Rockefeller's ninety-ninth birthday celebration). Valerie was the first chairman of the fifth generation, and also joined the board of directors of Rockefeller Philanthropy Advisors and the Council on Foreign Relations.

Stockholm Resilience Center

In May 2015, as part of the Rockefeller Foundation's anniversary, the Rockefeller Foundation Fellowship Program for Social Innovation was established at the Stockholm Resilience Center. The program's focus was on developing nations and vulnerable populations, with an objective of transforming suboptimal systems—politically, economically, education-ally, legally, environmentally, and socially—to a resilient society.[2]

This was part of the Great Transformation. Just like in previous decades, Stockholm and Sweden played a very central role.

Around the same time, countdown to the Paris Summit in December started, with high hopes for the signing of the Paris Agreement. Efforts to secure this outcome intensified.

The Stockholm Resilience Center (SRC) had been founded in 2007 by the Swedish governmental research foundation MISTRA (the Foundation for Environmental Strategic Research), the Stockholm Environment Institute (SEI), and the Beijer Institute. SRC was largely a continuation of the research of Beijer and SEI. Its purpose was to become a world-lead-ing institute for "sustainable governance and management of ecological and social systems" and be an important player in the UN-led process of implementing the sustainability goals.[3]

The center also became a vital node in the efforts to coordinate the spreading awareness of the "nine planetary boundaries" identified under the leadership of Johan Rockström (executive director of SEI 2004–12).

The concept of ecological resilience was introduced by Canadian ecologist C. S. Holling's article, "Resilience and the Stability of Ecological Systems" (1973), in which he mixed ecology with systems theory.[4] Holling also developed the theory of *panarchy* (from the Greek nature god, Pan), which provided a framework for describing the devel-opment of hierarchical systems with a number of interrelated elements—in this case it meant the complex interaction between man and nature. Holling identified the cycle of growth–collapse–rebirth, followed by new growth, etc.[5]

The theory also had parallells to anthropologist Joseph Tainter's *The Collapse of Complex Societies* (1988), describing how increasing complexity (in terms of increased bureaucracy, larger political entities, and increased energy consumption) inevitably leads to a social collapse, with examples such as the Maya culture and the fall of Rome.[6] Just like Barbara Marx Hubbard, Holling saw the inevitable collapse as something positive, offering new opportunities. In the case of the Roman Empire, however, the transformation phase from a high-energy to a low-energy system had not been handled properly. Part of the mission was therefore to create an effective control system in order to manage the critical phase.

As always, this would require global governance with a planetary institutional management. Society would proactively be rebuilt to respond to humanity's increasing crises (resource depletion, climate change, economy, etc). Holling (who had for a time been executive director of IIASA in Laxenburg, Austria) worked with the Beijer Institute and in 1996 initiated the network Resilience Alliance. This collaboration was formalised in 1998, with funding from the MacArthur Foundation and the Rockefeller Foundation.[7] The goal was to adapt, develop, and reorganise the current system into a resilient society within the nine planetary boundaries.

This self-appointed mission became an increasingly pressing priority for the Rockefeller family. Since 2006, they had also funded the Architecture 2030 initiative in order to build a CO_2-neutral urban environment.

In 2013, the Rockefeller Foundation and Rockefeller Philanthropy Advisors created the network 100 Resilient Cities.[8] The network received $31,350,000 from the Rockefeller Foundation during its first year.

The following year, the Rockefeller Foundation, in partnership with Swedish foreign aid agency SIDA, American USAID, and the Stockholm Resilience Center, also initiated the Global Resilience Partnership, specifically aimed at building resilient cities in Sahel, Horn of Africa, and in Southern and Southeast Asia.

The vision for the impending transformation also included smart globalization, which had been a priority for the RF since 2007. Smart information technology for monitoring and gauging all human activity had been identified as a crucial part of the solution to humanity's problems.

Global Challenges Foundation

In March 2013, the Swedish initiative Global Challenges Foundation (GCF) was announced, founded by the Hungarian-Swedish financier László Szombatfalvy, who had made a fortune on the stock market.

Its board of directors included Johan Rockström, with Swedish minister for foreign affairs, social Democrat Margot Wallström as spokesperson.

The purpose of GCF was to increase the knowledge base on global threats to humanity and to accelerate the advent of a strong Global Government to deal with these threats. According to Szombatfalvy, "Global challenges can only be solved through global action, but global action requires global decisions, and global decisions can only be made by supranational decision-making bodies, but today there are no effective, supranational, decision-making bodies."[9]

Collaboration with the Stockholm Resilience Center, the Future of Humanity Institute at Oxford (led by transhumanist philosopher Nick Bostrom), Kennette Benedict (*Bulletin of the Atomic Scientists*), and GLOBE International, was initiated. Rockström soon became known as a climate guru and media personality along with his partner, Hans Joachim Schellnhuber (whom he would later succeed as head of the Potsdam Institute for Climate Impact Research, PIK).

Earth League

In 2013, the research network Earth League was created, which included both Rockström (as chairman) and Schellnhuber. As scientific advisor to the world's top politicians and religious leaders, the network worked diligently to anchor the agenda. They were aided in this mission by the World Bank report, *Turn Down the Heat*, written at PIK in November 2012, under the leadership of Schellnhuber.[10] The doomsday message in the report (including dramatic changes, extreme heat waves, reduced food stocks, and sea level rise) was more alarmist than the more restrained IPCC reports (the alarmist approach followed the 2005 recommendations from psychologist David Wasdell which included "declaring a planetary emergency").

Just like Rockström, Schellnhuber was a strong advocate for global supranational solutions. In February 2013, he described his vision in an article in the *Center for Humans and Nature*:

> Let me conclude this short contribution with a daydream about those key institutions that could bring about a sophisticated—and therefore more appropriate—version of the conventional "world government" notion. Global democracy might be organized around three core activities, namely (i) an Earth Constitution; (ii) a Global Council; and (iii) a Planetary Court.[11]

In 2009, Schellnhuber had participated in the Great Transformation—Climate Change as Cultural Change conference in Essen along with a number respected profiles such as Obama's chief of staff, John Podesta, and professors Ottmar Edenhofer, Frank Biermann, and Stefan Rahmstorf. In a session led by Dr. David Held from the London School of Economics, the question was raised whether democratic solutions were compatible with the measures that were now "required."

> Technological innovation and political regulation can only be effective if "the people" participate in their various roles as polluters, producers, citizens and voters. Democratic regimes are not well prepared for the level of participation that is required: Can free democratic societies cope with the effects of grave changes in the global climate, or might authoritarian regimes possibly be better placed to enforce the necessary measures?

Schellnhuber's and the Rockefellers's dream of Global Governance seemed to be moving ever closer to being realised.

G20—A Global Politburo?

In 2009, in an op-ed in the *New York Times*, Mikhail Gorbachev asked about what place the newly formed G20 had within the system of global institutions. Was it a "global politburo," a "club for the powerful," or a prototype for a world government? Gorbachev felt that the G20 group should assume collective leadership in world politics but that, in order to achieve success and gain legitimacy, it needed to work closely with the United Nations.

He also suggested that the annual summit should be held at the UN headquarters in New York (instead of in the current system of rotating schedule and chairmanship of member states).[12]

On September 5 and 6, 2013, Vladimir Putin hosted the G20 Summit in St. Petersburg, Russia, which also included the G20 "engagement groups" (Business 20, Civil 20, Labor 20, Think 20, and Youth 20) "to make contributions such as drafting recommendations on their areas of interest."

The Group of Twenty had thus rapidly evolved into a growing global power factor with an ever closer relationship with UN institutions, even though the group's meetings had not yet moved into the UN headquarters.

UN Secretary-General Ban Ki Moon, who had attended all G20 meetings since the first one in Washington 2008, called for a joint responsibility in the emerging Syrian crisis. The year before, he had praised the collaboration that had developed between the UN and the G20. World leaders supported the UNFCCC and the secretary-general's efforts to mobilize the political will to introduce a legally binding agreement at COP21 in the conference communication.[13] Major decisions lay ahead.

High-level Political Forum 2013

On September 24, 2013, the High-level Political Forum on Sustainable Development (HLPF) held its first session at the UN headquarters.

The theme was "Building the future we want: from Rio+20 to the post-2015 development agenda." This new agency within the United Nations was to replace the Commission on Sustainable Development as coordinator of the sustainability agenda.

Now, the conclusions from the 2012 Rio Summit were to be turned into practical action and the new sustainability goals for 2015 and onwards established. This was to be executed through the Open Working Group on Sustainable Development Goals, formed on January 22, 2013, following a decision at the Rio Summit.[14] UN ambassadors Csaba Kőrösi from Hungary and Macharia Kamau from Kenya were appointed to lead the work.

Besides representatives from member states and the UN system, a large number of NGOs participated. At the 2012 Rio Summit, Ban Ki Moon had appointed an advisory group called the High Level Panel of Eminent Persons on the Post-2015 Development Agenda. Expert opinions were also submitted by the Leadership Council of the Sustainable Development Solutions Network, which included Johan Rockström, who inserted the "nine planetary boundaries" into the agenda.[15] A first report would be presented to the UN General Assembly and HLPF 2014.[16]

The HLPF acted as a guardian of sustainable development. Drafting the goals for global governance, strengthened interaction between the G20 and the UN was proposed.[17] It was imperative to create an effective mechanism for implementation that could take collective leadership.

From June 30 to July 9, 2014, HLPF held its second session, calling on the Ministerial Declaration to accelerate progress towards the Millennium Development Goals and "revitalise global partnership for sustainable development." Two thousand fifteen was to be a pivotal year, with both the establishing of the new Sustainability Goals and the Climate Summit in Paris.

Global Climate Legislation

In January 2013, GLOBE International had launched the Climate Legislation Initiative, in collaboration with the London School of Economics, with funding from the Zennström Foundation and the British Foreign Ministry.[18] Its purpose was to help lawmakers promote climate legislation around the world.

The summit, with one hundred delegates from twenty-six countries, took place at the British Foreign Ministry (with the blessings of conservative Foreign Minister William Hague) and was led by GLOBE President John Gummer (Lord Deben). One of the speakers was Christiana Figueres from the UNFCCC.

The following year, February 27–28, 2014, Partnership for Climate Legislation was launched at a meeting of the World Bank's premises and the Senate in Washington, again with the support of UN and the World Bank and the participation of Figueres (UNFCCC), Achim Steiner (UNEP), and Jim Yong Kim (the World Bank).[19]

This was followed by World Summit of Legislators in Mexico, June 6–8, 2014, which gathered three hundred parliamentarians from eighty countries and representatives from UN agencies such as UNEP, UNFCCC, the World Bank, and Global Environment Facility (GEF).[20]

The result was a resolution which the participants would bring back to their respective parliaments. They swore their allegiance to the climate agenda and the new sustainability goals and undertook to try to implement legislation that harmonized with them in their home countries. The goal was to create an international climate law.[21] Ban Ki Moon counted on parliamentary support and announced at the conference that on September 23 he was going to call for a climate summit for governments, business leaders, and representatives of the world of finance and civil society, and that he also intended to invite GLOBE representatives.[22]

From then on, GLOBE began holding annual meetings in connection with the UNFCCC's climate summits. These meetings were funded by the Global Challenges Foundation, who had partnered with GLOBE. There were close ties between these organizations and GLOBE was becoming an increasingly integrated part of the UN system.

The 2015 Climate March

During the months leading up to the COP21 Paris Climate Summit there was an increasing number of well-coordinated manifestations across the world aiming to spread climate awareness and pressure world leaders into action. This became a top priority for both the RBF and the Rockefeller Family Fund (RFF).

On September 21, 2014, two days before the preparatory UN Climate Summit in New York, three hundred thousand people filled the streets of New York in the mass demonstration People's Climate March. Participants included UN Secretary-General Ban Ki Moon, Al Gore, primatologist Jane Goodall, author Naomi Klein, and actor Leonardo DiCaprio.

There were 2,646 similar manifestations held in 125 other countries, including London, Paris, and Melbourne. Despite the name, however, these marches were not spontaneous expressions of grassroots climate anxiety, but had been initiated and orchestrated from above. They were organised by 350.org and funded by both the RBF and the RFF, following Pieter Winsemius's recipe for mobilizing "grassroots" organizations.

The People's Climate March was just one major climate milestone of 2014. Two months later Barack Obama and Xi Jinping jointly announced ambitious new targets to cut greenhouse gas emissions in the United States and China. RBF partners in China helped pave the way for this historic announcement.[23]

RBF and 350.org Call for Divestment

On September 22, the Rockefeller Brothers Fund announced that they would divest from fossil energy investments. RBF and 350.org led a call together with eight hundred organizations across the world, who all pledged to divest:

> Given the RBF's deep commitment to combating climate change, the Fund is now committing to a two-step process to address its desire to divest from investments in fossil fuels. Our immediate focus will be on coal and tar sands, two of the most intensive sources of carbon emissions. We are working to eliminate the Fund's exposure to these energy sources as quickly as possible.[24]

The New York Climate Summit

On September 23, the UN Climate Summit was held in New York, where a hundred world leaders met to put pressure on the climate negotiations at the upcoming 2015 climate summit in Paris and push for concrete measures against climate change. Al Gore presented the call for divestment.[25]

The president of the Rockefeller Foundation, Judith Rodin, presented the foundation's investment of more than half a billion dollars in creating climate resilience and sustainable cities and communities. She also announced other RF initiatives, such as the Global Resilience Partnership and the Asian Cities Climate Change Resilience Network.[26] At this summit, Leonardo DiCaprio was appointed a United Nations Messenger of Peace, with special focus on the climate.

Naomi Klein Becomes a Climate Activist

The week of the great climate march, Naomi Klein's new book, *This Changes Everything: Capitalism vs. the Climate*, was published. The book was part of the Message project initiated by the Rockefeller-founded organization Sustainable Markets Foundation (see 350.org in chapter 9) and was funded by some prominent members of the "billionaire's club" such as the Rockefeller Brothers Fund, the Rockefeller Family Fund, the Energy Foundation, the Wallace Global Fund, and the Tides Foundation.

The project also included a film with the same name, produced by director Susan Rockefeller's Louverture Films. The film took a year to make and was released in 2015.[27] The message in the book and the film was that the climate threat posed a chance to build a better world.

Naomi Klein (1970–), author, social activist and critic of capitalism, now helping leading capitalists spread their climate change shock doctrine.

Susan Cohn Rockefeller (1959–), filmmaker and conservationist, wife of David Rockefeller Jr., (photo by Samira Bouaou).

In November 2014, the United States and China signed an agreement on climate change—a subgoal that the RBF and their partners had actively worked towards for a couple of years.[28]

The Road to Paris

Two thousand fifteen marked a new anniversary. The RBF celebrated its seventy-fifth anniversary, David Rockefeller turned one hundred, while the family's "baby," the United Nations, had been founded seventy years earlier. The climate threat which for decades had been a priority for both the family and the UN became central in the celebration.

That same year, the family office relocated to One Rockefeller Plaza (in the Time-Life Building). "We decided to start again at One Rock" stated the new family patriarch, David Rockefeller Jr.[29]

During the World Economic Forum meeting in Davos, January 21–24, 2015, where the Rockefeller Foundation was a partner, Al Gore presented the project Live Earth: Road to Paris, which was to whip up expectations for the upcoming Paris meeting and gather "a billion voices with one message to demand climate action now."[30] Musician and producer Pharrel Williams was enlisted to organize the project. The idea was to let one hundred performers perform on seven continents on June 18. UN General Secretary Ban Ki Moon praised the initiative. However, the concerts were cancelled and Gore announced that a free concert instead would be held in Paris in the fall. This resulted in the project "24 Hours of Reality," a twenty-four-hour live broadcast from Paris and eight other countries on November 13–14, featuring prominent performers such as Duran Duran, Elton John, Jon Bon Jovi, and Peter Gabriel. Organizers were the Climate Reality Project in collaboration with 350.org, the WWF, Coalition Climat 21, and UNEP. The planned event would, however, take an unexpected and dramatic turn.

At the World Economic Forum, which that year had the theme "The New Global Context," Johan Rockström presented the "planetary boundaries" framework to the world elite. Just like his fellow "planetary stewards" Al Gore and Hans Joachim Schellnhuber, Rockström had a busy schedule in 2015.

Earth Day 2015

On Earth Day, April 22, the Global Challenges Foundation and The Earth League (both including Rockström) launched the "Earth Statement" campaign with the goal of influencing world leaders to commit to "8 essential elements," including keeping global warming within the two-degree target, 100 percent renewable energy and zero emissions by 2050.[31]

The petition was signed by prominent names such as Al Gore, Desmond Tutu, Richard Branson (Virgin Airlines), Gro Harlem Brundtland (former prime minister of Norway), Mary Robinson (Ireland's former president), Paul Polman (Unilever), actor Russell Brand, Hans Vestberg (Ericsson), and Swedish archbishop Antje Jackelén.

Religious Leaders Become Climate Activists

Fifty years after the Rockefeller family helped put the climate threat on President Lyndon Johnson's agenda, it was now time for the world's religious leaders to get involved.

> Religious Voices are a crucial block pushing for climate action. "Creation Care" has inspired many faithful Christians to understand that God calls upon them to be stewards of the planet, which includes supporting efforts to address climate change (RBF, *Sustainable Development Program Review 2005–10*, November 2010).[32]

On May 24, the papal encyclical *Laudato Si' of the Holy Father Francis on Care for Our Future Common Home* (note the reference to the Brundtland Report (*Our Common Future*) was issued. It supported commitments to combat climate change:

> A very solid scientific consensus indicates that we are presently witnessing a disturbing warming of the climatic system. In recent decades this warming has been accompanied by a constant rise in the sea level and, it would appear, by an increase of extreme weather events, even if a scientifically determinable cause cannot be assigned to each particular phenomenon. Humanity is called to recognize the

need for changes of lifestyle, production and consumption, in order
to combat this warming or at least the human causes which produce
or aggravate it.[33]

The goal was to create a global consensus and a plan for the world. The
crisis could only be managed if all nations joined together and acted as a
single unit.

On June 17, 2015, five days after David Rockefeller's one hundredth
birthday, Hans Joachim Schellnhuber (PIK) was elected as scientific advisor
to Pope Francis together with, among others, Veerabhadran Ramanathan
(Scripps Institution of Oceanography) and Nobel Laureate Paul Crutzen
(atmospheric chemist who popularised the term *anthropocene*).

In 2014, Schellnhuber had participated in a workshop at the Vatican,
discussing his favorite topic, "tipping points" (based on the theory developed with David Wasdell and advice from James Lovelock).[34]

The Pope, supposedly God's spokesperson on Earth, now received his
guidance from a more mundane source.

Two months later, the Muslim world followed. Religious leaders of
Islam announced that it was a religious obligation for the world's 1.6 billion Muslims to fight global warming.[35] Behind this proclamation we find
the British organization IFEES and its founder, Fazlun Khalid. A similar
message came from other religious leaders.

Big Oil Pleading for a Global Framework

On June 1, 2015, leading oil and gas companies such BG Group, BP, Eni,
Royal Dutch Shell, Statoil, and Total sent ta letter to France's Foreign
Minister Laurent Fabius and Christiana Figueres (UNFCCC), acknowl-
edging climate change as a "critical challenge for our world" and asking
governments attending the upcoming Paris Summit to:

- Introduce carbon pricing systems where they do not yet exist at the
  national or regional level;
- Create an international framework that could eventually connect
  national systems.[36]

The 2015 GLOBE Summit

During the countdown to the Paris Summit, GLOBE held its annual meeting in Los Angeles on from July 19 to 24, in close collaboration with UNEP and funding from the Global Challenges Foundation, to prepare the agenda for the implementation of the agreement. Attendees included Christiana Figueres (UNFCC), Laurent Fabius (coming chairman of COP21), and Margot Wallström (Swedish foreign minister, former spokesperson for the Global Challenges Foundation and candidate as Sweden's representative in the UN Security Council, to which she was elected on June 28, 2016).

GLOBE had become the UN bureaucracy's instrument for implementing the sustainability agenda from above.

> [T]he Paris Agreement must be a starting point for a profound paradigm shift that will make sustainable development possible and will lead to restructuring our economic models to achieve the decarbonisation of our economies by 2050.[37]

At this meeting, Anders Wijkman from the Club of Rome also presented his soon-to-be-published report on circular economy.[38]

Lovesong to the Earth

Finals were coming up. Now it was time to enlist top music artists. On September 15, "Lovesong to the Earth" premiered—a soundtrack to the battle against climate change and for the radical transformation required. According to Paul McCartney, "The climate crisis is near a global tipping point, we hope everyone who hears this anthem takes action to encourage our political leaders to keep our planet safe, by keeping fossil fuels in the ground and moving toward 100 percent renewable energy."[39]

The goal was to reach out to a new and wider audience with the message that it was time to act for the climate. The listeners were asked to share the song and sign a petition to sway world leaders. Twelve top artists, including McCartney and John Bon Jovi, participated. Organizers were Friends of the Earth and Ted Turner's UN Foundation. In the background we find organizations such as 350.org, UNEP, and Live Earth,[40] all connected to Al Gore's initiative "Live Earth: Road to Paris."

Agenda 2030

On September 25, 2015, the UN Summit on Sustainable Development gathered 150 world leaders in New York. Opening speakers were Ban Ki Moon and Pope Francis. Here, the Agenda 2030 framework was adopted, with seventeen sustainability goals (SDGs) for completely "transforming our world."[41]

The wording of the goals were grand and utopian and included total eradication of poverty and hunger in the world. The world was to be rebuilt from scratch and made fair, inclusive, and healthy for both man and nature. The links (sometimes verbatim) to the sixteen "commandments" of the Earth Charter (see appendices E and F) were obvious.[42] The framework applies to *all* nations, requiring each and every one to achieve the same results regardless of national or local legislation, traditions, or resources—no one is to be left behind.

The Global Goals.

According to Agenda 2030 special advisor David Nabarro, "The 17 Goals represent an indivisible tapestry of thinking and action that applies in every community, everywhere in the world. They are universal."[43]

In the media this historic agreement was hardly mentioned—despite the fact that is was designed to have a profound impact on the future of humanity, covering virtually every aspect of human activity, which was now to be micromanaged from above. All attention was focused on the upcoming Paris Summit.

The Paris Terror Attacks

On Friday, November 13, just as Al Gore had started the broadcast of "24 Hours of Reality: The World is Watching," terror struck Paris. A total of 153 people were killed in coordinated attacks in six locations in central Paris and in the suburb Saint Denis (including a concert at Bataclan with Eagles of Death Metal and outside the Stade de France sports stadium). The Islamic State of Iraq and the Levant (ISIL) later claimed responsibility for the attacks.

Al Gore immediately postponed the remaining broadcast, with only Duran Duran having had a chance to play. The eyes of the world and an intense wave of sympathy were now directed at Paris. The world needed to unite against such shocking threats.

The G20 Summit in Antalya

Only two days later, a G20 meeting was held in Antalya, Turkey, with President Tayyip Erdoğan as host. Global terrorism—which was already on the agenda—became the main theme, along with the war in Syria, the refugee crisis, and the climate threat. In his opening speech, UN Secretary-General Ban Ki Moon said that 2015 constituted a "watershed year for international cooperation" and praised the willingness to come together to solve humanity's greatest problems.[44]

> Climate change is one of the greatest challenges of our time. We recognize that 2015 is a critical year that requires effective, strong and collective action on climate change and its effects. We reaffirm the 2 [degrees] C goal as stated in the Lima Call for Action. . . . Agenda 2030, including the Sustainable Development Goals (SDGs) and the Addis Ababa Action Agenda, sets a transformative, universal and

ambitious framework for global development efforts. We are strongly committed to implementing its outcomes to ensure that no one is left behind in our efforts to eradicate poverty and build an inclusive and sustainable future for all. (G20 Leaders' Communiqué, Antalya Summit, November 15–16, 2015)[45]

World leaders of the largest economies were now prepared to unite against terrorism and support the implementation of the new sustainability goals and the creation of a global utopia. Every country had to meet its targets. They also pledged allegiance to the climate agenda, with the goal of signing a binding agreement in Paris.

The Global Climate March

From November 28 to 29, 2015, the Global Climate March was organised by 350.org, with Avaaz and Coalition Climat 21, with a call to keep fossil fuels in the ground and to implement a fair transformation of the energy system to 100 percent renewable by 2050. Many environmental organizations, including Greenpeace, Friends of the Earth, and the WWF, joined

Avaaz's shoe protest at Place de la Republique in Paris, December 2015.

in. Due to the terrorist attacks in Paris, however, the French government banished demonstrations during the Summit. Instead, Avaaz arranged a symbolic silent protest with twenty thousand shoes placed on Place de la Republique.[46]

Avaaz also organised an illegal poster campaign in central Paris before the Paris meeting with faux WANTED posters of persons who had expressed varying degrees of climate skepticism; from lobbyists such as Marc Morano (CFACT) and lawyer James Taylor (Heartland Institute) to Fiona Wild (vice-president of BHP Billiton's Department for Environment and Climate Change) and Danish statistician Bjørn Lomborg, who had not questioned the actual threat, but only how effective and economically justifiable the proposed measures would be in relation to the funds invested.[47]

Avaaz's WANTED posters in central Paris, December 2015.

Avaaz

Avaaz is a progressive NGO, founded in 2007, that channels public opinion through petitions and public campaigns aimed at world leaders. Anyone can suggest petitions for worthy causes such as human rights, animal rights, and protecting the environment, but its main focus has often been climate change action, challenging Monsanto, and support for refugees. Avaaz founding president and CEO Ricken Patel has previously been involved with the UN, the Rockefeller Foundation, the Bill & Melinda Gates Foundation, and the Carnegie Council for Ethics in International Affairs.

The Paris Climate Summit

On November 30, 2015, under intense media coverage, COP21 began (almost exactly two hundred years after the Treaty of Paris 1815, which marked the end of the Napoleonic wars in Europe). The conference was held at the Le Bourget airport north of Paris, with heavily armed police officers everywhere and making use of the airport security checks even for the public part of the conference. Security had been significantly increased after the terrorist attacks. Expectations were high. *This* time, nothing must go wrong.

Entrance to the COP21 Climate Summit at La Bourget, Paris, December 2015.

In his opening speech, Prince Charles connected the Syrian crisis with climate change, while the otherwise critical Vladimir Putin adhered to the G20 summit communiqué about the climate, arguing that, "climate change has become one of the gravest challenges humanity is facing."

On December 12, 2015, after lengthy, drawn-out negotiations, the representatives of 196 nations finally signed the Paris Agreement.

The agreement was, however, not binding and had many loopholes, necessary concessions in order to get all member states on board. Johan Rockström and Hans Joachim Schellnhuber viewed it as toothless and, just like a multitude of environmental organizations, wanted to see a full reduction to zero emissions by 2050, even if the target agreed upon had been tightened to not exceed a mean global temperature rise of 1.5°C.[48]

The Bulletin of the Atomic Scientists expressed similar dissatisfaction and refused to adjust Doomsday Clock back from "3 minutes to midnight" despite the Paris Agreement and the Iran nuclear agreement in April 2015.

257 Ecstatic world leaders after the signing of the Paris Agreement, December 12, 2015. Laurence Tubiana (European Climate Foundation), Christiana Figueres (UNFCC), Ban Ki-moon (UN Secretary-General), Laurent Fabius (Foreign Minister of France and president of the COP21 Climate Summit), and Francois Hollande (President of France).

The news coverage of the Paris Climate Summit finale was interspersed with special features from CNN Weather Center explaining the background in a simplistic way, with renderings of major capitals such as London and Sydney submerged to illustrate what a horrible fate the historic Paris Agreement would now be saving us from.

> The voluntary pledges made in Paris to limit greenhouse gas emissions are insufficient to the task of averting drastic climate change. These incremental steps must somehow evolve into the fundamental change in world energy systems needed if climate change is to ultimately be arrested. (Sivan Kartha, climate change expert at SEI and member of the Science and Security Board of *The Bulletin of the Atomic Scientists*)[49]

The Rockefeller Brothers Fund and its chief executive Stephen Heintz, however, saw it as a great victory. Now, efforts to make the agreement tougher could begin. The Fabian strategy continued. Paris was only one subgoal of many. The RBF was ready to finance organizations with the aim of both implementing and strengthening the goals. Governments would be held accountable for living up to their commitments and businesses forced to make necessary changes.[50]

During the conference, Stephen Heintz and May Boeve from 350.org announced that more than five hundred institutions, with $3.4 trillion in assets, had made divestment commitments.[51] In an interview in May 2016, Heintz later stressed the RBF's efforts to realize the Paris Agreement and that it "exceeded all their expectations."[52] This was obvious, not least through the good contacts with the White House through Chief of Staff John Podesta, and the lobby group Presidential Climate Action Project. President Obama praised the United States' leadership in the battle against climate change that had made the deal possible:

> Today, the American people can be proud—because this historic agreement is a tribute to American leadership. Over the past seven years, we've transformed the United States into the global leader in fighting climate change.[53]

The president of the Paris Conference, Laurent Fabius, and Christiana Figueres (executive secretary of the UNFCCC) were also pleased. The process of rebuilding the world could now proceed.

> This is the first time in the history of mankind that we are setting ourselves the task of intentionally, within a defined period of time to change the economic development model that has been reigning for at least 150 years, since the industrial revolution. That will not happen overnight and it will not happen at a single conference on climate change, be it COP 15, 21, 40—you choose the number. It just does not occur like that. It is a process, because of the depth of the transformation (Christiana Figueres, February 3, 2015).[54]

All that was needed now was to anchor the agenda more firmly with individual nations and to ultimately create binding commitments from each country.

Under the direction of Figueres, the UNFCCC had formed a partnership with the Rockefeller Foundation to demonstrate the essential role women could play in addressing climate change.[55] Figueres, the daughter of the former president of Costa Rica, José Figueres Ferrer, was herself a living example of this.

Before she started working at the UNFCCC, she had been a board member of Winrock International (founded in commemoration of Winthrop Rockefeller in 1985), together with Rockefeller family members Neva Goodwin, Peter O'Neill, and David Kaiser.[56]

Christiana Figueres (1956–), UNFCCC.

Figueres was now aiming higher. On July 6, 2016, she resigned as head of UNFCCC, to being nominated for the position as the new secretary-general of the United Nations. She felt that it was time for a woman to lead the organization. She was also a strong believer in Global Governance and thought the UN needed more muscles: "I am very convinced that society as a whole, global society, is moving to a point where we are going to need more and more global governance muscle than we have had in the past." She did, however, not receive enough support in the Security Council and later withdrew her candidacy.

So, what type of society was to be created and how would the economy be restructured? What did "smart globalization" mean, which the Rockefeller Foundation had declared in its 2007 annual report? How could humanity survive without degrading the environment?

The answer would come shortly after the signing of the Paris Agreement: a new revolution would begin, where mankind and the physical environment were to be digitized, upgraded, and monitored—all for the good of humanity and the environment.

Henry Kissinger
Special Studies Project 1956–58

The alternative to a new international order is chaos.
—Henry Kissinger

1956–1958	**RBF's Special Studies Project**	Coordinator
1957	Bilderberg	Member
1958–1969	Harvard Center for International Affairs	Director
1960–1968	Governor Nelson Rockefeller's presidential campaigns	Foreign Policy Advisor
1969–1975	The Nixon Administration	National Security Advisor
1973	Nobel Peace Prize	Laureate
1973	**Trilateral Commission**	Member
1973–1975	**Nelson Rockefeller's Critical Choices for Americans**	Member
1973–1974	The Nixon Administration	Secretary of State
1974–1977	The Ford Administration	Secretary of State
1977–1981	**Chase Manhattan International Advisory Committee**	Chairman
1977-1981	**Council on Foreign Relations**	Board of directors
1977–1987	**Rockefeller Brothers Fund**	Board member
1977	Aspen Institute	Board member
1982	Kissinger Associates	Founder
1984–1990	The President's Foreign Intelligence Advisory Board	Advisor
2001–2016	Defence Policy Board	Advisor

HENRY WAS BORN in 1923 in Fürth, Germany, as the eldest son of Louis Kissinger and Paula Stern. In 1949, he married Ann Fleischer. They had their children Elisabeth and David. He divorced Paula in 1964 and married Nancy McGinnes a decade later. Henry arrived in the United States in 1938 as a Jewish refugee. He quickly got adjusted to his new country. In 1943, he returned to Germany to fight the Hitler regime. After the war he got a PhD in political science at the Harvard Department of Government. In 1956, he was recruited by Nelson Rockefeller as head of RBF's Special Studies Project and became a close friend and ally of the Rockefeller family. After founding the Harvard Center for International Affairs with Robert Bowie he became advisor to Nelson Rockefeller and joined his Commission for Critical Choices for Americans. Both found themselves at the nexus of power when Nelson was vice president and Henry was secretary of state under Gerald Ford. Kissinger also helped found David Rockefeller's Trilateral Commission. After his sojourn in the White House, Henry became chairman of the Chase Manhattan International Advisory Committee and board member of RBF, CFR, and Aspen Institute. In 1984, he founded Kissinger Associates and has since been frequently hired by the White House for strategic counsel. Kissinger died in his home on November 29, 2023, at the age of one hundred.

The Smart City in which everyone and everything are connected.

Chapter Eleven

The Fourth Industrial Revolution

We stand on the brink of a technological revolution that will fundamentally alter the way we live, work, and relate to one another. In its scale, scope, and complexity, the transformation will be unlike anything humankind has experienced before.
—Klaus Schwab, 2016[1]

BRAVE NEW WORLD ECONOMIC FORUM

The World Economic Forum was founded in 1971 as the European Management Forum (see chapter 5) and defines itself as "the international organization for public–private cooperation."

Members and partners include many of the world's leading multinational corporations in virtually all sectors (e.g., IT, banking, oil and gas, automobile, aviation, freight, chemistry, biotechnology, pharmaceuticals, food, consumer goods, media, and entertainment).

Founder Klaus Schwab is executive chairman and Børge Brende (former minister of foreign affairs of Norway, Norwegian Red Cross, Mesta, Statoil) has been president since 2017.

The 28 members of the Board of Trustees (2023) included Peter Brabeck-Letmathe (Nestlé, Roche, Credit Suisse, L'Oréal, ExxonMobil), Laurence D. Fink (BlackRock), Al Gore (Climate Trace), André Hoffman (Roche, Masselaz), Christine Lagarde (IMF), Jack Ma (Alibaba), Yo-Yo Ma (cellist), Ngozi Okono-Iweala (World Trade Organization, GAVI), David M. Rubenstein (Carlyle Group, Council on Foreign Relations), Zhu

Min (Bank of China, People's Bank of China, IMF, the World Bank), and Queen Rania al Abdullah of Jordan.

In January 2016, only a month after the Paris Accord, at the World Economic Forum's annual summit in Davos, Schwab, a disciple of Henry Kissinger, proclaimed the start of the Fourth Industrial Revolution (4IR). This was the theme of the conference, which gathered the world's top economic and political players.

The world was about to be transformed in an unprecedented way. Digital, physical, and biological systems would be merged. The Internet of Things (IoT), nanotechnology, robots, artificial intelligence, brain-computer interface, and smart cities—a post-human world. Reality was to be blended with science fiction into something eerily like Aldous Huxley's *Brave New World*. Both the planet and the environmentally destructive humanity needed an upgrade. Schwab warned that this development could completely redefine what it means to be human!

> In its most pessimistic, dehumanized form, the Fourth Industrial Revolution may indeed have the potential to 'robotize' humanity and thus to deprive us of our heart and soul. But as a complement to the best parts of human nature—creativity, empathy, stewardship—it can also lift humanity into a new collective and moral consciousness based on a shared sense of destiny. It is incumbent on us all to make sure the latter prevails.

This dystopian vision was presented to the financial, political, cultural, and scientific superclass assembled after arriving in 1,700 private aircraft and helicopters and paying an admission fee of around $19,000 each for their participation.

Among speakers were Al Gore, actor Leonardo DiCaprio (National Resources Defense Council), UN Secretary-General Ban Ki Moon, Christiana Figueres (UNFCC), Naomi Oreskes (author of *Merchants of Doubt*), and Johan Rockström (the Stockholm Resilience Center, and soon-to-be appointed to the Swedish government's delegation for the implementation of Agenda 2030). Rockström emphasized that the transformation

World Economic Forum founder Klaus Schwab (1938–) delivering his 2016 version of "future shock."

needed to take place within the "nine planetary boundaries" but at the same time saw how a prosperous future could be created if resilience and justice were linked to the Fourth Industrial Revolution—visions which he felt must be rapidly implemented.[2]

The proposals from the Davos Summit soon found their way into policy documents around the world, including at the G20 Summit in

Hangzou, China, held on September 4 and 5, where a New Industrial Revolution Action Plan was presented.[3]

The following year, at the G20 Summit in Hamburg, Germany, July 7–8, 2017, under the presidency of Angela Merkel, the 4IR became the main focus, under the motto "Shaping an Interconnected World."

The technologies and ideals of the Fourth Industrial Revolution were thereafter introduced into the intellectual and political debate in member states and around the world, spawning a sudden flood of public relations articles, panels, lectures, and TV and radio shows about AI, robotics, transhumanism, and smart cities. There was, however, very little critical debate on potential consequences. Few asked if 4IR was realistic, desirable, financially justifiable, or safe. If mentioned at all, risks tended to be downplayed or presented as manageable, often through even more intrusive or disruptive technology. The general public, soon about to be monitored, controlled, and upgraded—and often paying for it through consumer products or taxes—was not consulted on their opinion.

The Fourth Industrial Revolution

Just in time for the Davos Summit 2016, Klaus Schwab's book *The Fourth Industrial Revolution* (ghostwritten by WEF employee Nicholas Davis) was published, outlining the vision of the impending revolution.

The tone of the book is chillingly rational and paints a picture of a perfect society where both man, nature, and the earth's weather system are programmable cogs in a global machinery that only needs calibrating to be perfect. The 4IR is presented as having the potential to change the course of history with a direct impact on all aspects of our existence.

After an initial disruptive transformation, a better and more efficient system is to be born: a sustainable future in the form of a global panopticon, where both human life and nature are to be supervised, controlled, and transformed. In this vision, both humans and the earth system must be merged with the technological system into a single controllable unit—a combination of Big Brother and Big Mother, where everyone is taken care of and guided safely from the cradle to the grave.

At the end of the book, twenty-three deep shifts are listed (including implants; portable internet; internet of things; smart cities; big data; driverless cars; artificial intelligence, robotics; blockchain; sharing economy, 3D printing; and design creatures)—each with their respective advantages, disadvantages, and unpredictable consequences, as well as an estimated timeline for their introduction.

Despite the serious risks outlined in the book, the conclusion is still that these new technologies are both inevitable and necessary for the implementation of the United Nations Sustainable Development Goals (SDGs), and for regulating carbon dioxide emissions.

Shaping the Fourth Industrial Revolution

Two years later, during the 2018 Davos Summit, Johan Rockström and Christiana Figueres declared their intention to gather a coalition of the large tech companies in order to turn them into "planetary stewards" with the task of creating a stable climate with zero emissions.[4] The new smart technologies were presented as a warranty for saving the world.

Just before the conference, the follow-up book, *Shaping the Fourth Industrial Revolution*, was published, with a foreword by Microsoft CEO Satya Nadella.[5] With the help of experts from the World Economic Forum's focus groups and conferences, graduates, senior executives, decision makers, and 240 leading thinkers, the agenda of the Fourth Industrial Revolution is outlined more in detail. Just like the SDGs, the 4IR aims at controlling every aspect of life, and not just human life but the whole earth system, by

- Extending digital technologies
- Reforming the physical world
- Altering the human being
- Integrating the environment

Schwab's two books read like a Bible of Alchemy—a combination of futurism and transhumanism, seemingly aiming at a global technocracy. Yet, in order to save the planet, world leaders, in cooperation with Big

Tech and international organizations such as the World Economic Forum, Trilateral Commission, G20, and the Rockefeller foundations, are rolling out the 4IR at breakneck speed, with the goal of reaching every corner of the world.

> The fourth industrial revolution is creating unprecedented opportunity for human advancement. Technologies such as AI, robotics, nano and bio tech offer opportunities for advancement in health, education, labor, jobs and massive increases in productivity. (Rockefeller Foundation, 2019)[6]

All that is needed to minimize the identified risks of this global technological quantum leap is—yes, *again*—that it is overseen and regulated by a global authority.

TECHNICAL CLIMATE SOLUTIONS

Smart Cities

The smart city concept is promoted as a new paradigm in urban planning and the expansion of "smart cities" is presented as a crucial part of the development of the future sustainable society. A smart city is generally based on ICT (information and communication technology) and a more or less developed network of sensors to collect data, monitor, and enable centralized control and management of the city's resources (e.g. traffic and transport systems; energy supply; water, sewage, and waste; street lighting; justice and information; schools, libraries, hospitals; buildings; goods, services, and human capital) with maximum efficiency. For citizens, it can signify anything from internet access and online services to full e-governance with systems for digital ID and payment, apps for communication with authorities, online income tax declarations, and even digital voting.

In 2012, the United Nations' new Green Climate Fund was established in the world's first smart city, Songdo, outside Seoul in the first free economic zone in South Korea, due to the city's environmental and climate profile, advertised as a car-free city with 40 percent green space and plenty of bicycle lanes.

The smart city Songdo, South Korea, is built in the International Style, following the outdated urban ideals of Le Corbusier (note the similarity between Posco Tower to the right and the new One World Trade Center in New York).

The planning of this experimental city began in 2003 and Songdo is a typical example of the new borderless global economy. It was a public-private partnership project between the steel giant Posco and Gale International, designed from scratch by American architects Kohn Pedersen Fox Associates (who had led the renovation and expansion of MoMA and Standard Oil's former headquarters at Rockefeller Center). Gale supplied the whole city with everything in it. All buildings are connected to the internet and to each other, with information and monitoring technology built into the infrastructure. Traffic flow and citizen behaviour is monitored in real-time via five hundred surveillance cameras. Household waste is automatically transported via the pneumatic system under the city and converted into energy.

All apartments have smart locks, with smart cards which can also be used for loaner bikes, parking, subway, and movie tickets. All apartments have smart meters (enabling residents to compare their energy consumption with that of their neighbors) and built-in cameras everywhere. Floor sensors detect pressure changes and automatically alert an alarm service of a suspected fall. Systems are tested where residents via the TV screen can receive language lessons or communicate with their physician as well as neighbors and relatives, and bracelets for locating children via GPS.[7] In other words, a futuristic dream straight out of the World Future Society's 1970s vision—or Orwell's *1984*. And this is *South* Korea.

How successful, environmentally friendly, and inclusive Songdo really turned out to be has been questioned. It was built primarily for an affluent middle class expected to be able to afford the higher standard and the new technology. The electricity comes from coal-fired power plants and the buildings are completely glazed with windows that cannot be opened, which requires air conditioning all year round.[8] Also, the pneumatic waste disposal system does not always work properly. As of March 2018 there was still no cultural life, no street vendors or old people, public transport systems was described as "a nightmare," and three quarters of the homes were still empty.[9]

There is actually very little evidence that a high-tech mega-city is the same as an ecologically, economically, or socially sustainable city. The specified goals in the New Urban Agenda and Agenda 2030 of biodiversity, health, inclusion, equality, security, and proximity are likely better met by small-scale New Urbanism and traditional towns than by high-tech high-rises.

This fact, alas, does not prevent a growing number of nations from racing to build new smart cities or implement ICT-based smart city programs in existing urban areas. India planned to build twenty new smart cities by 2021.[10] China has around five hundred smart city projects underway and intended to build one hundred new smart mega cities by 2020.[11]

After the 2016 summit, a flood of conferences and trade fairs across the world soon began peddling the smart city concept. With lucrative contracts up for grabs, leading tech consulting companies like Sweco, Cisco, IBM, and CGI stood ready to assist. The market for smart cities was expected to increase from $563.36 billion in 2016 to astronomical $2.57 trillion by 2025.[12]

Internet of Things

A central component of the smart city, and vital part of the vision for the global transformation to a sustainable society, is the Internet of Things (IoT).

> The IoT is not just a new paradigm, it is a new world order, not so much in the political sense but in the nature of the term: 'order' as in 'hierarchy,' reciprocity and communicative relations. We are entering a world in which the environment becomes the interface, and there will be no more dual relations (me and you, me and an object), but there will be always a third party (sensor–database) involved. (Internet of Things Council)[13]

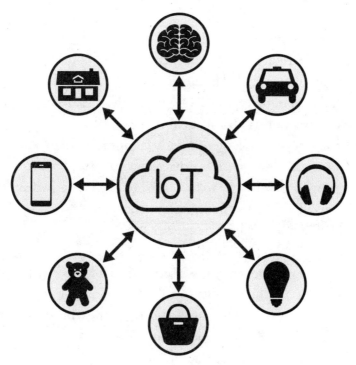

Internet of Things.

Internet of Things is a cyber-physical system where not just cell phones but everyday items such as household appliances, clothing, accessories, lighting, machines, vehicles, and buildings are equipped with passive or active (transmitting) chips and sometimes sensors and actuators that allow them to be tracked, exchange data, or be controlled over the network.

The network is intended to connect at least 80 billion devices and products, and can also include humans (via RFID chips in ID cards, credit cards, access cards, or inserted into the body). Blockchain and a digital currency can also be connected to the system.

All products and individuals included in the system are to be in continuous communication with each other and nothing is to pass under the radar. The data sets will be analysed in real time using artificial intelligence.

The European Union want to follow the test roll-outs in Hong Kong and Singapore and install millions of "smart lampposts" with sensors, cameras, speakers, Wi-Fi, advertising/information displays, e-vehicle charging, and many other features.[14]

For humans, Internet of Things is expected to help influence and improve our decision-making and behavior "in both open and subtle ways" (nudging) to promote "health and longevity" (e.g., by enabling one's physician or an app to keep track of one's medical status in real time).

Crime prevention is also to become more effective. Data collection via sensors and cameras equipped with facial recognition software, processed by artificial intelligence, is expected to lead to reduced crime. In the future, some police work may to be carried out by robots, where "criminal elements" may be rendered harmless by robots and drones—or by being identified before the crime is committed and "offered preventive councelling," as has been suggested in the UK.[15]

According to proponents, the losers in this brave new world will be those who don't see the benefits of the new innovative business models being developed, or who don't have access to it. The ethical challenges will be addressed by promoting development of digital solutions with "respect for human rights"—as long as they are not overemphasised and subject to "overly rigid regulations."

5G

In order to manage the dramatic increase in data flow required for a fully developed IoT (with smart houses, smart electricity grids, digital assistants, intelligent transport systems, etc.) and the increasing use of online, cloud, and streaming services, this new type of mobile network is now underway.

The 5G network is a complementary mobile network that transmits at higher frequencies with millimeter waves. The shorter range and weaker penetration requires more base stations, placed closer together (e.g., on lampposts). The directional high frequency waves can also be directed in real time towards specific receivers (e.g., in self-driving vehicles). Jon Markman wrote in a 2019 *Forbes* article,

> 5G isn't just the next generation of a wireless connectivity. It is the foundation for the first generation of truly smart things. Fast, low latency networks will support billions of connected devices communicating at a machine level.[16]

Advocates promise faster internet, increased efficiency in manufacturing processes, and enabling "circular economy" with cradle-to-grave tracking of every product—and citizen.

> Companies can refine their monitoring of projects by using drones and embedded sensors to enable real-time communication and to track people, machines, components and the construction process itself. (Shaping the Future of Construction, World Economic Forum)[17]

Despite being controversial, 5G is currently being implemented in a growing number of regions across the world. Potential risks to wildlife, the environment, human health, and personal integrity tend to be downplayed by proponents and public concerns either ignored or ridiculed.

Blockchain

Blockchain is a digital distributed ledger shared in a network, where each transaction is encrypted and added to a block in the ledger, forming a blockchain of coded information which is difficult to manipulate. The technology is mainly used for digital crypto currencies such as Bitcoin but can also be used for automated and decentralised management of existing currencies, royalty payments, contracts other uses requiring trust, without (but not excluding) the need for an authorized third party such as a bank, law firm, or government agency.

In *Shaping the Fourth Industrial Revolution*, blockchain technology is presented as safe, transparent, democratizing, and inclusive. One of the problems, however, is that the energy and bandwidth used by blockchain technology to process and store all transactions far exceeds the environmental benefit. According to an analysis by the Bank of International Settlements (BIS), the exponentially increasing adding of transaction blocks could soon cause a system overload.

> A big part of the appeal of many cryptocurrencies to their supporters is that they are decentralized rather than tied to a central bank like the US Federal Reserve. Records of transactions are kept on a digital ledger. But because every single transaction is added to the digital ledger, the report said using a cryptocurrency like bitcoin for retail transactions around the world would quickly swell the ledger beyond the capacity of computer servers to store it.[18]

Artificial Intelligence and Robotics

AI and robots, including self-driving vehicles and remote-controlled drones, are described by Schwab as such a rapidly growing presence in our everyday life that what was previously only science fiction is now fast becoming reality. Advantages, disadvantages, and risks are addressed, such as the fact that the robotisation risks replacing increasingly qualified professions and competing with developing countries which would otherwise attract investments through cheap labor, as well as the risk of hacking and the use of robots, drones, and AI for warfare, terrorism, and crime. It can also be used for "pre-crime" law enforcement.

> AI is already monitoring data from sensor networks and video streams and can alert security officials to suspicious patterns. Meanwhile, police have deployed robots for search and rescue, and have also used them to kill an armed gunman. (Klaus Schwab, *Shaping the Fourth Industrial Revolution*)[19]

The very first AI conference was held at Dartmouth College in 1956 and the first industrial robot was launched in 1961. As always, the Rockefellers were at the forefront (see chapter 2):

> At The Rockefeller Foundation, we know that supporting communities of dedicated engineers can change the world. In 1956, we funded the Dartmouth conference that coined the phrase "Artificial Intelligence" and launched a new way of thinking about computation.[20]

Energy Production

In *Shaping the Fourth Industrial Revolution*, great hopes are placed on the renewable energies—especially for developing countries with poor or unreliable energy supply—if only strategic investments are made. Smart electricity grids controlled by AI, nanotechnology, biotechnology, and fusion are proposed as promising future avenues of development for greener, cheaper and more efficient energy supply.

The intermittency, distribution and storage shortcomings of solar and wind power are recognized by Schwab, but hopes are placed on batteries with better storage capacity solving some of the problems. The potential harmful effects on health or the environment (mining, land use, transportation, and danger to birds, bats, and insects) are not considered.

That biofuels can have both a positive and negative net effect depending on how they are produced is also not mentioned. Biogas (e.g., from sewage or food waste) can have a more positive net effect than ethanol (created by fermenting sugar-containing crops) which may require use of agricultural land needed for food production.[21]

The most controversial biofuel is probably biodiesel, often made from palm oil, which contributes to rainforest degradation, eutrophication, and pollution of waterways.[22]

Nuclear fission, so popular with the early anti-coal and -oil activists, is not viewed by Schwab as an option, only hopes for future fusion reactors. The issue is still controversial. Climate scientist James Hansen, for example, has earlier expressed a certain realism when it comes to energy production but his position on fourth-generation nuclear power is not welcomed by the environmental movement:

> I think it's unfortunate that so many environmentalists are just assuming that these renewables will be able to satisfy all of our requirements. Renewables, the "soft renewables," are only providing between one and two percent. Hydropower provides a significant amount of electricity. But that's limited as to how much of that we can have. The hope that sun and wind and geothermal can provide all of our energy is a nice idea, but I find it unlikely that that is possible. The environmental community is basically asking the governments to reduce their emissions and subsidizing clean energy. Well, that simply doesn't work, we don't get enough energy from renewables to make a difference. That then forces any government to approve expanded oil drilling, hydro fracking to get more gas, mountain top removal to get more coal. . . . We're not going to turn the lights out. No government, no president, no governor is going to turn out the light, there has to be energy. And if renewables are not providing it, then it's fossil fuels.[23]

Geoengineering

Geoengineering is the human attempt to control Earth's complex biosphere and atmosphere. Theoretical geoengineering techniques for counteracting global warming include carbon capture and storage (CCS), marine fertilizers, artificial islands, large-scale tree planting, cloud seeding with aerosols to create rain or artificial smog, large mirrors, and nanotechnology.

Human intervention on the planet's weather system in order to mitigate challenges such as air pollution, drought, and global warming are also outlined in *Shaping the Fourth Industrial Revolution*. The potential risks of such large-scale experiments are addressed by the book's co-authors, who believe that relying solely on reactive technical methods will not suffice, but need to be combined with emission reducing strategies.

In their *Global Risks Report* (2019), the World Economic Forum points out the additional risk of individual nations using weather manipulation to induce drought and flooding as weapons against each other.

> Weather manipulation tools—such as cloud seeding to induce or suppress rain—are not new, but deploying them at scale is becoming easier and more affordable. As the impacts of climate-related changes in weather patterns intensify, the incentives to turn to technological fixes will increase in affected areas.[24]

To manage these risks, supranational regulations are again called for. These ideas are very similar to Hans Joachim Schellnhuber's futuristic-technocratic ideas outlined in his 1988 article, "Geocybernetics: Controlling a Complex Dynamical System under Uncertainty," proposing a future geocybernetic control system for managing both nature, climate and man.

> Global change, i.e., the mega-process radically transforming the relationship between nature and human civilization since the end of World War II, is investigated from the point of view of systems analysis. It is argued that this unbridled process should rather be domesticated by planetary control strategies transpiring from a new science called "geocybernetics."[25]

This concern echoes the conclusions in *Rockefeller Panel Reports* (1958):

> If it becomes possible to interfere actively in the big processes with
> the atmosphere, the results are likely to transcend national boundar-
> ies. The problems that will then arise must be handled on an interna-
> tional basis. They may well be insoluble if the development leading
> up to weather control has been carried out by uncorrelated national
> efforts.[26]

Space Technology

Shaping the Fourth Industrial Revolution also includes space technology,
such as Elon Musk's SpaceX space program. High hopes are placed on
new or improved technologies such as microsatellites, nano materials, 3D
printing, virtual reality, robotics, and enhanced space telescopes making
space travel cheaper and easier to implement. The awaiting future promises to become "a whole new era of understanding how man fits into a
global and cosmic context." Space technology will of course not save the
climate but enthusiasts view it as a long-term Plan B for humanity.

ECONOMIC CLIMATE SOLUTIONS

Several models for reducing carbon dioxide emissions and environmental degradation have been tested, proposed or are being introduced, from
CO2 taxes to carbon quotas to an overhaul of the entire economic system
into a New International Economic Order.

Carbon Tax

Carbon tax is a tax on non-renewable fuels, based on their carbon content,
producing carbon dioxide at combustion. Among the first to introduce
a carbon tax were the Netherlands (1990), Sweden and Norway (1991),
Denmark (1992), and Great Britain (1993), followed by a growing number
of countries around the world.

Using an incrementally but exponentially increasing carbon tax has
been advocated by, among others, the World Bank (after guidance from
Maurice Strong in the 1990s).[27]

[T]he expected effect of a carbon tax is not to decrease emissions immediately or brutally, as that would be a costly shock to the economy. Instead, a carbon price is expected to first progressively reduce the pace at which GHG emissions are growing until that growth stops and emissions finally start to decrease. Also, the best design for a carbon price is to make it grow exponentially over time. (*Decarbonizing Development: Three Steps to a Zero-Carbon Future*, World Bank Group, 2015)[28]

Carbon Quotas

Cap and trade is a system where national governments set a quota of greenhouse gas emissions for companies. Any excess can be sold to other companies needing more. The system has been used in various regions for several years, including in the European Union.

The proposal for emission quotas originates with William W. Kellogg (RAND and NCAR) and Margaret Mead at the 1975 Endangered Atmosphere conference, where they suggested that various nations be assigned polluting rights to keep carbon dioxide emissions below a globally agreed-upon standard.[29] The proposal was later revived by Al Gore (in *Earth in the Balance*, 1992) and David Blood, Mark Ferguson, and Peter Harris (Goldman Sachs Asset Management), seeing profits to be made on the scheme. In a 2010 *Rolling Stone* article, Matt Taibbi reported,

The feature of this plan that has special appeal to speculators is that the "cap" on carbon will be continually lowered by the government, which means that carbon credits will become more and more scarce with each passing year. Which means that this is a brand new commodities market where the main commodity to be traded is guaranteed to rise in price over time. The volume of this new market will be upwards of a trillion dollars annually.[30]

Personal Energy Quotas

Tradeable Energy Quotas (TEQs) is a system where each individual is allocated a quota of energy (or greenhouse gas emissions) to consume. Any surplus can be sold to others.

The idea of personal or domestic energy quotas was developed by British David Fleming (1940–2010), with a past in the Green Party, Transition Towns, and New Economics Foundation. It was first published in June 1996.

However, the idea originally dates back to the technocracy movement of the 1930s, which advocated—and still does—even more extreme measures such as replacing the price-based economic system with an energy-based system calculated on the distribution and consumption of energy for each product or service.[31] Just like M. King Hubbert (one of the founders of Technocracy Inc.), Fleming warned that peak oil was imminent, but combined this threat with the threat of climate change.

In 2010, Fleming presented his concept at a seminar in the Swedish parliament with Johan Rockström, Anders Wijkman and parlament member Per Bolund, current leader of the Swedish Green Party.[32] According to Fleming's plan, "Every adult is given an equal free Entitlement of TEQs units each week. Other energy users (Government, industry etc.) bid for their units at a weekly Tender, or auction."[33]

The energy unit is measured in carbon dioxide—one kilogram per unit. The individual's values and behavior should also be modified by being able to track in real time how one's lifestyle gives rise to the greenhouse gas emissions said to be destroying our planet.

> TEQs have long been Green Party policy, as we believe that we need a fair and transparent system to reduce energy demand and give each person a direct connection to the carbon emissions associated with their lifestyle. (Caroline Lucas, Green Party and GLOBE)[34]

The idea is also that we should "help each other" by keeping track of our neighbors' emissions—just like in Songdo. A Climate Change Committee, independent of national governments, is to determine the size of the annual budget. The ration is then meant to decrease year by year and each individual gets less and less non-renewable energy to use, until reaching the desired target.[35] The European Union, or example, has set the target of an 80–95 precent reduction by 2050. Given how negligible the portion of renewable energy so far is, and how dependent on fossil fuels it is to be produced, the concept appears diabolical. The rich can buy their way out while the poor end up getting less and less each year.

This closely resembles the dystopian film *In Time* (2011), where a time quota (to stay alive) is the currency and the ration keeps decreasing—except for the ruling class, which has unlimited time rations.

The infrastructure for managing TEQs is based on smart grid networks and each individual being part of the Internet of Things. When connected to home electronics these could be automatically shut down if the allocated quota is exceeded.[36] This opens up for extensive surveillance and mapping to determine who is sufficiently "sustainable," as well as for hacking and abuse of power.

According to Swedish Smartgrid, "The development of smart electricity networks means that data and information will be collected with ever higher resolution and with ever shorter time intervals. When the data on the individual's electricity consumption increases, the possibilities of mapping persons and companies' movements are also increasing."[37]

A recent unexpected spokesperson for this solution is the young climate activist Greta Thunberg (daughter of the Swedish opera singer Malena Ernman and descendant of Svante Arrhenius on her father's side).

In January 2019, she was invited to the World Economic Forum Summit in Davos, where she urged world leaders to introduce a new global energy currency:

> No other current challenge can match the importance of establishing a wide, public awareness and understanding of our rapidly disappearing carbon budget, that should and must become our new global currency and the very heart of our future and present economics.[38]

Most likely she is unaware of the technocratic roots of the idea, or the Orwellian dystopia it might lead to if it was actually implemented.

Carbon Currency

Carbon currency is a system where each product or service gets its carbon footprint fixed. With blockchain technology or special apps collecting and storing each consumer's transactions, it is possible to calculate that person's total carbon consumption footprint.

In March 2017, the Stockholm Environment Institute and the WWF launched its digital climate calculator.[39] A year later, Swedish supermarket chain ICA offered a digital tool to enable their customers to follow their own climate footprint month by month.[40] Although initially voluntary, such registration opens up for it becoming mandatory and no longer private in the future.

Carbon Offset

Carbon offset is a means of compensating for emissions (e.g., from aviation) where the company or the consumer pays a fee towards projects assumed to reduce CO_2 emissions, e.g., solar panel projects or tree planting programs.

For companies, carbon offset schemes is a convenient way for a polluting or carbon-intensive company to greenwash their image by investing in offset projects, often in developing nations. Now, concerned consumers can also ease their assumed climate conscience by buying such climate indulgences.

A growing number of companies are now offering various online calculators or mobile apps for calculating and compensating for one's carbon emissions, including MyClimate.org,[41] SAS,[42] and Poseidon (a Swiss foundation).[43]

Offset schemes such as REDD (Reducing Emissions from Deforestation and Forest Degradation) have also become a growing business for opportunistic entrepreneurs and NGOs—and for developing countries in need of investments. However, they are often fraught with problems, including monitoring and assessing success rate.[44]

In some cases, projects have led to ruthless land grabs where the original inhabitants get violently evicted from land they've inhabited for generations, and/or prohibited from using land earmarked for the project.[45]

Circular Economy

An essential component of the vision of sustainability is circular economy—a utopian model based on the idea of recycling and more efficient management and use of resources, ideally leading to a world without waste.

In 2017, the World Economic Forum initiated the public-private collaboration Platform for Accelerating the Circular Economy (PACE), chaired by the CEO of Philips, Frans van Houten.[46] Connected to this project is also British solo long-distance sailor, Dame Ellen MacArthur. In 2000, she founded the Ellen MacArthur Foundation (in partnership with Cisco, the BT Group, B&Q, Renault, and National Grid). In 2013, she became a member of the Club of Rome and has become a passionate champion for circular economy.

To be effective over time, proponents insist that circular economy requires the Internet of Things, where all products and components are tracked in real time throughout their life cycle.

> To be sustainable, a system must be responsive; actions and behaviours must be connected via data and knowledge. With the embedding of intelligence in almost every object, we can imagine systems that adapt and respond to change in order to remain fit for purpose. (Tim Brown, IDEO)[47]

The origin of the idea of circular economy can be traced to the view of earth as a closed system with finite resources which need to be carefully used and recycled, like on a spaceship. This idea was spread through Barbara Ward's book *Spaceship Earth* (1966),[48] made into a film just before the Stockholm Conference; Kenneth E. Boulding's essay "The Economics of the Coming Spaceship Earth" (1966);[49] and Buckminster Fuller's *Operating Manual for Spaceship Earth* (1969).[50] "Spaceship Earth: The Life Support System" was also the title of chapter 7 in the RBF's *The Unfinished Agenda* (1977). It has since been marketed by the Club of Rome and the New Economics Foundation.

In 2016, former European Parliament member Anders Wijkman (GLOBE, Club of Rome, World Future Council) made a "study" for the

Club of Rome, based on models predicting positive effects on climate, environment, and economy and pointed out the engagement from the European Commission resulting in the Circular Economy Package:

> The 'circular economy' is an industrial system that is restorative by intention and design. The idea is that rather than discarding products before the value are fully utilized, we should use and re-use them.[51]

David Rockefeller Jr.'s wife, Susan, a film director, has also been enthusiastic about the vision of a zero-waste Utopia:

> My greatest hope is for a worldwide spiritual transformation simultaneous with advanced technology and global empathy in the way we approach all our production and processes—in essence, to have a circular economy where there is zero waste.[52]

Sharing Economy

Closely related to circular economy is the sharing economy (or access economy). It is an informal peer-to-peer business model where individuals can "share" (free of charge or for a fee) personal assets like their homes, vehicles, tools, or time with strangers, often via a mobile app, which charges a commission for offering a platform for booking and payment. Such digital platforms are said to make it easier for entrepreneurs, freelancers, and private citizens to market themselves and rent goods or services, resulting in lower prices and greater accessibility for the consumer.

However, this "informal" rental market has rapidly become dominated by giants such as Airbnb and Uber, resulting in hotels and taxi companies facing unfair competition from amateur operators without professional experience, education, trade union membership, fixed costs, or tax registration.[53]

This is also true of the closely related "gig economy" where easy access to global labor for small temporary jobs (e.g., TaskRabbit or Amazon's Mechanical Turk), which have been criticized for creating a "race to the bottom" for disenfranchised workers desperate for even the smallest income.

The grand vision is that eventually all of our everyday products (including clothes, furniture, lighting, household appliances, and means of transport), should be rented rather than owned. This is said to increase the incentive for making products more durable instead of today's planned obsolescence and price wars which often result in substandard products.

Already, more and more of our entertainment is consumed as streaming services, and many software products are sold as subscription services rather than as apps that you buy and own (to the dismay of many users).

In a futuristic article for the World Economic Forum, with the remarkable title "Welcome to 2030. I Own Nothing, Have no Privacy, and Life Has Never Been Better," Danish parliamentarian and WEF contributor Ida Auken describes a future where the sharing economy and the circular economy have gradually developed into a total relinquishing of private property rights. Even homes are shared and used for other purposes, e.g., as offices, when occupants are not home.[54] The vision presupposes universal digital connection, and that the items leased and reused are included in the Internet of Things. The goal is for us all to become "one happy family," sharing everything. The downside is that there is no longer any privacy or private property.

In some cases, it can certainly be more practical to rent than to buy, e.g., a tuxedo, ski equipment, machines, or vehicles, needed only for a specific occasion; or to share seldom-used power tools or garden equipment with one's neighbors. But forced collectivization and proletarization has already been tried, with well-known horrendous results.

The Green Climate Fund

The Green Climate Fund (GCF), in Incheon, South Korea, was established in 2010 within the framework of the UNFCCC. It has set a goal of raising $100 billion per year by 2020 from UN member states for supporting projects, programs, policies, and other activities in developing countries using "thematic funding windows"—in other words, another wealth redistribution scheme with vague guidelines, no transparency, and no oversight, opening up endless possibilities of fraud and corruption at taxpayer expense.

Technocracy

The sustainable utopia as described by, among others, the World Economic Forum, entails a total transformation of society, replacing it with a new social and economic system; a synthesis between socialism and capitalism. "The future demands that we reinvent capitalism for the sake of the planet and the life it sustains," noted Rockefeller Brothers Fund CEO Stephen Heintz in the RBF 2016 annual report.

This coincides with the vision outlined by W. Warren Wagar at the World Future Society Conference 1980, where technocracy was presented as the final phase of capitalism—a merging of state bureaucracy and big business, forming a monolith that can act as a single unit (see chapter 5). This was one of the goals of the early futurists, initially using social justice as means towards this end, now the environment and climate change.

There are several types and degrees of technocracy:

- Bureaucratic technocracy, where experts ("technocrats") are appointed by an elected government as advisors, administrators, or reviewing authority.
- Political technocracy, a hypothetical system of governance run by scientists instead of elected officials (related to the meritocracy state outlined in Plato's *Republic*).
- Economic technocracy, a conceptual planned resource allocation system, where money is replaced by energy credits (based on the amount of energy used to produce goods or services). The goal is a circular economy, with automated production and distribution systems, managed by specialized engineers, resulting in minimal waste and efficient use of resources. The inhabitants of a technate are guaranteed a basic income of energy credits and more leisure time (at least in theory).

In his books, *Technocracy Rising: The Trojan Horse of Global Transformation* (2015) and *Technocracy: The Hard Road to World Order* (2018), author and researcher Patrick M. Wood, editor-in-chief of *Technocracy News & Trends*, has mapped out in detail the development of the technocratic movement and its impact on world politics and sustainable development—to a large extent through the Trilateral Commission.[55]

It started with Technical Alliance, formed in 1919 by a group of scientists, engineers, economists, and educators to study the effects of technology on our social structure. Out of this grew Technocracy Inc., a research and educational organization founded by Howard Scott at Columbia University in 1933, advocating economic technocracy. As Wood has brought to public attention, Technocracy Inc. still exists and the agenda is the same; transforming the current economic system into a new global economic technocracy.

This also relates to the dream of Trilaterals David Rockefeller, Henry Kissinger, and Zbigniew Brzezinski, of a New International Economic Order.

> The post-industrial society is becoming a 'technetronic' society: a society that is shaped culturally, psychologically, socially, and economically by the impact of technology and electronics—particularly in the area of computers and communications. (Zbigniew Brzezinski, *Between Two Ages*, 1970)[56]

Once the technocratic system is in place, it is absolute and cannot be revoked by popular vote. In such a world organism, man is only a sub-component, subordinate to the collective, guided by common values and governed by the central planning of a small scientific elite. According to proponents, this is the road to the lost paradise.[57]

In 2008, a New Age version of technocracy was spread to a wider alternative audience through the film *Zeitgeist Addendum*, based on Jacque Fresco's visionary Venus Project.[58]

The World Economic Forum's version of this high-tech utopia is, however, devoid of any New Age imagery or terminology and focuses solely on outlining the promising technology, science, politics, and business opportunities.

A full political-economic technocracy does not yet exist, but China is rapidly moving ever closer to becoming an autocratic political technocracy, while leading international organizations and corporations use Agenda 2030 as a tool for implementing softer versions of economic technocracy in the West.

Social Credits

Social credits is a top-down ranking system, developed and tested in China, where citizens and companies are scored according to reliability, credit-worthiness, law-abidingness, and behaviour. Scoring can be done manually (in rural villages) or automatically (in cities). The most advanced system includes ubiquitous camera surveillance with AI facial recognition software for scoring of each individual in real time, based on behaviour, consumption patterns, lifestyle, opinions, friends, and activity in social media.

The social credit system was developed from Sesame Credits, created by ANT Financial Services Group, a subsidiary of Alibaba (whose founder, Jack Ma, was a board member of the World Economic Forum) to rank the credit-worthiness of their customers.

In the extensive surveillance system now being implemented in China, all human activity is to be monitored and rated in real time. Conscientiousness, loyalty, obedience, and "wise" lifestyle choices are rewarded with VIP service on hotels and airports, favorable loans, prime schools, and attractive jobs and housing. Criminal offenses, criticizing the regime, and undesirable personal choices are punished with slower internet connection, travel bans, and difficulties getting home loans or access to certain products and services. Once blacklisted (e.g., for displeasing authorities or being associated with a low-score person; for minor misdemeanours such as littering or jaywalking; or even *by mistake*) it is not possible to appeal. No warning is given. One only notices the blacklisting through tangible restrictions in everyday life—or by being named and shamed on public billboards.

In 2014, Chinese authorities planned to have the system operational in all of China by 2020, but as of 2023 it is still under development. However, by 2019 more than ten million Chinese people in the ten first test cities had already met restrictions such as not being able to buy a train or airplane ticket and efforts to expand the system nationwide continues.[59]

Stockholm 2040

Meanwhile in the West, official authorities and media have been working hard at normalizing surveillance as a natural part of modern life.

In September 2018, the city council of Stockholm published an astonishing brochure (also available online in English) depicting a vision of city life in the year 2040, featuring, among other things, special offline "tin foil hat" zones for citizens to meet and interact without supervision, facial recognition, virtual reality, and digitalized commercials.[60]

"The tin foil hat has become a natural meeting place and I like to be completely offline for a while." ("Elin Zakholy, analogian," fictitious 2040 citizen.)

Other features in this 2040 vision are "firefly" mini drones to light up parks, streets, and squares. They can "follow you when you are out jogging or taking the dog for a walk in the woods at night" and "supply communication, data, and navigation services." Another ficticious 2040 citizen states, "I use my personal Firefly to film, document and broadcast my life in realtime on social media."

The All-Seeing Eye

In its *Global Risks Report* (2019), the World Economic Forum actually warns that a digital panopticon with an "all-seeing eye" (AI) is now being implemented:

> Facial recognition, gait analysis, digital assistants, affective computing, microchipping, digital lip reading, fingerprint sensors—as these and other technologies proliferate, we move into a world in which everything about us is captured, stored and subjected to artificial intelligence (AI) algorithms.

Examples include Moscow, which in 2017 had a total of 160,000 surveillance cameras at the entrances to about 95 percent of the city's res-
idential buildings and other buildings. When the system is upgraded
with facial recognition software, it can automatically compare with law
enforcement databases and identify wanted and suspects.[61] This surveillance state is not decades into the future but is already being gradually introduced. Camera drones for both civilians and corporations are
suddenly an everyday thing, even if there are some legal restrictions on
how they may be used. Smart TVs and the AI function in cell phones
and computers may both watch, listen and share content with a third
party—which can be used by law enforcement and other government
authorities.[62]

Sometimes, though, there are limits to what citizens will accept. In
August 2019, the city council of Norrköping in Sweden announced that
they had purchased bracelets with bluetooth tracking chips, to be tested
on twelve to fifteen children in a kindergarten. The bracelets would alert
staff if children attempted to move outside the kindergarten area (known
as *geofencing*). This pilot project was said to be part of the Department of
Education's efforts to "increase school attendance."[63] The announcement
elicited intense reactions from both the public, the political opposition,
and the Swedish Data Protection Authority, leading to the project being
stopped.[64]

Information Control

RBF chief executive Stephen Heintz was concerned that the new technology with its high-speed information flow could be manipulated and abused
to achieve political goals. This had become apparent during the United
States presidential election in 2016 when real estate magnate and reality
show star Donald Trump was elected. This sparked concerned debates on
filter bubbles, fake news, foreign influence, and people's behavior online.
According to Heintz, the information shared in social media often lacked
"a basic standard of accuracy and documentation." Trump's nationalist
and populist rhetoric required countermeasures. A more effective control of the information flow would be needed to help people distinguish

between "deliberate manipulation, unfounded lies, lively debate, and evidence-based knowledge."[65]

By 2019, popular resistance to the technocratic globalization agenda had grown into a real obstacle for the Rockefeller sphere. In their report *Democracies under Stress*, the Trilateral Commission expressed concern that social media was dividing people and making developed democracies turning inward. Their new approach was to initiate "domestic dialogues."

> Trilateral Commission domestic dialogues will bring together "coastal elites" and individuals from rural and other areas. The two-day dialogues, taking place in different locations in the heart of the North American continent, will each be organized around a concrete issue—such as urban renewal, manufacturing, or various aspects of the energy industry.[66]

In April 2019, a Rockefeller family-sponsored event was held at the Columbia Journalism School, with the stated objective of changing how climate change is reported in the media. Panelists included Naomi Klein, Bill McKibben, Chris Hayes, Kyle Pope, and Alexandria Ocasio-Cortez, discussing such topics as how to sell the New Green Deal to the public. Shockingly, many panelists agreed that the journalistic aim of *neutrality* (which Klein calls "fetish for centrism") needed to be replaced by journalistic *activism* in order to save the planet![67]

A Global Surveillance State

In 2010, the EU project FuturICT was launched, with the aim of using information technology and data analysis to be able to understand and control complex, global, socially complex systems and achieve sustainability and resilience by anticipating crises and future opportunities. The project included a large number of universities and received financial support from both the European Commission and private philanthropists such as George Soros. The head of FuturICT, Professor Dirk Helbing, later warned that a new global fascism based on surveillance was being introduced through the technological platform he himself had helped build.

We are faced with the emergence of a new kind of totalitarianism
of global dimensions that must be stopped immediately. "An emergency operation is inevitable, if we want to save democracy, freedom,
and human dignity," I warned. "Arguments such as terrorism, cyber
threats and climate change have been used to undermine our privacy,
our rights, and our democracy."[68]

The implications of the emerging digital society could become devastating to
man. Helbing mentioned British security service system Karma Police which
analyzes what you are watching and listening to and noted that all the features of fascism have been already been implemented digitally and could be
utilised on a society-wide scale at any time. The features of fascism include

- Mass surveillance
- Unethical experiments with humans
- Social engineering
- Forced conformity
- Propaganda and censorship
- "Benevolent" dictatorship
- (Predictive) policing
- Different valuation of people
- Relativity of human rights
- And even euthanasia for the expected times of crisis in our unsus-
  tainable world.

Helbing poses the question if the sustainability agenda can be seen as
totalitarianism clad in nice wording. To stay within the planetary boundaries while maintaining growth, the world's population may need to be
reduced by one-third. Systems such as Social Credits can, in the face of a
"digital doomsday" where artificial intelligence makes decisions about life
and death, be used to evaluate the benefit of each citizen and determine
who should have access to food and resources. Despite these risks, he is
nevertheless hopeful, believing that an alternative and democratic digital
society with a new economic system could solve the problems of humanity.

Global Governance

In order to manage the new global technological challenges, both the WEF and the RBF keep hammering home the message of transnational institutions.

> Humankind faces unprecedented challenges from global warming, nuclear proliferation, terrorism, declining trust in government, eroding faith in democracy, and extreme economic inequality, as well as profound questions arising from advances in technology, social media, and artificial intelligence—to name just a few. The institutions and systems on which we have relied, in some cases for centuries, and in others for decades, seem increasingly anachronistic and, therefore, unable to manage the nature and pace of global developments. (Stephen Heintz, Rockefeller Brothers Fund, 2016)[69]

The problems are said to be just too many and too severe to be handled by individual nations.

But how should an effective global governance be implemented? What problems could arise? In 2016, World Economic Forum policy advisor Olivier Woeffray had analyzed the risk of public rule and the "tyranny of the majority" in relation to the implementation of the seventeen sustainable development goals:

> The changes brought about by the Fourth Industrial Revolution will hopefully be largely positive, but empowerment can bring about unintended consequences. For these we'll need new governance models that are more effective, accountable and inclusive. New questions will arise. Can we trust the crowd? How can we manage the risk of the tyranny of the majority? How do we ensure reflexivity and long term-thinking in a fast-paced environment? How can we ensure effective collaboration while including more actors?[70]

It was clear that decision-making should not be left to the "ignorant" masses, incapable of making the "correct" decisions.

A Global Despotic Council

According to the Global Challenges Foundation, one way of dealing with the problems was to "raise awareness of global, catastrophic risks and accelerate the emergence of a global governance that can handle them." The GCF also warned of "future bad global governance," i.e., either failing to solve major problems and instead creating worse outcomes, or the development of a world dictatorship with total surveillance. The latter risk should, however, be weighed against the risks of *insufficient* global governance which could result in "several billion victims or a total system collapse"—a difficult choice, according to report's author, Dennis Pamlin.[71] (Comments had been obtained from experts such as Johan Rockström, László Szombatfalvy, and Nick Bostrom and GLOBE International helped promote the report.)[72]

During the 2010s, the demands from proponents of global governance would keep increasing. In 2010, James Lovelock had concluded that the climate change could be "an issue as serious as war" and that it might be "necessary to put democracy on hold for some time."[73]

In April 2018, the organization Democracy Without Borders published the book *A World Parliament: Governance and Democracy in the 21st Century*, by Jo Leinen and Andreas Bummel.[1] In it they wrote,

> More than at any time in history, all the people in the world are linked together in a shared civilization, encompassing the entire planet. Their multiple interconnections generate mutual dependencies and affinities. Humanity now has a common destiny. Global challenges such as war, poverty, inequality, climate change and environmental

---

1   Andreas Bummel is co-founder and director of Democracy Without Borders, co-founder of the Campaign for a United Nations Parliamentary Assembly, council member of the World Federalist Movement's Institute for Global Policy in New York, a fellow of the World Academy of Art and Science, and an honorary member of the Society for Threatened Peoples. Jo Leinen (Social Democratic Party), member of the Advisory Council of Democracy Without Borders, member of GLOBE EU, and president of the Union of European Federalists since 1997.

destruction are overwhelming nation-states and today's international institutions. Doing the right thing requires more than having the right policies; it requires having the right political structures to implement them.[74]

In December 2018, Swedish philosopher Torbjörn Tännsjö, author of *Global Democracy: A Case for World Government* (2008), took the even more extreme position of calling for an actual coup d'état and establishing a global enlightened despotic council, forcing nation states to cease to exist, in order to halt the disaster highlighted by Johan Rockström and colleagues.[75]

Rockström, who in 2018 succeeded Schellnhuber as head of the Potsdam Institute, had also on several occasions called for authority to be transferred from national to global level in order to effectively implement the transition to a sustainable development within the nine planetary boundaries. In a 2015 interview, he said, "I cannot see any other way than 200 nations having to surrender some of their decision-making sovereignty to a planetary institutional management. We have to work with the institutions we have, and there is only one institution that is global, the UN."[76]

This echoed the visions of sixteenth-century philosopher Thomas Hobbes, with a *Leviathan* to monitor the technological system and provide security to the people. In 1946, professor Oliver Reiser wrote,

If society is not to collapse from unresolved conflicts and resulting failures at integration, the nations of the world must surrender some measure of their sovereignty and begin to function within the texture of a world whole. The social nervous system, center of intellectual–social unification, is called the world brain.[77]

This vision was also reflected in the "singleton" concept developed by Nick Bostrom, founder of the Future of Humanity Institute (see chapter 12).

Once formed, a future singleton might be perpetually stable. This could happen if surveillance, mind control, and other security technologies develop in such a way as to enable a singleton to effectively prevent the emergence of internal challenges.[78]

This was the vision of the all-seeing eye. The efforts to provide the World Brain with an effective leadership would continue.

G20 as an Emerging World Government

In 2016, during the Chinese chairmanship in Hangzhou, the G20 group committed to the implementation of Agenda 2030 with its seventeen UN Sustainable Development Goals (see appendix F). Efforts to find new solutions intensified. The ambition was now, among other things, for G20 to develop from a management discussion group into an executive body for "the great transformation."

With the 2019 G20 Summit in Japan, the Sustainable Development Goals became even more closely linked with the Fourth Industrial Revolution, under the motto "Society 5.0 for SDGs." The G20 also got a more solid foundation with its complementary meetings for interest groups (see appendix C)—all with the stated objective of influencing the G20 leaders into creating "a New International Economic Order."[79]

Despite the SDG ideals of inclusiveness, none of these engagement groups, with the possible exception of mayors, have been given the mandate by popular vote to represent their particular segments of society (such as all the world's youth or all the world's women). The whole setup appears rather like a way of circumventing traditional democratic procedures and institutions.

The use of external engagement groups happens to coincide precisely with the strategy outlined by RBF in 2010.

> As the Fund began to seriously pursue the goal of securing climate policy at the federal level in 2005, staff recognized that meaningful climate policy at the federal level would only be possible when the majority of those calling for action were from outside the environmental community and, therefore, set out to diversify the voices calling for action on climate change.

The RBF has supported "allied voices for climate action" that include businesses, investors, evangelicals, farmers, sportsmen, labor, military leaders, national security hawks, veterans, youth, and governors and mayors. Each of these constituencies has an important role to play.

In addition to the grantmaking aimed at supporting individual constituency groups, a core piece of the RBF's strategy throughout this period has been to enable coordination among organizations to generate the necessary pressure to encourage a strong national policy response. (Rockefeller Brothers Fund, "Building Constituency Support for Policy Action," *Sustainable Development Program Review 2005–10*)[80]

In June 2019, the Rockefeller Foundation in Bellagio gathered leaders from politics, business, and civil society "to explore how technological developments will impact the future of the state, the future of capitalism and the future of international cooperation. We will look to answer some of the most pressing and difficult questions in a bid to figure out how to ultimately upgrade the system."[81]

However, it was not just the global system that was about to be upgraded. This Brave New World would require both a technologically improved human and a reduction of the population to sustainable levels.

Barbara Marx Hubbard
Foundation for Conscious Evolution

There is truly the quantum disruptive technology, if you talk about technology. Take robotics—think of Ray Kurzweil's work—and add biotechnology, nanotechnology, robotics, space development, zero-point energy research, and put that together with spiritual and social evolution. We become what I call a "universal species." Homo universalis.

—Barbara Marx Hubbard

1964	Salk Institute	Fundraiser
1966	World Future Society	Co-founder, board member
1970	Committee for the Future	Founder
1975	SYNCON	Initiator
1973	New Dimensions Radio	Founder
–	International Committee for the Future	Chairman
1984	U.S. Democrat Party	Vice-President candidate
1988	Global Family	Founder
1992	Foundation for Conscious Evolution	Founder and Manager
–	Emerson Institute	Ph.D.
1993	Woman of Vision and Action	Board member
1993	Club of Budapest	Honorary member
1996	Foundation for the Future	Advisor
1996	Association for Global NewThought	Founding member
2008	Evolutionary Leaders	Member
2008	Worldshift Network	Member
2009	Worldshift 2012	Member
2011	Thrive Movement	"Pioneer"
2013	Occupy Love	"Visionary"

BARBARA MARX HUBBARD, PhD, was born in 1929, daughter of toy manufacturer Louis Marx and his first wife Irene Freda Salzman. Barbara married philosopher Earl Hubbard and had five children with him. In the early 1960s, Barbara found Pierre Teilhard de Chardin's evolutionary philosophy and the Omega Point theory which, along with the theories of philosopher Sri Aurobindo and architect Buckminster Fuller, strongly influenced her views on human evolution. She co-founded the World Future Society, advocated the establishment of space colonies and, through the International Committee for the Future, arranged the SYNCON conferences in the 1970s. In the early 1980s, she became a leading voice in the emerging New Age movement. A decade later, with financing from Laurance Rockefeller, she also founded the Foundation for Conscious Evolution and got a PhD in Conscious Evolution at Emerson Theological Institute. She was also active in Ervin László's Club of Budapest, Worldshift 2012, and several similar organizations. Barbara passed away on April 10, 2019.

Homo noosphericus—the new cybernetic human (artwork by Kimmie Fransson).

Chapter Twelve

Homo Universalis

There is truly the quantum disruptive technology, if you talk about technology. Take robotics—think of Ray Kurzweil's work—and add biotechnology, nanotechnology, robotics, space development, zero-point energy research, and put that together with spiritual and social evolution. We become what I call a "universal species." Homo universalis.

—Barbara Marx Hubbard[1]

POPULATION CONTROL

Perfectly in line with the Neo-Malthusian, transhumanist, and eugenics views of John D. Rockefeller Jr. and John D. Rockefeller III in the early 1900s, modern-day solutions to the planet's problems still include population control and a radical transformation of the human species. All under the watchful eye of an artificial intelligence. The Fourth Industrial Revolution, with its rapid advances in automation and AI, had once again brought to the fore the Rockefeller family's key concern: how many people will actually be needed in the Brave New World of our near future.

ATCA and Philanthropia

In 2011, the futurist British elite think tank ATCA raised the question, "What is globalised human society going to do with the mass of under-employed or unemployed human beings that are rendered irrelevant or redundant by the fast approaching Super Convergence of the Bio–Info–Nano Singularity?"[2]

The ATCA (Asymmetric Threats Contingency Alliance) was founded in 2011 as a philanthropic expert initiative aiming to solve global problems and build a "wisdom-based" global economy through Socratic dialogue. The organization has more than five thousand select members, including politicians, scholars, business leaders, and NGO representatives. It was founded by Indian engineer and IT guru D. K. Matai (founder of the London-based security consultancy mi2g), Mark Lewis (lawyer at Berwin Leighton Paisner), and the vice president of the Trilateral Commission Europe, Hervé de Carmoy (a protegé of David Rockefeller's and earlier CEO of Chase Manhattan).[3]

Connected to ATCA is also the global network Philanthropia with one thousand philanthropists (ultra-high-net-worth individuals, foundations, private banks, and NGOs) focusing on global challenges such as climate change, poverty, and supporting young global leadership through science and technology. The Philanthropia network includes government leaders from the G10 group, British MPs, American senators, and European MEPs, as well as David Rockefeller's daughter Peggy Dulany (chairman of Synergos), Michael Northrop (head of the RBF's Sustainability Program), Deepak Chopra (Alliance for a New Humanity), and Edward Goldsmith (*The Ecologist*).[4]

ATCA's question was a chilling example of an elitist and misanthropic worldview shared and propagated by numerous NGOs, think tanks, scientists, and prominent influencers.

The Sixth Mass Extinction

According to Stanford University professor Paul Ehrlich, if we do not choose the right path away from anthropogenic climate change and stay within the nine planetary boundaries, we will face the sixth mass extinction of life on earth.[5] Ehrlich, who has been spreading the view of mankind as a cancer on the planet since 1968, has argued for an ideal population of 1.5 to 2 billion to ensure the well-being of the planet and mankind.[6] This clearly means a very sharp reduction in the world's population.

Similar views were expressed by Hans Joachim Schellnhuber in 1998:

> It is conceivable, on the other hand, that geocybernetics will follow completely different (or complementary) courses that lie more in the realm of social management. Here the demographic issue overrides other themes: Is there an optimal number of human beings to be supported by the ecosphere?[7]

A decade later, Schellnhuber made a controversial statement at the Copenhagen Climate Summit 2009, where he said that the planet's planetary boundaries can only tolerate a population of less than 1 billion people: "In a very cynical way, it's a triumph for science because at last we have stabilized something—namely the estimates for the carrying capacity of the planet, namely below 1 billion people."[8]

He later claimed that this statement has been misinterpreted and only applies to a scenario where we fail to gain control over the climate through more effective management of the earth's resources and people, and where national sovereignty is relinquished to a global governance under the UN. Whichever path we take, however, the world, according to Schellnhuber, is

"Alchemists" Isaac Newton, Hans Joachim Schellnhuber, and Johan Rockström at the Royal Academy of Sciences, Stockholm, May 25, 2016.

facing a complete transformation: "Whatever we do or don't do, the world as we know it will soon cease to exist."[9]

The Global Challenges Foundation, with board member Johan Rockström (Schellnhuber's successor as director of the Potsdam Institute), also view population growth as a crucial problem causing environmental degradation and higher CO_2 emissions.[10]

The Unfinished Agenda

In a 1994 speech before the business council for the United Nations, David Rockefeller said, "The negative impact of population growth on all of our planetary ecosystems is becoming appallingly evident."[11]

John D. Rockefeller III's old agenda from the 1950s, a global plan for controlling population growth, has always been lurking in the background to the climate threat that has been used to sell the bitter medicine. It seems that the time has come to finish the Rockefeller Brothers Fund's *The Unfinished Agenda* (which, in 1977, with panelists such as David Brower, had recommended far-reaching approaches to lowering population growth, and underlying Jimmy Carter's presidential report *Global 2000*).

Apart from the activities of Population Council, population is no longer an open priority for the Rockefeller Foundation's development program. Instead, the foundation's close partner Bill & Melinda Gates Foundation has taken over and become a leader in this field with its "soft" fertility-reducing methods in the form of vaccination programs, strengthening of women's positions, and the spread of contraceptives in developing countries.

The Rockefeller Foundation is, however, still very active in the field of agriculture, including food safety and the development of GMO (including the controversial "golden rice"). Continuing the old pursuit of reforming traditional agriculture over the world into biotech business, the RF and Bill & Melinda Gates Foundation in 2006 founded the Alliance for a Green Revolution in Africa (AGRA), with Kofi Annan as chairman.[12] The initiative was criticized in *Voices from Africa* for imposing quick-fix biotech solutions without the involvement of local representatives, and for forcing farmers to buy their seeds from large corporations each year.[13]

The Rockefeller-initiated petrochemical, later biotechnical, agricultural practices have increased crop yield by 250 percent. At the same time, the RF points out that food production needs to increase by another 70 percent in order to feed the growing world population in the future.

Agriculture is also classified as having a larger climate impact than all the world's transports taken together.[14] Man's need for sustenance is deemed unsustainable.

This issue has been investigated by Johan Rockström, with Jonathan Fowley at the University of Minnesota and others. In a 2011 article in *Nature*, they suggested that it *is* possible to stay within the planetary boundaries with a transition to a climate-smart agriculture.[15] This is to be achieved through a more efficient agriculture with reduced waste, no expansion of agricultural land use, strategic use of pesticides, reduced harvest loss and changed dietary habits from meat to vegetable crops.[16] Rockström stated that "if humanity continues on its current trajectory, it will likely be unable to meet the needs of a world population that is expected to reach at least nine billion by 2050."

In 2015, GMO biotechnology was also presented as a "resilient solution" to the threat of climate change.

> While agricultural biotechnology remains controversial, these techniques provide an especially promising set of tools that have produced dramatic improvements in yield and reductions in production costs and input use intensity. Examples of new crops that have benefited agriculture and reduced emissions include genetically modified crops with pest resistance and herbicide tolerance. (Travis Lybbert and Daniel Sumner, Agricultural and Resource Economics, University of California, Davis)[17]

None, however, address the question of how the world's population will be fed with an industrial agriculture totally dependent on fossil fuels, if all fossil energy is to be phased out completely by 2050. Just as Barbara Marx Hubbard had announced, the Sustainable Utopia does not seem to be for everyone.

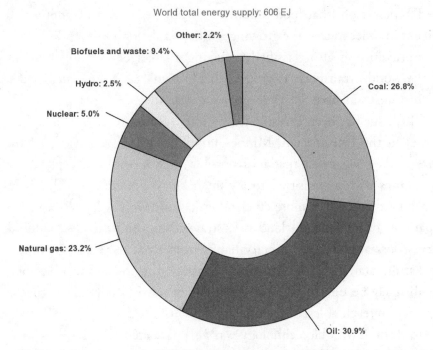

World total energy supply 2019 (chart from International Energy Agency iea.org). Note that wind and solar is included in the Other 2.2%.

Earth Overshoot Day

The message from the WWF, the Club of Rome, and the Global Footprint Network (with a network of interconnected elite-funded organizations) is clear: for every year, Earth Overshoot Day—the day of the year (e.g., August 2)—when the year's total production is consumed—occurs sooner and sooner. During the rest of the year, we either have to live on our savings or grant ourselves an advance on future consumption. According to these organizations' calculations, we live on borrowed capital, increasing our debt each year. In other words, an extensive reduction of the global population is required to create a sustainable society.

Christiana Figueres, closely associated with the Rockefeller sphere, has made chilling statements about what the United Nations should do: "Really, we should make every effort to change those numbers because we are already, today, already exceeding the planet's planetary carrying capacity."[18]

The Worldwatch Institute has also called for drastic measures.

> Looking past the near-term concerns that have plagued population
> policy at the political level, it is increasingly apparent that the long-
> term sustainability of civilization will require not just a levelling-off
> of human numbers as projected over the coming half-century, but a
> colossal reduction in both population and consumption.[19]

Population Matters

Population reduction, originating with Malthus in the eighteenth century,
is also still actively advocated by the British elite.[20] This, despite the horrendous effects of Malthusian views in British colonies in the nineteenth
century.[21]

In 1991, David Willey founded Population Matters, with patrons
such as Sir Crispin Tickell, Dame Jane Goodall, Sir David Attenborough,
Professor Paul Ehrlich, and James Lovelock. According to Population
Matters, the most effective national and global climate strategy to keep
global mean temperature rise below 1.5 degrees Celcius is limiting the
number of child births.[22]

> Those who fail to see that population growth and climate change
> are two sides of the same coin are either ignorant or hiding from the
> truth. These two huge environmental problems are inseparable and
> to discuss one while ignoring the other is irrational. (James Lovelock,
> 2009, Population Matters)

UN High-Level Panel on the Post-2015 Agenda

Population reduction is also a concern for the United Nations' new sustainable development goals. In a 2015 discussion paper from the UN High-
Level Panel on the Post-2015 Agenda, Johan Rockström and Jeffrey Sachs
pointed out that fertility must, among other things, be reduced in sub-Saharan Africa. They proposed "soft" and voluntary measures to achieve
"sustainable fertility." If this fails, a "Malthusian catastrophe" awaits.
Rockström believes that the curves need to be turned down quickly.[23] So

far, however, the population issue has not received sufficient response as a possible climate policy measure.

In an August 2018 article in *Science*, John Bongaarts, vice-president of the Population Council, and Brian O'Neill (of National Center for Atmospheric Research and main author of IPCC's Fourth and Fifth Assessment reports) say that population policy and family planning as a solution to the climate issue have so far been met with a great resistance from conservative interests. This now had to be changed. Like Rockström, the authors highlight the great challenge from sub-Saharan Africa population being expected to increase from one billion to four billion by 2100. By lowering the birth rates in southern Africa, they claim, the quality of education will improve and crime, terrorism, and unemployment can be kept in check, as well as eliminating poverty, reducing the burden on the environment, and reducing CO_2 emissions. They therefore propose that the IPCC incorporate population policy as part of possible measures to manage climate change and that these be linked to the UN's sustainability goals.[24]

Birth Strike

Even more drastic measures are proposed by the philosopher Travis Rieder at the Berman Institute of Bioethics at Johns Hopkins University: "In order to help avert a 2°C increase in global average temperatures this century, we must reduce population growth faster than choice-enhancing policies are capable of doing on their own."[25]

Now voluntary or "soft" methods are off the table. The end justifies the means. Rieder proposes a global population engineering program to make women have fewer children with a variety of propaganda techniques, so-called "nudging," in order to save the world from the looming climate catastrophe outlined in Schellnhuber's World Bank Report. Reider believes that fertility needs to be lowered to 0.5 children per woman to avoid the "catastrophic tipping points" of Schellnhuber and Wasdell, saying "[W]e need to investigate the defensibility of additional fertility-reducing population engineering interventions."[26]

This should be done, e.g., with moral arguments about what a burden each new individual will have on the planet. Having many children should not be rewarded. Rieder believes that it is easier to reduce the population than to change people's living habits

> even if we are able to make the kind of radical cuts to our emissions hoped for by the IPCC, the total CO2 emissions saved by refraining from having one additional child is larger than the summed lifetime savings from six common "green activities" (such as lowering one's transportation related GHG emissions, increasing the energy efficiency of one's home, etc.).

Since the beginning of 2019, the British organization BirthStrike, with predominately young fertile women as members, has begun advocating these ideas.[27]

TECHNOLOGICAL TRANSHUMANISM

The most far-reaching change is planned for man himself, who is to be modified at a fundamental level. After centuries of scientific and technological development, transhumanists finally see real opportunities for realising the old Hermetic dream of creating an upgraded superhuman with the help of technology. Now using the threat of climate change as an excuse.

Klaus Schwab and the World Economic Forum's expert panel describe man as an object that can be altered and improved to perfection, applying both bio- and neurotechnological methods. It is stated matter-of-factly that the future will "challenge our perception of what it means to be human." When technology moves into our body, the question arises of the boundary between machine and human, recognising that the new technology can be used "to manipulate our worldview and influence our behavior."

Humanity+ and Future of Humanity Institute

In 1998, Nick Bostrom founded the World Transhumanist Association (from 2004 renamed Humanity+) to make transhumanism more respectable and scientific. H+ is currently represented in 120 countries and very active in promoting the transhumanist agenda. In 2005, Nick Bostrom also founded the Future of Humanity Institute (FHI) under the Faculty of Philosophy, Oxford Martin School. Bostrom is director of the institute and the team includes Anders Sandberg, among others. Together with the Global Challenges Foundation, they are also policy advisors to the World Economic Forum. The institute analyzes risks and opportunities with future "human enhancement" technologies such as gene therapy, life extension, brain implants and brain-computer interfaces, and population control.

In 2015, the Global Challenges Foundation and the Future of Humanity Institute issued the report, twelve risks that threaten human civilization, which analyzed catastrophic threats such as extreme climate change, nuclear war, pandemics, ecological disasters, global system collapse, asteroids, super-volcanoes, synthetic biology, nanotechnology, artificial intelligence, unknown consequences, and a substandard global governance. The hope was that a powerful AI would be able to solve all problems—while at the same time posing a danger in itself by potentially coming to view humanity as redundant.[28] A *Terminator* scenario echoing the warnings from ATCA and others.

In a speech to the Swiss Civil Society Association on November 11, 2017, German Hans Ulrich Gumbrecht, professor of literature at Stanford University, declared an ambition to create a God-like being with superhuman knowledge to guide mankind—and deliver us from our sins. There was, however, no guarantee that this entity would turn out to be benevolent.[29]

This potential danger was also pointed out by Stephen Hawking, Elon Musk (Tesla Motors), and Bill Gates[30]—for which they received the 2015 Luddist Award by the Information Technology and Innovation Foundation for being "AI alarmists."[31] This is even more astonishing considering the fact that they all supported Bostrom's work and that Musk has invested heavily in creating a brain-computer interface, Neuralink.

Human Engineering

In 2012, Anders Sandberg and Rebecca Roache (Future of Humanity Institute, Oxford Martin School), with Matthew Liao (Center for Bioethics,

New York University), proposed biomedical modifications of humans, "so that they can mitigate and/or adapt to climate change (presented as less risky option, than for example, geoengineering)." The team of transhumanists offered several outlandish examples of human engineering fantasies:

1. "Pharmacological meat intolerance" by creating a bovine protein intolerance–inducing meat patch (akin to a nicotine patch) and encouraging people to use them.

2. "Making humans smaller" (reducing birth weight and height) by using hormone therapy or genetic diagnosis and modification before implantation (in this science fiction–like vision it is taken for granted that you visit a fertility clinic to procreate) where you will be given the "liberty-enhancing" choice of having a greater number of smaller children or a smaller number of larger children to meet your allocated greenhouse gas emissions quota, instead of the strict limit of a maximum of two children per family suggested by Guillebaud and Hayes.

3. "Lowering birth-rates through cognitive enhancement" by education and cognition enhancement for women, which can lead to them choosing to have fewer children.

4. "Pharmacological enhancement of altruism and empathy" through hormone treatment, in order to make them care more about the environment increase people's willingness to assist victims of climate change.

5. Other bioengineering possibilities mentioned include "increasing our resistance to heat and tropical diseases" and "reducing our need for food and water."[32]

Humanity 2.0

The ideas of a human-machine fusion were starting to spread a few years before Schwab declared that the dawn of the Fourth Industrial Revolution. In 2014, James Lovelock stated that instead of trying to save the planet, humanity should live in cities with regulated temperature and focus on evolving from biological creatures to merging with technology.[33]

According to Lovelock, "If we can somehow merge with our electronic creations in a larger scale endosymbiosis, it may provide a better next step in the evolution of humanity and Gaia."[34]

This solution, however, is not something he would prefer for himself.

> I must admit an empathetic dread for some unfortunately future person whose body becomes connected to one of more of the ubiquitous social networks. I can imagine no punishment more severe than having my still comparatively clear mind overtaken by the spam of hucksters and the never-ceasing gossip of the Internet.

Elon Musk promotes this vision with another motive. In order to avoid an evil and autocratic AI, humans should be merged with technology through a neural link between the cerebral cortex and the digital world, forming a symbiosis between man and machine. According to Musk and the Global Challenges Foundation, by merging with AI, we don't have to worry about a malicious AI because then we will collectively constitute the AI ourselves. The long-term goal is full integration with the technology and becoming one with the internet through brain–computer interface (BCI). We will all be a part of the Internet of Us.

The Trilateral elite network ATCA/Philanthropia has also embraced the ideas that humans should increasingly be merged technology: "What's the Q-BRAIN Singularity about? Simply put, Quantum-Blockchain-Recursion-Artificial-Intelligence-Nano (Q-BRAIN) smart technologies coming together in our global civilisation to synthesise man and machine as one in a hybrid formulation where man becomes part machine and machine becomes part man."

In March 2016, ATCA predicted that this brave new world would be implemented by 2020:

> Everything that you see happening today between man and machines will change and metamorphose beyond recognition, in the coming 4–5 years. Expect total disruption via new Q-BRAIN enabled

products and applications in terms of challenging legacy technology solutions; societal behaviours, habits and norms; global trading,
finance and economics; and absolutely everything including the way
we live, work and play![35]

This neurotechnological revolution entails influencing our brains with
microelectrodes, which proponents claims will expand our abilities, and
change our behaviors and our interaction with the external world. The
boundaries between the real world and virtual reality will become blurred
through augmented reality where virtual objects, information and data
are merged with the physical world. In *Shaping the Future of the Fourth
Revolution*, Klaus Schwab wrote,

> Influencing the brain in more precise ways could change our sense of
> self, redefine what it means to have experiences and fundamentally
> alter what constitutes reality. By affecting how we govern ourselves,
> the system management of human existence, brain science encour-
> ages a huge step for humans beyond natural evolution.[36]

As positive benefits, proponents hope the technology will be able to cure
neurological disorders and motor disabilities. At the same time there are
warnings that this development can lead to employers starting to use the
technology to vet job applicants and monitor employees.

Following controversies around the use of RFID identification and
workplace tracking, according to the World Economic Forum, the monitoring of employee brains is expected to be the next ethical dilemma. This
also includes the risk of judicial systems starting to use the technology in
order to analyse the likelihood of criminal activity, assess guilt, and extract
memories directly from human brains and that security risks can be identified at border controls through brain X-ray.

The European Commission's HIVE Project, which ran between
2008 and 2012 with the aim to develop a noninvasive Brain–Computer
Interface (BCI), warned of the ethical implications of the new technology:
"The project can open the door to breakthrough technologies that could

be used in negative ways, such as (conceivably) mind control and BCI-related military applications."[37]

Oxford Martin School's Program on Mind and Machine expresses similar concerns.

> Advances in understanding how the brain works are rapidly leading to new possibilities for intervention in brain function. The ability of brains and machines to talk to each other directly is fast becoming a very real possibility. This raises profound ethical issues related to understanding behaviour and potentially manipulating it, so called "mind control."[38]

The perfect society risks developing into an electronic prison where our perception of reality is manipulated and our behavior controlled.

Transhumanism Goes Mainstream

The old dreams of superhuman abilities and immortality have now moved out from the occult secret societies and small futurist groups to being launched to a wider audience, not just as science fiction but as a real option, through a growing stream of articles, panel discussions, conferences with leading transhumanists, philosophical radio programs, and science TV shows. What began in 1998 with Professor Kevin Warwick's tests on himself with implanted chips has now evolved into a growing biohacker movement.

In 2012, Swedish piercer Jowan Österlund started Biohax International, offering chip implants to companies and individuals. In 2014, the Swedish biohacking organization BioNyfiken (BioCurious) began hosting "chipster parties" where biohackers could be "upgraded" with chip implants to "become digital super humans" (meaning at that early stage only simple things such as using the chip for digital doorlock systems and later participating in a 2017 train ticket chip trial). The chipping spectacles received a lot of media attention in Sweden and internationally, and reached a fever pitch in 2017. Biohax claimed that their technology contributes to a more sustainable future through the reduced need for plastic

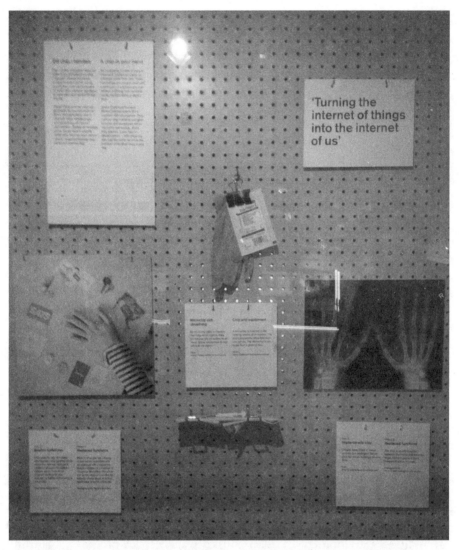

Exhibition with chipping propaganda at the Visualisation Center in Norrköping, Sweden, March 2017, displaying Biohax International's motto "Turning the internet of things into the internet of us."

cards, in accordance with EU's circular economy and the UN Sustainable Development Goals.[39] Meanwhile, biohacking conferences have been organized by enthusiastic pioneers in Finland, Sweden, and Estonia. The Scandinavian countries have been pioneers and countless seminars and lectures on transhumanist subjects have been held, influencing both pop culture and intellectual debate.[40] However, contrary to how the World

Economic Forum portrays it, the young Scandinavian tech enthusiasts who volunteered as pioneers were only a few thousand and Biohax is now out of business. But this was only the initial test phase before rolling out chipping on a larger scale for the general public, once the necessary infrastructure is in place.

Transhumanism in Media
During the last decades, there has been constantly growing flood of science fiction books and magazines, Japanese manga and anime, TV series and films with transhumanist themes and dystopian visions of our near and far future.

Two of the more astonishing examples are the sci-fi series *Black Mirror* and National Geographic's futuristic docudrama series *Year Million* (based on Damien Broderick's 2008 book *Year Million: Science at the Far Edge of Knowledge*.[41] The *Year Million* TV series featured a mix of live action sequences, high-quality 3D renditions, and interviews with leading futurologists and transhumanists. With overtly religious references, the series promised that by using genetic engineering, nano-robots, implants, and robotics, we could become genetically "perfect" and super-intelligent; merge with AI; connect with others in a swarm consciousness and become telepathic; rebuild the Tower of Babel; conquer the galaxy; and finally be uploaded to the internet and live forever in a "digital Nirvana." How National Geographic's respectable TV channel could be transformed into a simple propaganda channel for futurism and transhumanism, with specific focus on projects which happen to be under development by Elon Musk (SpaceX space program,[42] Starlink satellite system,[43] Neuralink brain implants,[44] and Tesla Motors electric cars) may perhaps be explained by National Geographic in September 2015 partnering with 21st Century Fox, whose CEO, James Murdoch (son of media mogul Rupert Murdoch, 20th Century Fox and Fox News), was elected to Tesla's Board of Directors in 2017.

The Omega Point
The development of Teilhard de Chardin's transhumanist notion of an "Ultra-human" and the journey towards the Omega Point is now being

pushed with an increasing intensity, with alluring promises of "immortality" in a technological "heavenly kingdom" under the watchful eye of a Digital God (AI). The question is who really understands what this mystical event will mean for the survival of mankind?

We are faced with difficult choices and it is up to each individual to accept or reject this future as our ultimate destiny. It is my belief that this alchemical quest to transform the world and mankind in order to achieve lasting peace, harmony, and balance is based on false premises. True harmony cannot be achieved if lies, manipulation, and control are part of the recipe.

The path towards becoming a better person is something uniquely individual and cannot be applied to humanity as a whole. It is not possible to take shortcuts with the help of technology and become a godlike super-human. Real human development can only be achieved through the personal experiences, challenges, and lessons we all face during the course of our lives, through the hard work of acquiring true knowledge, maturity,

Prometheus with the fire stolen from the gods, sculpture outside Rockefeller Center. The inscription reads: "Prometheus, teacher in every art, brought the fire that hath proved to mortals a means to mighty ends."

and wisdom. Only we ourselves are responsible for building and caring for the temple.

Isn't it time to save our humanity and to stop listening to these false prophets and self-appointed "planetary custodians"?

> *For whatever is hidden is meant to be disclosed, and whatever is concealed is meant to be brought out into the open.*
>
> —Mark 4:22

Peter O'Neill and Neva Rockefeller.

Epilogue

Some even believe we [the Rockefeller family] are part of a secret cabal working against the best interests of the United States, characterizing my family and me as "internationalists" and of conspiring with others around the world to build a more integrated global political and economic structure—One World, if you will. If that's the charge, I stand guilty, and I am proud of it.
—David Rockefeller, *Memoirs*, 2003[1]

The Divestment

So what happened after the family's smaller foundation, the Rockefeller Family Fund, in March 2016 announced that they would divest from all fossil energy holdings and sell their shares in Exxon Mobil after pointing out how corrupt and irresponsible the family's old crown jewel was?

In January 2016, the RFF had held a secret meeting in Manhattan with green NGOs (including Bill McKibben from 320.org) discussing how to establish in the public's mind that their old crown jewel Exxon was "a corrupt institution that has pushed humanity (and all creation) towards climate chaos and grave harm."[2]

In July 2016, it was announced that Rockefeller Financial Services Inc. had bought up 43,568 shares in Exxon Mobil Corporation during the second quarter of 2016, thereby increasing its holding by 4.2 percent. As a result, the company had a total of 1,074,179 shares in Exxon Mobil Corporation worth $100,983,568. Additionally, they continued to increase their holdings in oil companies such as Chevron, BP, ConocoPhilips, and Cabot Oil. The various organizations within the family empire apparently had very different priorities! It all, however, appears to have been a carefully planned and coordinated campaign, executed with the blessings of the family office at One Rockefeller Plaza. The chairman of the 5600 Operating Committee for the Rockefeller Family, Peter O'Neill, was also a member of the board of directors of the RBF, Winrock International, and chaired the Rockefeller Family Fund's finance committee.[3] He was thus deeply involved in the Rockefeller's divestment initiative. As director of Rockefeller Financial Services and a member of its finance committee, he was also responsible for expanding its holdings in the "corrupt institution" Exxon Mobil.

Ariana Rockefeller (1982–), member of Next Generation Advisory Council at Rockefeller Capital Management, daughter of David Rockefeller Jr.

At the end of 2017, these ambitions were expanded when Rockefeller Financial merged with Viking Global Investors, forming the company Rockefeller Capital Management with Wall Street superstar Greg Fleming as CEO. The goal was to grow from $18.3 billion to $100 billion within five years. As of June 30, 2018, more than 8 percent of the holdings was invested in the energy sector and oil companies such as BP, ConocoPhilips, ExxonMobil, Chevron, Cabot Oil & Gas, Royal Dutch Shell, and Total. According to RBF's Justin Rockefeller, "Given that we fight climate change, to us, investing in fossil fuels is somewhat akin to a cancer-fighting foundation investing in tobacco."[4]

So, how much were high morals worth in the world of the Rockefellers? With one hand—the one appearing in the media—they would divest "for the future of climate and the planet" while the other hand continued to expand their holdings in fossil energy. In addition, ExxonMobil, Chevron, BP, and Total continued as members and donors to the Council on Foreign Relations (which received $25 million from David Rockefeller upon his death in 2017).

There were also family members still working within the oil industry who did not support the public actions against their old company. David Jr.'s daughter Ariana Rockefeller said the campaign was "deeply misguided."[5] This opinion, however, did not prevent her from distributing money to climate marches as a board member of the David Rockefeller Fund.

The Mitsubishi Deal

From 1989 to 1991, the family, under David's leadership, sold 80 percent of the shares in Rockefeller Group Inc. to Japanese Mitsubishi for $1.3 billion as ownership was no longer "in the Rockefeller family interest." This gave $800 million after tax to the family's "1934 Trust."

Four years earlier, David and the Rockefeller family had created the company Rockefeller Center Properties Inc., with David as chairman, which lent $1.3 billion of raised share capital for investments in Rockefeller Center. The Rockefeller Group, with Mitsubishi as the new owner, was to repay the loan with rent revenues from the Rockefeller Center. A year later, the overheated property market crashed, resulting in dwindling revenues from Rockefeller Center[6] After the Rockefeller family refused to assist with capital in this crisis, in May 1995, Mitsubishi withheld a mortgage of $20 million and was forced to hand over Rockefeller Center to its creditor, Rockefeller Center Properties Inc. The chairman of Rockefeller's family trust, William Bowen, stated coldly, "We were willing to do more than our pro rata share, but the terms of that additional investment had to make business sense."

Mitsubishi was then forced, through a clause in the original agreement, to purchase the remaining 20 percent of the Rockefeller Group from the Rockefeller family in 1997, adding an extra $160 million to the family fortune. After the loan defaulted, Rockefeller Properties Inc. was near bankruptcy. David then gathered a new ownership group (including Gianni Agnelli) who bought Rockefeller Properties Inc. for $306 million and settled the debt to the shareholders. Then, in 2000, Rockefeller Center was sold to Jerry Spreyer for $1.85 billion. David Rockefeller personally made $45 million on the deal, tripling his investment in four years.[7] David had clearly inherited his grandfather John D. Rockefeller's talent as a ruthless businessman.

David's Passing and Legacy

On March 20, 2017, David Rockefeller, 101 years old, passed away. He had helped to change the world in a way that few others have done in world history. Valerie Rockefeller wrote in RBF's annual report,

> The entire Rockefeller family mourns the passing of our Uncle David, who has guided the family as a whole and shaped our individual work in philanthropy, all the while carrying himself with a sense of humility that will endure in us and our work.[8]

His longtime friend Henry Kissinger also praised his life achievements:

> When David Rockefeller left us, all over the world, lives became emptier. Over the decades, we had come to think of David as a custodian of our aspirations, who would see to it that basic issues affecting

freedom and governance, health and art would be appropriately defined and attended to.

In David's 1941 dissertation, *Unused Resources and Economic Waste,* he had discussed what motivated businessmen and argued that entrepreneurship was not just about maximising profit, an opportunity to satisfy man's creativity, power-seeking, and gambling instincts. It also had a higher meaning.

> In other words, part of the joy of business is achieving what one has set out to do, accomplishing goals that are important, and building something that has permanence and value beyond itself. (David Rockefeller, *Memoirs,* 2003)[9]

In the case of David and the Rockefeller Family it was all about "conspiring with others around the world to build a more integrated political and economic structure—a unified world."

The long-term pursuit of power and dominance in a technocratic world system of central planning would continued after his death, with the aim set for 2020 and the implementation of the Paris Agreement in order to create the post-human Utopia. In the *Rockefeller Panel Reports* (1961), the Rockefeller brothers had written that they "could not escape the task that history had assigned" to them—a task that meant "helping to shape a new world order in all its dimensions."

Now, this assignment was left to a new generation both within and outside the family—especially the Bill & Melinda Gates Foundation and the World Economic Forum—to bring to fruition.

The Pension Funds Campaign

On March 13, 2018, the Rockefeller Brothers Fund donated $70,000 to the Greenpeace Fund and their Swedish divestment project. Three months later, Greenpeace Sweden initiated an activist campaign against one of the largest pension funds in Sweden (with a total investment capital of more than SEK 1.404 billion) demanding that it would divest its holdings in fossil energy. This was part of a global effort in which RBF worked with organizations such as 350.org and Greenpeace to persuade pension funds and other institutions around the world to create "a fossil-free world."

RBF had also become early members of the Divest-Invest Philanthropy (created by Wallace Global Fund).[10] As of July 2020, the most recent update at the time of writing, 1,246 organizations have joined this global network, with a staggering $14.1 trillion in assets. The question is, who will acquire these huge holdings in the world's most important commodity? Oil is still the bloodstream of the globalised economy which the Rockefeller family has helped create.

World Economic Forum Partners with United Nations

In June 2019, the World Economic Forum and United Nations signed a strategic partnership "to accelerate the implementation of the 2030 Agenda for Sustainable Development" with an attached digital agenda to "meet the needs of the Fourth Industrial Revolution."[11] This partnership was then manifested in WEFs report *Unlocking Technology for the Global Goals*, issued for the Davos Summit in January 2020, which outlines in detail how technocratic technologies are to be used to solve each of the seventeen sustainability goals.[12]

The 2020 Coronavirus Crisis

Only a few of months into the new decade, the crisis occurred that was to kickstart the Fourth Industrial Revolution and the implementation of the UN Global Goals. Just like when the climate threat entered the global political arena in the 1980s, following the Tjernobyl disaster and the threat of a nuclear holocaust, this new threat was also an invisible enemy. This time, however, it was not CO2 or radioactivity but a coronavirus, COVID-19.

As soon as the World Health Organization (WHO) upgraded the contagion to a global pandemic on March 11, 2020, governments around the world reacted with drastic authoritarian measures of varying degrees of swiftness and severity. Martial Law was declared, borders were closed, gatherings were banned or limited, and some countries ordered partial or total curfews.[13] Some countries used cellphone data to track potential carriers and drones to inform or disperse crowds. People advised or ordered to stay at home were suddenly forced to do most of their schooling, work, business meetings, shopping, and social gatherings online. As the economy took a nosedive and the stock market collapsed, the smart surveillance technology was rolled out en masse over the world.

What happened was very similar to the scenario "Lock Step" in Rockefeller Foundation's report *Scenarios for the Future of Technology and International Development*, written in 2010 with the objective of investigating "what new or existing technologies could be leveraged to improve the capacity of individuals, communities, and systems to respond to major changes, or what technologies could improve the lives of vulnerable populations around the world."[14]

These same scenarios were also predicted by the World Economic Forum working groups and in Bill & Melinda Gates Foundation's search for solutions for how pandemics can be managed through public-private partnership, as exemplified in the Event 201 Pandemic Exercise in October 2019, hosted by the Johns Hopkins Center for Health Security in partnership with the World Economic Forum and the Bill & Melinda Gates Foundation.[15]

Meanwhile, the Rockefeller-funded organization ID2020 Alliance had been working on a global digital ID, necessary for shopping, travelling, handling finances, storing medical data, and interacting with authorities. Partners included Microsoft, Accenture, and GAVI—the Vaccine Alliance.[16] Big Tech was now teaming up with Big Pharma for mutual profit.

Never letting a good crisis go to waste, the Club of Rome also pointed to COVID-19 as a golden opportunity to usher in the high-tech Green Deal.

> The COVID-19 crisis shows us that it is possible to make transformational changes overnight. We have suddenly entered a different world with a different economy. Governments are rushing to protect their citizens medically and economically in the short term. But

there is also a strong business case for using this crisis to usher in global systemic change.[17]

In their view it was the same agenda. The world would never be the same.

> What if a small group of world leaders were to conclude that the principal risk to the Earth comes from the actions of the rich countries? . . . In order to save the planet, the group decides: Isn't the only hope for the planet that the industrialised civilizations collapse? Isn't it our responsibility to bring that about? (Maurice Strong, 1992)[18]

The Great Reset

For the World Economic Forum, the COVID-19 crisis was the perfect trigger event to implement the long-hatched grandiose plan for a global technocracy, with Big Tech coming to the "rescue." In June 2020, WEF chairman Klaus Schwab, backed up by luminaries such as Prince Charles and UN General Secretary Antonio Guterres, Microsoft president Brad Smith, Mastercard CEO Ajaypal Singh Banga, and IMF director Kristalina Georgieva, declared the need for a Great Reset.

> The COVID-19 crisis has shown us that our old systems are not fit anymore for the 21st century. It has laid bare the fundamental lack of social cohesion, fairness, inclusion and equality. Now is the historical moment in time, not only to fight the real virus but to shape the system for the needs of the Post-Corona era. We have a choice to remain passive, which would lead to the amplification of many of the trends we see today. Polarisation, nationalism, rasism, and ultimately increasing social unrest and conflicts. But we have another choice. We can build a new social contract, particularly integrating the next generation, we can change our behavior to be in harmony with nature again, and we can make sure the technologies of the Fourth Industrial Revolution are best utilized to provide us with better lives. In short, we need a Great Reset.[19]

Conclusions

Unhindered by the restraints of traditional liberal values, this elite would not hesitate to achieve its political ends by using the latest modern techniques for influencing public behavior and keeping society under close surveillance and control.
—Zbigniew Brzezinski, *Between Two Ages*, 1970

THE ROCKEFELLER FAMILY has had two overarching and intertwined goals: power and dominance over a "perfect world" (by their own definition).

The postwar scientific and political developments in the fields of environment and climate have been permeated by the family's ambitions towards economic monopoly and power, and on creating a New International Economic Order with a united world, One World.

In this context, the climate threat has been identified as an international problem requiring increased global cooperation and the strengthening of supranational organizations—often in conjunction with other global threats, such as nuclear war, pandemics, and terrorism.

This has been orchestrated from a position of extreme privilege, in a highly elitist project in which the Rockefeller family has mobilized and collaborated with a super-rich clique of billionaires and their multinational corporations, as well as with socialist utopians and green idealists. They have cast a very wide net and recruited some of the world's most prestigious scientists, respected leaders, and prominent activists, as well as some outright maniacs, to work for their vision for the world. Their identification of climate change and CO_2 as crucial for the survival of mankind was done long before the environmental movement started engaging in the issue.

Much of the background to the climate issue can be traced to Neo-Malthusian notions of an overpopulated planet, and ideas of genetic improvement of humans. These are areas in which the Rockefeller family has played a leading role internationally through its foundations and organizations. Mankind, our activities and behavior, have been identified as the great enemy and a burden on the planet.

These views have thereafter been very effectively disseminated to both legislators and a wider audience by

- Founding and/or helping to establish of a large number of (ostensibly independent and unconnected) foundations, institutes, NGOs, and think tanks, to give the impression of a wide interest in and support for their ideas;
- Coordinating and controlling these organizations by having the same clique of loyal agents on their board of directors.
- Creating informal but powerful behind-the-scenes networks in the arenas of international politics, in order to realize their ambitions without being hindered by the democratic demands for openness and transparency.
- Funding activists and organizations to spark public debate.
- Conducting orchestrated and carefully crafted media campaigns to further anchor the impression of a serious threat.
- Taking advantage of triggering events (energy crises, financial crises, hurricanes, forest fires, oil spills, etc.) in order to influence policymakers.

In short, the strategy has been to focus attention on a problem and then offering the solution. This has required long-term planning, careful strategic thinking, as well as a global philanthropic network with financial muscles.

In the planning of this scheme, strategists such as Henry Kissinger, Zbigniew Brzezinski, Graham T. T. Molitor, and Peter Winsemius stand out, in collaboration with the Rockefeller Brothers Fund, the Rockefeller Foundation, Trilateral Commission, the German Marshall Fund, etc. The solution to the "climate chaos" offered is the implementation of a global institutional management, where both population growth and the use of natural resources are regulated in order to reduce carbon dioxide emissions and achieving a highly efficient and resource-efficient circular economy.

These aspirations were formulated in the Trilateral Commission's version of the New International Economic Order (NIEO) in the 1970s, and are now a part of the UN Sustainable Development Goals and the Paris Agreement on Climate, with the G20 Group emerging as the leading and executive global council.

Another central cornerstone of the agenda is the development of digital smart solutions where all human activity must be carefully documented and their CO_2 emissions calculated. This involves a refined technological surveillance system built around technocratic ideals and "fair distribution" through the application of artificial intelligence (AI).

There are also far-reaching visions of a transhumanist "upgrading" of humanity, creating a World Brain (Internet of Us), and geocybernetic control of the natural processes of the earth system. Grandiose examples are Elon Musk's Starlink and Neuralink systems, which seem like a direct implementation of Oliver Reiser's Project Prometheus and Krishna.[1]

The ideas of a hyper-technological and transhumanist world civilization, referred to as the Fourth Industrial Revolution and Society 5.0, have been inspired by, among others, Pierre Teilhard de Chardin, Buckminster Fuller, Oliver Reiser, H. G. Wells, the 1930s technocracy movement, and the World Future Society. However, its roots can be traced even further back, to alchemy, hermeticism, and theosophy.

In transhumanism, the spiritual evolution in occultism is combined with Darwinism and the techno-optimist aspirations of futurism into a new techno-religion (evolutionary humanism) where man, using technology and biotechnology, assumes control of his own evolution and ultimately refines himself to perfection.

The offered solutions to the climate threat, however, risks becoming a very costly experience for mankind and is quite far from the utopian visions of the 1970s environmental movement. The social engineering of technocracy, with detailed regulation and behavioral modification, entails alarmingly far-reaching restrictions on human liberty. In addition, this control is now about to literally get under our skin—and even inside our skulls.

> *If you want a vision of the future, imagine a boot stamping on a human face—forever.*
>
> —George Orwell

Appendix A

Rockefeller Foundation

Bold = Rockefeller Family or Rockefeller-initiated organizations

ROCKEFELLER FOUNDATION PRESIDENTS		
PERIOD	**PRESIDENT**	**OTHER POSITIONS**
1913–17	**John D. Rockefeller Jr.**	**Standard Oil**
1917–29	George E. Vincent	**General Education Board**
1929–36	Max Mason	**University of Chicago** (President)
1936–48	Raymond Fosdick	**League of Nations** (Undersecretary)
1948–52	Chester Barnard	Chairman of National Science Foundation
1952–61	Dean Rusk	Secretary of State
1961–72	J. George Harrar	**Rockefeller Foundation's Division of Natural Sciences and Agriculture**
1972–79	John Knowles	Massachusetts General Hospital (Director)
1980–88	Richard Lyman	Stanford University (President)
1988–97	**Peter Goldmark Jr.**	International Herald Tribune (CEO)
1998–2004	Gordon Conway	Royal Geographical Society (President)
2005–17	Judith Ronin	University of Pennsylvania (President)
2017–	Rajiv Shah	Bill & Melinda Gates Foundation USAID

ROCKEFELLER FOUNDATION CHAIRPERSONS		
PERIOD	**CHAIR**	**OTHER POSITIONS**
1917–40	**John D. Rockefeller Jr.**	**Bureau of Social Hygiene** **Museum of Modern Art (MoMA)** **League of Nations** **Council of Foreign Relations**
1941–45	Walter W. Stewart	Institute for Advanced Study
1946–49	John J. McCloy	**Chase Manhattan Bank** **Council of Foreign Relations** Ford Foundation World Bank

(Continued on next page)

ROCKEFELLER FOUNDATION CHAIRPERSONS		
PERIOD	**CHAIR**	**OTHER POSITIONS**
1950–52	John Foster Dulles	Secretary of State Carnegie Endowment for International Peace **Council of Foreign Relations**
1952–71	**John D. Rockefeller III**	**Bureau of Social Hygiene Institute of Pacific Relations Population Council General Education Board Asia Society; Rockefeller Brothers Fund Council on Foundations Council of Foreign Relations**
1972–75	Douglas Dillon	Secretary of the Treasury Brookings Institution **Chase Manhattan Bank Council of Foreign Relations**
1976	Cyrus Vance	Secretary of State **Council of Foreign Relations Trilateral Commission**
1977–82	Theodor Hesburgh	Congregation of the Holy Cross **Chase Manhattan Bank Council of Foreign Relations Trilateral Commission**
1983–86	Clifton Wharton	Deputy Secretary of State Aspen Institute **Asia Society Council of Foreign Relations**
1987–95	John R. Evans	World Bank Rhodes Scholar
1996–99	Alice Stone Ilchman	Assistant Secretary of State **Council of Foreign Relations**
2000–10	James Orr III	Mellon Financial Corporation
2010–16	**David Rockefeller Jr.**	**Rockefeller Brothers Fund Rockefeller Family Fund Rockefeller Family & Associates Rockefeller & Co Council of Foreign Relations Museum of Modern Art** Bohemian Club
2016–21	Richard R. Parsons	**Citigroup** Time Warner World Trade Center Memorial Foundation **Rockefeller Brothers Fund Council of Foreign Relations Museum of Modern Art**

Appendix B

Models

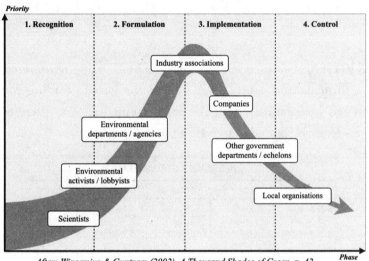

After: Winsemius & Guntram (2002), A Thousand Shades of Green, p. 42

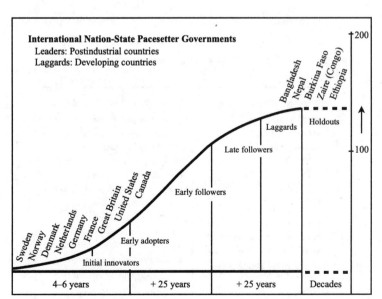

After: Molitor, Graham T. T. (2003), "Molitor Forecasting Model: Key Dimensions for Plotting the 'Patterns of Change'", Journal of Futures Studies, Vol. 8 No. 1

Appendix C

G20

THE G20 GROUP, formed in 1999 out of G7/G8, is an international forum for governments and central banks from the world's leading economies and invited international organizations, countries, and interest groups. The presidency is rotated between the member states.

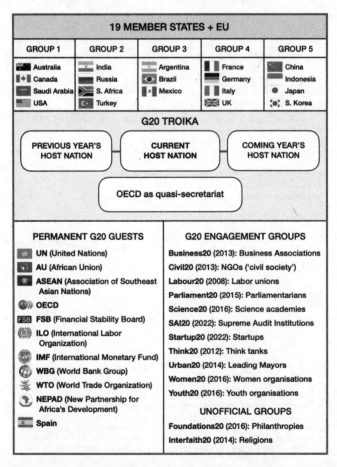

Global Summits

OVERVIEW OF MAJOR global summits during 2019. Each organization's annual summit is usually held around the same time every year (not necessarily on that the exact date). Some have fixed meeting places, other summits (in parentheses in the table) are hosted by different countries. Themes can vary from year to year, and between summits, but often there is a common major theme for any given year. The discussions and resolutions of one summit can affect subsequent meetings that year, and may influence both global and regional policy, as well as what gets highlighted by the media.

GLOBAL SUMMITS 2019			
DATE	**SUMMIT**	**PLACE**	**THEME**
Jan 22–25	**World Economic Forum**	Davos, Switzerland	Globalization 4.0
Feb 10–12	**World Government Summit**	Dubai, UAE	Shaping the Future of Governments
March 6	**S20 Summit**	(Tokyo, Japan)	Threats to Marine Ecosystems and Conservation of the Marine Environment—with Special Attention to Climate Change and Marine Plastic Waste
March 14	**UNEP One Planet Summit**	(Nairobi, Kenya)	Africa Pledge
March 14–15	**B20 Summit**	(Tokyo, Japan)	Towards the Realisation of Society 5.0 for SDGs
March 17–19	**Global Solutions Summit**	Berlin, Germany	Recoupling Social and Economic Progress— Towards a new International Paradigm

(Continued on next page)

GLOBAL SUMMITS 2019			
DATE	**SUMMIT**	**PLACE**	**THEME**
March 21–22	**Europeiska rådet**	Brussels, Belgium	
April 9–13	**Africa Climate Week**	(Accra, Ghana)	
April 2–123	**World Bank/IMF**	Washington, USA	
April 21–23	**C20 Summit**	(Tokyo, Japan)	
May 12–14	**Petersberg Climate Dialogue**	Berlin, Germany	
May 26–27	**T20 Summit**	(Tokyo, Japan)	Seeking a Sustainable, Inclusive and Resilient Society
May 28–29	**R20 Austrian World Summit**	Wien, Austria	To Step up Climate Ambition
May 30–June 2	**Bilderberg**	(Montreux, Switzerland)	A Stable Strategic Order
June 13	**F20 High Level Forum**	(Tokyo, Japan)	Heading towards a New Climate and Sustainable Economy—Shifting the Trillions for a Just Transition
June 14–16	**Trilateral Commission** (Rockefeller)	(Paris, France)	Democracies Under Stress
June 17–27	**UNFCCC Session**	Bonn, Germany	
June 20–21	**European Council Summit**	Brussels, Belgium	Brexit
June 22–23	**ASEAN Bi-annual Summit**	(Bangkok, Thailand)	
June 28–29	**G20 Leaders Summit**	(Osaka, Japan)	Society 5.0
July 9–18	**High-level Political Forum on Sustainable Development (HLPF)**	New York, USA	Empowering people and ensuring inclusiveness and equality
July 11–12	**NATO**	Brussels, Belgium	
Aug 19–23	**Latin America and Caribbean Climate Week**	(Salvador, Brazil)	
Aug 25–27	**G7 Leaders Summit**	(Biarritz, France)	A G7 fighting inequality
Sep 2–6	**Asia-Pacific Climate Week**	(Bangkok, Thailand)	
Sep 17–30	**UN General Assembly**	New York, USA	
Sep 21	**UN Youth Climate Summit**	New York, USA	
Sep 23	**Climate Action Summit**	New York, USA	Climate Action Summit 2019: A Race We Can Win. A Race We Must Win
Nov 11–13	**Paris Peace Forum**	Paris, France	Dive deeper with the experts; Global solutions for global issues; From past to present
Nov 13–14	**BRICS Summit**	(Brasilia, Brazil)	Economic growth for an innovative future
Dec 2–13	**COP 25 (UNFCC)**	(Madrid, Spain)	The Time for Action is Now

Earth Charter

THE EARTH CHARTER, drafted 1994–2000, is an international declaration of values and principles, "to guide the transition towards a more just, sustainable, and peaceful world."

The Earth Charter

I. Respect and Care for the Community of Life

1. Respect Earth and life in all its diversity.
2. Care for the community of life with understanding, compassion, and love.
3. Build democratic societies that are just, participatory, sustainable, and peaceful.
4. Secure Earth's bounty and beauty for present and future generations.

II. Ecological Integrity

5. Protect and restore the integrity of Earth's ecological systems, with special concern for biological diversity and the natural processes that sustain life.
6. Prevent harm as the best method of environmental protection and, when knowledge is limited, apply a precautionary approach.
7. Adopt patterns of production, consumption, and reproduction that safeguard Earth's regenerative capacities, human rights, and community well-being.
8. Advance the study of ecological sustainability and promote the open exchange and wide application of the knowledge acquired.

III. Social and Economic Justice

9. Eradicate poverty as an ethical, social, and environmental imperative.

10. Ensure that economic activities and institutions at all levels promote human development in an equitable and sustainable manner.

11. Affirm gender equality and equity as prerequisites to sustainable development and ensure universal access to education, health care, and economic opportunity.

12. Uphold the right of all, without discrimination, to a natural and social environment supportive of human dignity, bodily health, and spiritual well-being, with special attention to the rights of indigenous peoples and minorities.

IV. Democracy, Nonviolence, and Peace

13. Strengthen democratic institutions at all levels, and provide transparency and accountability in governance, inclusive participation in decision-making, and access to justice.

14. Integrate into formal education and lifelong learning the knowledge, values, and skills needed for a sustainable way of life.

15. Treat all living beings with respect and consideration.

16. Promote a culture of tolerance, nonviolence, and peace.

Source: earthcharter.org

Appendix F

Agenda 2030

On September 25, 2015, the global framework Agenda 2030 was signed by United Nations member states. It includes seventeen Sustainable Development Goals (SDGs) and 169 targets.

The Sustainable Development Goals

1. End poverty in all its forms everywhere.
2. End hunger, achieve food security and improved nutrition, and promote sustainable agriculture.
3. Ensure healthy lives and promote well-being for all at all ages.
4. Ensure inclusive and equitable quality education and promote life-long learning opportunities for all.
5. Achieve gender equality and empower all women and girls.
6. Ensure availability and sustainable management of water and sanitation for all.
7. Ensure access to affordable, reliable, sustainable, and modern energy for all.
8. Promote sustained, inclusive, and sustainable economic growth, full and productive employment, and decent work for all.
9. Build resilient infrastructure, promote inclusive and sustainable industrialization, and foster innovation.
10. Reduce income inequality within and among countries.
11. Make cities and human settlements inclusive, safe, resilient, and sustainable.
12. Ensure sustainable consumption and production patterns.

13. Take urgent action to combat climate change and its impacts by regulating emissions and promoting developments in renewable energy.

14. Conserve and sustainably use the oceans, seas, and marine resources for sustainable development.

15. Protect, restore, and promote sustainable use of terrestrial ecosystems, sustainably manage forests, combat desertification, and halt and reverse land degradation and halt biodiversity loss.

16. Promote peaceful and inclusive societies for sustainable development, provide access to justice for all, and build effective, accountable, and inclusive institutions at all levels.

17. Strengthen the means of implementation and revitalize the global partnership for sustainable development.

Source: sustainabledevelopment.un.org

Appendix G

Timeline Milestones

YEAR	MILE STONES IN THE ENVIRONMENTAL AND CLIMATE AGENDA
1945	**UN Conference on International Organization** (UNCIO), San Francisco
1948	**International Union on the Conservation of Nature** (IUCN)—Founded by Julian Huxley
1948	**Conservation Foundation**—Founded by Laurance Rockefeller and Fairfield Osborn
1952	**The Conference on Population Problems**—John D. Rockefeller III
1956	**Special Studies Project** 1956—1960—Rockefeller Brothers Fund
1957	**International Geophysical Year** (IGY) 1957—58
1961	**WWF**—Founded by Julian Huxley and Prince Bernhard of the Netherlands
1963	*Implications of Rising Carbon Dioxide Content of the Atmosphere* (conference and report)—Conservation Foundation
1968	**The Club of Rome**—Founded by Aurelio Peccei and Alexander King
1970	**Earth Day** (the first Earth Day event)—Initiated by Gaylord Nelson
1970	*Study of Critical Environmental Problems* (MIT study)—Caroll L.Wilson
1971	*Study of Man's Impact on the Climate* (MIT study)—Caroll L. Wilson
1972	*Limits to Growth* (Report to the Club of Rome)—Meadows, Meadows, Randers, and Behrens
1972	**The Stockholm Conference**, UN Conference on the Human Environment, and the Stockholm Declaration (motto: "Only One Earth")
1974	**Declaration on the Establishment of a New International Economic Order** (NIEO)—United Nations
1975	**International Symposium on Long-term Climate Fluctuations** (CO_2 is defined as primary force driving climate change)
1975	**The Next 25 Years: Crisis and Opportunity** (conference)—World Future Society
1977	*Goals for Mankind* (report to the Club of Rome)—Ervin László
1980	**First Global Conference on the Future** (international conference)—World Future Society
1982	**One World Program** (project)—Rockefeller Brothers Fund
1987	*Our Common Future* (report to the United Nations)—The Brundtland Commission
1988	**Conference of the Atmosphere**, Toronto—Hosted by Canada

(Continued on next page)

YEAR	MILE STONES IN THE ENVIRONMENTAL AND CLIMATE AGENDA
1988	**Intergovernmental Panel on Climate Change** (IPCC) (UN body)
1991	*Beyond Interdependence* (report) The Trilateral Commission
1991	*The First Global Revolution* (report)—The Club of Rome
1992	**Earth Summit**, UN Conference on Environment and Development in Rio de Janeiro (motto: "In Our Hands")
1992	**United Nations Framework Convention on Climate Change** (UNFCCC) (UN body)
1995	*Our Common Neighborhood* (report by the UN Commission on Global Governance)—MacArthur Foundation and Ford Foundation
1997	**COP3, Kyoto Protocol** (climate conference and climate treaty)—UNFCCC
2000	**United Nations Millennium Declaration** (goals set for 2000–2015)—The United Nations
2000	*The Earth Charter*—UNESCO and the government of the Netherlands
2002	**UN World Summit on Sustainable Development** in Johannesburg (motto: "Building Partnerships for Sustainable Development")
2004	*A More Secure World: Our Shared Responsibility* (report)—UN High-level Panel on Threats, Challenges, and Change
2006	*An Inconvenient Truth* (climate documentary)—Al Gore
2009	**Club of Rome's global assembly** (conference)—Johan Rockström presents the Planetary Boundaries Framework
2009	**COP 15**, UN Climate Conference in Copenhagen, and The Copenhagen Accord (climate treaty)—UNFCCC
2012	**Rio+20**, UN Conference on Sustainable Development in Rio de Janeiro (motto: "The Future We Want")
2012	**Global Challenges Foundation** (Founded by László Szombatfalvy in Stockholm)
2015	*Confronting the Crisis of Global Governance* (report)—The Albright–Gambari Commission
2015	**Transforming Our World: The 2030 Agenda for Sustainable Development** (UN sustainability treaty)
2015	**COP 21**, UN Climate Change Conference in Paris, and The Paris Agreement)—UNFCCC
2019	**UN–WEF Strategic Partnership Framework** (official partnership)––Signed by the United Nations and the World Economic Forum
2020	**The Great Reset** (agenda declared)—United Nations and World Economic Forum
2020	**Designing a Roadmap to The Future We Want, The UN We Need** (UN conference)—UN75 Global Governance Forum
2021	*Our Common Agenda* (United Nations report)—António Guterres
2022	**Stockholm+50,** UN conference in Stockholm (motto: "A healthy planet for the prosperity of all")
2024	**Summit of the Future** (UN conference)

Appendix H

Timeline Organisations

Bold = Rockefeller Family or Rockefeller-initiated organizations

YEAR	ORGANISATION	FOUNDER	FINANCIERS
1870	**Standard Oil**	**John D. Rockefeller** **William A. Rockefeller Jr.**	
1890	**University of Chicago**	**John D. Rockefeller** Frederick T. Gates	**Rockefeller Foundation**
1901	**Rockefeller Institute of Medical Research**	**John D. Rockefeller** *Frederick T. Gates*	**Rockefeller Foundation**
1902	Carnegie Institution for Science	Andrew Carnegie	Carnegie Steel Company
1903	Scripps Institution of Oceanography	Fred Baker	E. W. Scripps Ellen Browning Scripps
1904	**General Education Board**	**John D. Rockefeller** Frederick T. Gates	**Standard Oil** **Rockefeller Foundation**
1910	Carnegie Endowment for International Peace	Andrew Carnegie	Carnegie Steel Company
1911	Carnegie Corporation of New York	Andrew Carnegie	Carnegie Steel Company
1911	**Equitable Trust Company**	**John D. Rockefeller**	
1911	**Bureau of Social Hygiene**	**John D. Rockefeller Jr.**	**Rockefeller Foundation**
1913	**Rockefeller Foundation**	**John D. Rockefeller** Frederick T. Gates	**Standard Oil**
1913	**International Health Division**	**Rockefeller Foundation**	**Rockefeller Foundation**
1914	**China Medical Board**	**Rockefeller Foundation**	**Rockefeller Foundation**
1914	MIT Sloan School of Management	Alfred P. Sloan	
1920	League of Nations		Member states **Rockefeller Foundation**

(Continued on next page)

YEAR	ORGANISATION	FOUNDER	FINANCIERS
1921	**Council on Foreign Relations**	**John D. Rockefeller Jr.** J. P. Morgan	**Rockefeller Brothers Fund** **Rockefeller Foundation** **Chevron** **ExxonMobil** **JPMorgan Chase** Carnegie Corporation German Marshall Fund of the U.S. Ford Foundation MacArthur Foundation Open Society Institute William and Flora Hewlett Foundation United Nations Foundation
1924	**International House of New York**	**John D. Rockefeller Jr.**	**Rockefeller Brothers Fund** **Rockefeller Foundation**
1927	**Colonial Williamsburg**	**John D. Rockefeller Jr.**	**Rockefeller family** Lila & DeWitt Wallace
1927	Kettering Foundation	Charles F. Kettering	General Motors
1929	**MoMA (Museum of Modern Art), New York**	**Abby A. Rockefeller**	**Rockefeller Brothers Fund** **David Rockefeller**
1929	**Chase Manhattan Bank**	**Equitable Trust Co.**	
1930	**Woods Hole Oceanographic Institution**	**Rockefeller Foundation**	**Rockefeller Foundation**
1930	Institute for Advanced Study (IAS)	Abraham Flexner	National Science Foundation
1931	International Council of Scientific Unions (ICSU)		Membership fees UNESCO United Nations
1931–1939	**Rockefeller Center**	**John D. Rockefeller Jr.**	**John D. Rockefeller Jr.**
1934	Alfred P. Sloan Foundation	Alfred P. Sloan	General Motors
1936	Ford Foundation	Henry Ford Edsel Ford	Ford Motor Company
1940	**Rockefeller Brothers Fund (RBF)**	**John D. Rockefeller III** **Nelson Rockefeller** **Laurance Rockefeller** **Winthrop Rockefeller** **David Rockefeller**	**Standard Oil**

YEAR	ORGANISATION	FOUNDER	FINANCIERS
1942	University of Chicago Department of Meteorology	Carl-Gustaf Rossby	
1944	W. Alton Jones Foundation	Alton Jones	Cities Service Company
1945	United Nations		**John D. Rockefeller Jr. Rockefeller Brothers Fund Rockefeller Foundation**
1945	UNESCO	Julian Huxley	**Rockefeller Brothers Fund Rockefeller Foundation**
1945	The World Bank		
1945	Bulletin of the Atomic Scientists	Eugene Rabinowitch Hyman H. Goldsmith	**Rockefeller family** Carnegie Corporation MacArthur Foundation
1946	Atomic Energy Commission	Vannevar Bush	U.S. Government
1946	Office of Naval Research	Vannevar Bush	U.S. Military
1946	Meteorological Institution, Stockholm University (MISU)	Carl-Gustaf Rossby University of Chicago US Weather Bureau	Swedish Government
1946	**Rockefeller Brothers Inc.**	**Laurance Rockefeller** *Harper Woodward* Ted Walkowicz	**Rockefeller family**
1948	**Conservation Foundation**	Fairfield Osborn **Laurance Rockefeller**	**Rockefeller Brothers Fund Rockefeller Foundation** Ford Foundation
1948	International Union for Conservation of Nature	Julian Huxley	**Rockefeller Brothers Fund Rockefeller Foundation**
1948	World Health Organisation (WHO)	Brock Chisholm	**Rockefeller Foundation** Member states
1949	Aspen Institute	Walter Paepcke Guiseppe Borgese Robert M. Hutchins (all from **University of Chicago**)	**RBF Rockefeller Foundation Exxon** Ford Foundation ARCO Carnegie Corp.
1952	**Population Council**	**John D. Rockefeller III** Frederick Osborn	**Rockefeller Brothers Fund Rockefeller Foundation**

(Continued on next page)

YEAR	ORGANISATION	FOUNDER	FINANCIERS
1954	Bilderberg Meetings	**David Rockefeller** Prince Bernard Jozef Retinger	
1955	International Meteorological Institute	Carl-Gustaf Rossby	
1955–1956	**Biological Effects of Atomic Radiation (BEAR)**	**Rockefeller Foundation** NAS	**Rockefeller Foundation**
1956	**Asia Society**	**John D. Rockefeller III**	**Rockefeller Brothers Fund**
1956	Atmospheric CO_2 Program of the IGY	Scripps Institution of Oceanography	US Weather Bureau
1957	Scientific Committee on Oceanic Research (SCOR)	ICSU	Membership fees
1958	**Am. Conservation Association**	**Laurance Rockefeller**	**Rockefeller Brothers Fund**
1958	Center for International Affairs, Harvard University (now Weatherhead Center for International Affairs)	McGeorge Bundy Henry Kissinger Robert Bowie	Ford Foundation
1960	National Center for Atmospheric Research (NCAR)	Walter Orr Roberts	
1961	WWF	Prince Bernhard Prince Philip Julian Huxley	Fundraising
1963	**Council of the Americas (COAS)**	**David Rockefeller**	
1963	**Asian Cultural Council**	**John D. Rockefeller III**	**John D. III Fund**
1964	**Harvard Center for Population Studies**	**Rockefeller Foundation** Roger Revelle	**Rockefeller Foundation**
1964	David and Lucile Packard Foundation	David Packard Lucile Packard	Hewlett–Packard Company (HP)
1964	UN Conference on Trade and Development (UNCTAD)	Raúl Prebisch	Member states
1965	National Endowment for the Arts	Lyndon B. Johnson	US Government
1966	William and Flora Hewlett Foundation	*William Hewlett* *Flora Hewlett*	Hewlett–Packard Company (HP)

YEAR	ORGANISATION	FOUNDER	FINANCIERS
1966	World Future Society	Edward Cornish Barbara Marx Hubbard	
1967	**Rockefeller Family Fund**	**John D. Rockefeller III Nelson Rockefeller Laurance Rockefeller David Rockefeller Martha Rockefeller**	**Standard Oil**
1967	Environmental Defence Fund	George Woodwell Ford Foundation	**Rockefeller Brothers Fund Rockefeller Foundation** Ford Foundation Bloomberg Philanthropies
1968	**Greenacre Foundation**	**Babs Rockefeller Mauzé**	
1968	Club of Rome	Aurelio Peccei Alexander King	Gianni Agnelli (FIAT) Volkswagen Foundation
1969	Friends of the Earth	David Brower	Robert O. Anderson
1969	**Venrock**	**Rockefeller Brothers Inc.**	
1969	**Chase International Advisory Committee**	**Chase Manhattan Bank David Rockefeller**	**Chase Manhattan Bank**
1970	Natural Resources Defence Council	John Bryson Gustave Speth George Woodwell	Ford Foundation **Rockefeller Brothers Fund**
1970	MacArthur Foundation	John MacArthur Catherine MacArthur	Bankers Life
1971	Climate Research Unit (CRU), University of East Anglia	Graham Sutton Solly Zuckerman	BP Royal Dutch Shell **Rockefeller Foundation**
1971	**Consultative Group for International Agricultural Research (CGIAR)**	**Rockefeller Foundation**	
1971	European Management Forum	Klaus Schwab	
1971	Planetary Citizens	Donald F. Keys	**Rockefeller Brothers Fund**
1971	International Institute for Environment and Development (IIED)	Robert O. Anderson Joseph Slater	ARCO Ford Foundation
1972	Lindisfarne Association	William Irving Thompson **Rockefeller Brothers Fund**	
1972	German Marshall Fund	Guido Goldman	The Government of West Germany

(Continued on next page)

YEAR	ORGANISATION	FOUNDER	FINANCIERS
1972	International Federation of Institutes of Advanced Study (IFIAS)	Aurelio Peccei Alexander King Arne Tiselius	**Rockefeller Brothers Fund** **Rockefeller Foundation** Lilly Endowment John Deere Wallenberg Foundation **Exxon Foundation**
1972	International Institute for Applied Systems Analysis (IIASA)	McGeorge Bundy Jermen Gvishiani Aurelio Peccei Solly Zuckerman	Ford Foundation
1972	United Nations Conference on Human Environment, Stockholm	United Nations	Member states
1972	UNEP	United Nations	
1973	**Trilateral Commission**	**David Rockefeller** Zbigniew Brzezinski	**David Rockefeller** **Rockefeller Brothers Fund** **Rockefeller Foundation** **ExxonMobil** William and Flora Hewlett Foundation. David and Lucille Packard Foundation. IBM Andrew W. Mellon Foundation Ford Foundation Kettering Foundation
1973- 1974	**Commission on Critical Choices for Americans**	**Nelson Rockefeller** *Richard Nixon*	
1974	Worldwatch Institute	Lester Brown **Rockefeller Brothers Fund** **Rockefeller Foundation**	**Rockefeller Brothers Fund** **Rockefeller Foundation**
1975	G6	6 member states	Member states
1976	G7	7 member states	Member states
1976	Congressional Clearinghouse on the Future	Annie Cheatham Charlie Rose	
1977	Beijer Institute	Kjell Beijer Märta Beijer	Kjell & Märta Beijers Stiftelse **Rockefeller Brothers Fund**

YEAR	ORGANISATION	FOUNDER	FINANCIERS
1978	Congressional Institute for the Future	Al Gore Newt Gingrich John Heinz	Carnegie Corp. C.S. Mott Foundation W. Alton Jones Foundation IBM Merck & Co Siemens
1978	**Group of 30**	**Rockefeller Foundation** Geoffrey Bell	**Rockefeller Foundation**
1979	**New York City Partnership**	**David Rockefeller**	**Rockefeller Brothers Fund**
1980	Brandt Commission	Willy Brandt Robert McNamara	Governments Ford Foundation German Marshall Fund
1982	Palme Commission	Olof Palme	Governments
1982	World Resources Institute	Gus Speth MacArthur Foundation	MacArthur Foundation **Rockefeller Brothers Fund** **Rockefeller Foundation**
1982	Kissinger Associates	Henry Kissinger	
1983–1987	Brundtland Commission	United Nations	Jacob Wallenberg Timothy Wirth Paul A. Volker
1985	**Environmental Grantmakers Association**	**Rockefeller Family Fund**	**Rockefeller Brothers Fund** **Rockefeller Foundation**
1985	**Winrock International Institute For Agricultural Development**	**Winrock International Research and Training Center** **Agricultural Development Council**	**Rockefeller Brothers Fund**
1986	**Synergos**	**Peggy Rockefeller Dulany**	**Rockefeller Foundation**
1986	Climate Institute	Crispin Tickell	**Rockefeller Brothers Fund** **Rockefeller Foundation** BP C. S. Mott Fund Shell Foundation World Bank World Resources Institute Governments United Nations
1987	World Economic Forum	Klaus Schwab	Transnational corporations **Rockefeller Foundation**

(Continued on next page)

YEAR	ORGANISATION	FOUNDER	FINANCIERS
1988	IPCC	WMO ICSU UNEP	Member states
1989	GLOBE	Congressional Institute for the Future Al Gore Hemmo Muntingh	German Marshall Fund Int. Fund for Animal Welfare W. Alton Jones Foundation Unilever
1989	**David Rockefeller Fund**	**David Rockefeller** **Peggy McGrath Rockefeller**	**David Rockefeller**
1989	**Global Climate** **Coalition**	**Exxon** **Mobil** **Chevron** Royal Dutch Shell	
1989	Stockholm Environment Institute (SEI)	Beijer Institute Gordon Goodman Bert Bolin Birgitta Dahl	**Rockefeller Brothers** **Fund** **Rockefeller Foundation** Formas Mistra SIDA United Nations
1991	**LEAD International**	**Rockefeller Foundation**	**Rockefeller Foundation**
1992	UNFCCC	United Nations	Member states
1992	Potsdam Institute for Climate Impact Research	H. J. Schellnhuber	Government of Germany The Federal State of Brandenburg
1993	Club of Budapest	Ervin László	
1993	Open Society Foundation	George Soros	George Soros
1994	Earth Charter	Queen Beatrix Maurice Strong Mikhail Gorbachev Steven Rockefeller	Dutch Government
1995	Global Scenario Group	SEI Tellus Institute	**Steven Rockefeller** **Rockefeller Foundation** Nippon Foundation UNEP
1996	International Human Dimensions Programme on Global Environmental Change	ICSU International Social Science Council	
1998	UN Foundation	Ted Turner	Time Warner
1998	Humanity+	Nick Bostrom David Pearce	
1999	Skoll Foundation	Jeffrey Skoll	eBay
1999	G20	20 member states	Member states

YEAR	ORGANISATION	FOUNDER	FINANCIERS
2000	Bill & Melinda Gates Foundation	Bill Gates Melinda Gates	Microsoft
2000	Foundation for the Law of Time	José Argüelles	
2001	ATCA 5000	D. K. Matai Hervé de Carmoy Mark Lewis	
2002	**Rockefeller Philanthropy Advisors**	**Eileen Rockefeller Growald**	**Rockefeller Brothers Fund** Bill & Melinda Gates Foundation MacArthur Foundation
2003	Global Footprint Network	Mathis Wackernagl Susan Burns	
2004	The Climate Group	**Rockefeller Brothers Fund** Steve Howard	**Rockefeller Brothers Fund** HSBC
2005	Future of Humanity Institute	Nick Bostrom University of Oxford	
2006	Bloomberg Philanthropies	Mike Bloomberg	Bloomberg LP
2007	Growald Family Fund	**Eileen Rockefeller Growald Paul Growald**	
2007	Presidential Climate Action Project	**Rockefeller Brothers Fund** University of Colorado Foundation	**Rockefeller Brothers Fund** Denver Foundation Aspen Community Foundation Prentice Foundation
2007	Stockholm Resilience Center	Mistra Beijer Institute SEI	Mistra **Rockefeller Foundation**
2008	Skoll Global Threats Fund	Jeffrey Skoll	eBay
2008	350.org	Sustainable Markets Foundation	**Rockefeller Brothers Fund** **Rockefeller Family Fund**
2009	**The Future We Want**	**Rockefeller Brothers Fund** Michael Northrop	**Rockefeller Brothers Fund** Surna Foundation Capstone Turbine Production
2009	Earth System Governance Project	International Human Dimensions Programme on Global Environmental Change	United Nations

(Continued on next page)

YEAR	ORGANISATION	FOUNDER	FINANCIERS
2009	Worldshift 2012	Ervin László	
2009	Noosphere Forum	José Argüelles	
2010	FuturICT	EU Future & Emerging Technologies *Dirk Helbing*	European Commission *George Soros*
2010	Stalliance	ICSU UNEP UNESCO WMO	
2011	**Climate Reality Project**	**Rockefeller Philanthropy Advisors** *Al Gore*	**Rockefeller Brothers Fund** Skoll Foundation
2012	Future Earth	Stalliance	Governments Foundations Universities
2012	Biohax International	Jowan Österlund	
2013	**100 Resilient Cities**	**Rockefeller Philanthropy Advisors Rockefeller Foundation**	**Rockefeller Foundation**
2013	Global Challenges Foundation	László Szombatfalvy	László Szombatfalvy
2014	BioNyfiken	Hannes Sjöblad	
2015	**Global Resilience Partnership**	**Rockefeller Foundation** Stockholm Resilience Center SIDA USAID	
2017	**ID2020 Alliance**	**Rockefeller Foundation** Accenture GAVI Microsoft IDEO.org	**Rockefeller Foundation** Bill & Melinda Gates Foundation
2018	**Rockefeller Capital Management**	**Rockefeller Financial Services Viking Global Investors**	

Bibliography

Rockefeller Family Primary Sources

Rockefeller, David. "From a China Traveler," *New York Times*, August 10, 1973.

———. *Memoirs*. New York: Random House, 2003.

———. *Unused Resources and Economic Waste: A Dissertation*. Chicago: University of Chicago Press, 1941.

Rockefeller, David Jr. "Does Happiness Come from Material Wealth?," *Financial Times*, February 16, 2018.

Rockefeller, John D. III. *Population and the American Future: The Report of the Commission on Population Growth and the American Future*, Rockefeller Commission Report, The Center for Research on Population and Security, July 27, 1972. www.populationsecurity.org/rockefeller/001_population_growth_and_the _american_future.htm.

Rockefeller, Steven. "Commentary on 'Meaning, Religion, and a Great Transition,'" Great Transition Initiative, December 2014. www.greattransition.org/commentary /steven-rockefeller-meaning-religion-and-a-great-transition-michael-karlberg.

———. "The Earth Charter at 15: A Spiritual Lens on Sustainability," interview by Allen White, Great Transition Initiative, December 2015.

———. "The Legitimacy of the Earth Charter," A Lecture Presented at Exeter College, Oxford University February 9, 2007. earthcharter.org.

Rockefeller, Steven, and John C. Elder, eds. *Spirit and Nature: Why the Environment Is a Religious Issue—An Interfaith Dialogue*. Boston: Beacon Press, 1992.

Rockefeller Archives Center, Subpanel II: International Security Objectives and Strategy, 1956–1964. dimes.rockarch.org, search: "International Security Objectives and Strategy."

———. The Population Council, World Population Problem, and Contraceptive Studies during the Early Postwar Era by Huang, Yu-ling, 2011. www.rockarch.org/ publications/resrep/huang2.pdf.

Rockefeller Brothers Fund. "75 Years of Engaged Philanthropy, Timeline: The One World Program 1983," 75.rbf.org/#!trigger=one-world-program.

———. "2014 Divestment Statement." www.rbf.org/sites/default/files/rbf divestment _statement-2017-oct.pdf.

————. "About." www.rbf.org/about.

————. *Annual Reviews* 1951–2017. www.rbf.org/about/annual-reviews-archive.

————. Grantees: 350.org. www.rbf.org/grantees/350.org.

————. Grantees: Climate Accountability Institute. www.rbf.org/grantees/climate
-accountability-institute.

————. Grantees: Rockefeller Philanthropy Advisors. www.rbf.org/grantees/rockefeller
-philanthropy-advisors.

————. Grants search: Presidential Climate Action Project. www.rbf.org/grantees
/natural-capitalism-solutions-inc.

————. Memorandum from Thomas Wahman to Russell Phillips, September 23,
1985, RBF Records, RG3 Project Files, Subgroup 2, Box 1163, Folder 7175,
FA005.

————. "On the Loss of David Rockefeller, 1915–2017." www.rbf.org/news/on-the
-loss-of-David-Rockefeller. March 20, 2017.

————. *Prospect for America: The Rockefeller Panel Reports.* Garden City, NY:
Doubleday, 1961.

————. "RFF's Decision to Divest." www.rffund.org/divestment. 2016.

————. "Solutions to Global Warming: A National Conversation We Desperately
Need to Have." 2006. www.rbf.org/news/solutions-global-warming-national
-conversation-we-desperately-need-have.

————. "Stephen Heintz Praises the Agreement Reached at the Paris Climate
Conference." December 12, 2015. www.rbf.org/news/stephen-heintz-praises
-agreement-reached-paris-climate-conference.

————. *Sustainable Development Program Review 2005–2010.* November 2010. www
.rbf.org/sites/default/files/sustainabledevelopmentprogramreview.pdf.

Rockefeller Financial. "RIT Capital Partners, Chaired by Lord Rothschild, and
Rockefeller & Co. Announce Strategic Partnership," press release, May 30, 2012.
www.ritcap.com/sites/default/files/RIT_-_Rockefeller_Press_Release_FINAL_5
-29-12.pdf.

Rockefeller Foundation Archives. Agriculture. rockfound.rockarch.org/agriculture.

————. Bureau of Social Hygiene. rockfound.rockarch.org/bureau-of-social-hygiene.

————. Computer Science. rockfound.rockarch.org/computer-science.

————. Evolution of a Foundation. rockfound.rockarch.org/evolution-of-a
-foundation.

————. Family Planning. rockfound.rockarch.org/family-planning.

————. Finding a Footing. rockfound.rockarch.org/finding-a-footing.

————. Kinsey Reports. rockfound.rockarch.org/kinsey-reports.

————. Simon Flexner. rockfound.rockarch.org/biographical/-/asset_publisher
/6ygcKECNI1nb/content/simon-flexner.

Rockefeller Foundation. Addressing the UN Climate Summit. 2014. www.rockefeller
foundation.org/blog/addressing-un-climate-summit.

———. "Advancing a More Nourishing and Sustainable Food System." 2018. www
.rockefellerfoundation.org/our-work/topics/agriculture-and-food-security

———. *Annual Reports.* 1956–2018. HYPERLINK "http://www.
rockefellerfoundation.org/about-us/governance-reports/annual-reports" www
.rockefellerfoundation.org/about-us/governance-reports/annual-reports.

———. "Bellagio 60, Innovative Frontiers of Development, Upgrading the System
Rethinking Capitalism, Digital State, Strengthening Multilateralism." June 5–9,
2019. "http://www.rockefellerfoundation.org/bellagio60/innovative-frontiers
-development" www.rockefellerfoundation.org/bellagio60/innovative-frontiers
-development.

———. Innovative Frontiers of Development Conference. June 5–9, 2019. www
.rockefellerfoundation.org/bellagio60/innovative-frontiers-development.

———. "Meet Africa's Inclusive AI Community Solving Global Problems." August
26, 2019. www.rockefellerfoundation.org/blog/meet-africas-inclusive-ai-community
-solving-global-problems.

———. "Reducing Loss from What We Grow and Harvest." 2018. www.rockefeller
foundation.org/our-work/initiatives/yieldwise.

Books

Adler, Mortimer J. *How to Read a Book: The Art of Getting A Liberal Education.* New
York: Simon and Schuster, 1940.

Argüelles, José. *Manifesto of the Noosphere: The Next Stage in the Evolution of Human
Consciousness* (Manifesto Series). Berkeley: North Atlantic Books, 2011.

Bacon, Francis. *New Atlantis.* UK, 1626.

Bailey, Alice. *The Externalisation of the Hierarchy.* New York: Lucis Publishing Co.,
1957.

Bailey, Christopher. *US Climate Change Policy.* London: Routledge, 2015.

Baran, Stanley, and Dennis Davis. *Mass Communication Theory: Foundations, Ferment,
and Future.* Boston: Cengage Learning, 2009.

Bayh, Birch. *One Heartbeat Away: Presidential Disability and Succession.* Indianapolis:
Bobbs-Merrill, 1968.

Beckwith, Burnham Putnam. *The Next 500 Years: Scientific Predictions of Major Social
Trends.* New York: Exposition Press, 1967.

Begich, Nick. *Angels Don't Play This HAARP: Advances in Tesla Technology.*
Anchorage: Earthpulse Press, 1995.

Bello, Walden. *Dark Victory: The United States and Global Poverty* (2nd edition).
Oakland: Food First Books, 1999.

Bishop, Matthew, and Michael Green. *Philanthrocapitalism: How the Rich Can Save the World*. New York: Bloomsbury USA, 2008.

Bolin, Bert. *A History of the Science and Politics of Climate Change: The Role of the Intergovernmental Panel on Climate Change*. Cambridge, UK: Cambridge University Press, 2007.

———. *The Atmosphere and the Sea in Motion: Scientific Contributions to the Rossby Memorial Volume*. New York: Rockefeller Institute Press, 1959.

Bolton, Kerry. *Revolution from Above*. Budapest: Arktos Media, 2011.

Boyer, John B. *University of Chicago: A History*. Chicago: University of Chicago Press, 2015.

Broderick, Damien. *Year Million: Science at the Far Edge of Knowledge*. New York: Atlas & Co., 2008.

Brzezinski, Zbigniew. *Between Two Ages: America's Role in the Technetronic Era*, Santa Barbara, CA: Greenwood Press, 1970.

Burris, Beverly. *Technocracy at Work*. Albany, NY: SUNY Press, 1993.

Byers, Horace. *Carl-Gustaf Arvid Rossby 1898–1957*. Washington, DC: National Academy of Sciences, 1960.

Campbell, Charles. *A Questing Life: The Search for Meaning*. Lincoln, NE: iUniverse, 2006.

Caro, Robert A. *The Power Broker: Robert Moses and the Fall of New York*. New York: Vintage Books, 1975.

Carson, Rachel. *Silent Spring*. Boston: Houghton Mifflin, 1962.

Chernow, Ron. *Titan: The Life of John D. Rockefeller, Sr.* New York: Random House, 1998.

Clinton, Bill. *My Life*. New York: Knopf Publishing Group, 2004.

Collier, Peter, and David Horowitz. *The Rockefellers: An American Dynasty*. New York: Holt, Rinehart and Winston, 1976.

Colomer, Josep M. *How Global Institutions Rule the World*. Basingstoke, UK: Palgrave Macmillan, 2014.

Constanza, Robert, and Ida Kubiszewski, eds. *Creating a Sustainable and Desirable Future: Insights from 45 Global Thought Leaders*. Singapore: World Scientific Publishing, 2014.

Council on Foreign Relations. *The War and Peace Studies of the Council on Foreign Relations 1939–1945*. New York, The Harold Pratt House, 1946.

Critchlow, Donald. *Intended Consequences: Birth Control, Abortion, and the Federal Government*. Oxford: Oxford University Press, 1999.

Ehrlich, Paul, and Ann Ehrlich. *The Population Bomb*. New York: Ballantine Books, 1968.

Engdahl, William. *A Century of War*. London: Pluto Press, 2004.

Engfeldt, Lars-Göran. *From Stockholm to Johannesburg and Beyond: The Evolution of the International System for Sustainable Development Governance and Its Implications.* Stockholm: Ministry of Foreign Affairs, 2009.

Feather, Frank. *Through the '80s: Thinking Globally, Acting Locally.* Washington, DC: World Future Society, 1980.

Fitzgerald, Michael. *Making Modernism: Picasso and the Creation of the Market for Twentieth-Century Art.* New York: Farrar, Straus and Giroux, 1995.

Flohn, Hermann. *Meterologie im Übergang: Erfahrungen und Erinnerungen (1931–1991),* Bonn: Dümmler, 1992.

Franks, Angela. *Margaret Sanger's Eugenic Legacy: The Control of Female Fertility.* Jefferson, NC: McFarland & Co., 2005.

Fuller, Buckminster. *Operating Manual for Spaceship Earth.* Carbondale: Southern Illinois University Press, 1969.

Gibson, Donald. *Environmentalism: Ideology and Power.* New York: Nova Science Publishers, 2002.

Glover, Leigh. *Postmodern Climate Change.* London: Routledge, 2006.

Gonzales, George. *Corporate Power and the Environment.* Lanham, MD: Rowman & Littlefield, 2001.

Goodspeed, Thomas W. *A History of the University of Chicago, Founded by John D. Rockefeller.* Chicago: University of Chicago Press, 1916.

Gore, Al. *Earth in Balance.* New York: Earthscan, 1992.

Gunderson, Lance H., and C. S. Holling. *Panarchy: Understanding Transformations in Human and Natural Systems.* Washington, DC: Island Press, 2001.

Harkavy, Oscar, et al. *Curbing Population Growth: An Insider's Perspective on the Population Movement.* New York: Population Council, 1996.

Harkness, Deborah. *John Dee's Conversation with Angels.* Cambridge, UK: Cambridge University Press, 1999.

Harman, Willis, and O. W. Markley, eds. *Changing Images of Man.* London: Pergamon Press, 1982.

Hatfield, Mark O. "Nelson Rockefeller" in *Vice Presidents of the United States, 1789–1993.* Washington, DC: US Government Printing Office, 1977, 505–512.

Heinze, Thomas, and Richard Münch. *Innovation in Science and Organizational Renewal: Historical and Sociological Perspectives.* Berlin: Springer, 2016.

Hironaka, Ann. *Greening the Globe: World Society and Environmental Change.* Cambridge, UK: Cambridge University Press, 2014.

Hobbes, Thomas. *Leviathan.* London: Andrew Crooke, 1651.

Howe, Joshua P. *Behind the Curve: Science and the Politics of Global Warming.* Seattle: University of Washington Press, 2014.

Hyman, Sidney. *The Aspen Idea.* Norman: University of Oklahoma Press, 1975.

Jacobs, Jane. *The Death and Life of Great American Cities*. New York: Random House, 1961.

Jarrett, Henry, ed. *Environmental Quality in a Growing Economy: Essays from the Sixth RFF Forum*. Baltimore: Johns Hopkins University Press, 1966, 3–14.

Kanninen, Tapio. *Crisis of Global Sustainability* (Global Institutions). London: Routledge, 2012.

Kert, Bernice. *Abby Aldrich Rockefeller: The Woman in the Family*. New York: Random House, 1993.

Keys, Donald. *Earth at Omega: Passage to Planetization*. Boston: The Branden Press, 1982.

King, Alexander. *The First Global Revolution* (see Reports).

Kinsey, Alfred, W. B. Pomeroy, and C. E. Martin. *Sexual Behavior in the Human Male*. Philadelphia: W. B. Saunders Co., 1948.

Klein, Naomi. *This Changes Everything: Capitalism vs. the Climate*. New York: Simon & Schuster, 2015.

Knudsen, Dino. *The Trilateral Commission and Global Governance*. London and New York: Routledge, 2013.

Kristiakis, Alexander. *How People Harness Their Collective Wisdom and Power to Construct the Future*. Charlotte: Information Age Publishing Inc., 2006.

Lamb, R. *Promising the Earth*. London: Routledge, 1996.

László, Ervin. *Goals for Mankind* (see Reports).

———. *Macroshift: Navigating the Transformation to a Sustainable World*. San Francisco: Berrett–Koehler Publishers, 2001.

———. *Quantum Shift in the Global Brain: How the New Scientific Reality Can Change Us and Our World*. Rochester, VT: Inner Traditions, 2008.

———. *WorldShift 2012: Making Green Business New Politics & Higher Consciousness Work Together*. Rochester, VT: Inner Traditions, 2009.

László, Ervin, Robert Baker Jr., Elliott Eisenberg, and Venkata Raman. *The Objectives of the New International Economic Order*. Oxford: Pergamon Press, 1978.

Legget, Jeremy. *Carbon War: Global Warming and the End of the Oil Era*. London: Routledge, 2001.

Leinen, Jo, and Andreas Bummel. *A World Parliament: Governance and Democracy in the 21st Century*. Berlin: Democracy Without Borders, 2018.

Lett, Donald Jr. *Phoenix Rising: The Rise and Fall of the American Republic*, Bloomington, IN: AuthorHouse, 2008.

Livingstone, David. *Transhumanism: The History of a Dangerous Idea*. Sabilillah Publications, 2015.

Lomborg, Bjørn. *Cool It: The Skeptical Environmentalist's Guide to Global Warming*. New York: Knopf Publishing Group, 2007.

Luterbacher, Urs, and Detlef F. Sprinz. *International Relations and Global Climate Change*. Cambridge, MA: MIT Press, 2001.

Malone, Thomas, Edward Goldberg, and Walter Munk. *Roger Dougan Revelle 1909– 1991, A Biographical Memoir*. Washington, DC: National Academy of Sciences, 1998.

Martinus. *Livets Bog: Det Tredje Testamente*, seven volumes. Stockholm: Verdisbild, 1932–69.

Marx Hubbard, Barbara. *The Book of Co-Creation: An Evolutionary Interpretation of the New Testament*. New Visions, 1980.

———. *The Book of Co-Creation—The Revelation: Our Crisis is a Birth*. Sonoma, CA: Foundation for Conscious Evolution, 1993.

———. *Conscious Evolution: Awakening the Power of Our Social Potential*. Novato, CA: New World Library, 2010.

Mazur, Laura, and Louella Miles, eds. "Sir Crispin Tickell," in *Conversations with Green Gurus: The Collective Wisdom of Environmental Movers and Shakers*. Hoboken, NJ: John Wiley & Sons, Inc., 2012.

McKibben, Bill. *The End of Nature*. New York: Anchor/Random House, 1989.

McNeill, Jim, et al. *Beyond Interdependence* (see Reports).

Meynard, Thierry, ed. *Teilhard and the Future of Humanity*. New York: Fordham University Press, 2006. www.jstor.org/stable/j.ctt13x04dq.

Mires, Charlene. *Absolutely Not New York: Finding a Home for the United Nations*. New York: New York University Press, 2013.

Mittal, Anuradha, and Melissa Moore. *African Farmers and Environmentalists Speak Out Against a New Green Revolution in Africa*. Oakland, CA: The Oakland Institute, 2009.

More, Thomas. *Utopia*. Habsburg, Netherlands, 1516.

Mourlon-Druol, Emmanuel, and Federico Romero, eds. *International Summitry and Global Governance: The Rise of the G7 and the European Council, 1974–1991*. London: Routledge, 2014.

Oreskes, Naomi. *Merchants of Doubt: How a Handful of Scientists Obscured the Truth on Issues from Tobacco Smoke to Global Warming*. New York: Bloomsbury Press, 2011.

Osborn, Fairfield. *Our Plundered Planet*. London: Faber and Faber, 1948.

Palmstierna, Hans, and Lena Palmstierna. *Plundring, svält, förgiftning*. Stockholm: Rabén & Sjögren, 1968.

Pearce, Fred. *The Coming Population Crash: And Our Planet's Surprising Future*. Boston: Beacon Press, 2010.

Quigley, Carroll. *Tragedy and Hope*. San Pedro, CA: GSG & Associates, 1966.

Ray, Dixy Lee. *Trashing the Planet: How Science Can Help Us Deal with Acid Rain, Depletion of the Ozone, and the Soviet Threat among Other Things*. Washington, DC: Regnery Publishing, 1990.

Raymond, Jack. *Robert O. Anderson: Oil Man/Environmentalist and His Leading Role in the International Environmentalist Movement: A Monograph.* Aspen, CO: Aspen Institute for Humanistic Studies, 1988.

Reed, Henry Hope. *Rockefeller, New York: A Tour.* New York: Greensward Foundation, 1988.

Reiser, Oliver. *Cosmic Humanism and World Unity.* New York: Gordon and Breach, 1975.

———. *The World Sensorium: The Social Embryology of World Federation.* Whitefish, MT: Kessinger Publishing, LCC, 1946.

Rifkin, Jeremy. *Entropy.* New York: Bantam, 1981.

Robertson, Thomas. *The Malthusian Moment: Global Population Growth and the Birth of American Environmentalism (Studies in Modern Science, Technology, and the Environment).* New Brunswick, NJ: Rutgers University Press, 2012.

Rosner, Lisa. *The Technological Fix: How People Use Technology to Create and Solve Problems.* New York: Routledge, 2004.

Samaan, A. E. *From a "Race of Masters" to a "Master Race": 1948 to 1848.* Charleston, SC: CreateSpace, 2013.

Sarnoff, David, ed. *The Fabulous Future: America in 1980.* Boston: E. P. Dutton & Co., 1955.

Satterthwaite, David. *Barbara Ward and the Origins of Sustainable Development.* International Institute for Environment and Development, London, 2006.

Schwarzenbach, Alexis. *Saving the World's Wildlife: The WWF's First Fifty Years.* London: Profile Books, 2011.

Schwab, Klaus. *The Fourth Industrial Revolution.* Geneva: World Economic Forum, 2016.

———. *Shaping the Fourth Industrial Revolution.* Geneva: World Economic Forum, 2018.

Shaplen, Robert. *Toward the Well-Being of Mankind: Fifty Years of the Rockefeller Foundation.* Garden City, NY: Doubleday & Company Inc., 1964.

Sklar, Holly. *Trilateralism: The Trilateral Commission and Elite Planning for World Management.* Boston, MA: South End Press, 1980.

Slee, Tom. *What's Yours Is Mine: Against the Sharing Economy.* Berkeley: OR Books, 2017.

Slichter, Charles. *Frederick Seitz 1911–2008.* Washington, DC: National Academy of Sciences, 2010.

Spekke, Andrew. *The Next 25 Years: Crisis and Opportunity.* Washington, DC: World Future Society, 1975.

Stonor Saunders, Frances. *Who Paid the Piper? The CIA and the Cultural Cold War.* London: Granta, 1999.

Sörlin, Sverker. *Science, Geopolitics and Culture in the Polar Region: Norden beyond Borders.* Oxford: Taylor and Francis, 2016.

———. *A Tribute to the Memory of Carl-Gustaf Rossby.* Stockholm: The Royal Swedish Academy of Engineering Sciences (IVA), 2015. www.iva.se/globalassets /info-trycksaker/iva/minnesskriften-2015.pdf.

Szombatfalvy, László. *Vår tids största utmaningar.* Stockholm: Ekerlids, 2009.

Tainter, Joseph. *The Collapse of Complex Societies.* Cambridge, UK: Cambridge University Press, 1988. wtf.tw/ref/tainter.pdf.

Tännsjö, Torbjörn. *Global Democracy: A Case for World Government.* Edinburgh: Edinburgh University Press, 2008.

Teilhard de Chardin, Pierre. *The Phenomenon of Man.* New York: Harper & Brothers, 1955.

Tickell, Crispin. *Climatic Change and World Affairs (Harvard Studies in International Affairs).* Cambridge, MA: Center for International Affairs at Harvard University, 1986.

Toffler, Alvin. *Future Shock.* New York: Random House, 1970.

———. *Creating a New Civilization: The Politics of the Third Wave.* Atlanta: Turner Publishing, 1995.

Toppin, Don, ed. *This Cybernetic Age.* New York: Human Development Corporation, 1969.

Train, Russell. *Politics, Pollution, and Pandas: An Environmental Memoir.* Washington, DC: Island Press, 2003.

Turque, Bill. *Inventing Al Gore: A Biography.* Boston: Houghton Mifflin, 2000.

Vogt, William. *Road to Survival.* New York: T. William Slodne Associates, 1948.

Wagar, W. Warren. *The City of Man: Prophecies of a World Civilization in Twentieth-Century Thought.* Boston: Houghton Mifflin, 1963.

Walsch, Neale Donald. *The Mother of Invention: The Legacy of Barbara Marx Hubbard and the Future of YOU.* New York: Hay House, 2011.

Ward, Barbara. *Spaceship Earth.* New York: Columbia University Press, 1966.

Ward, Barbara, and René Jules Dubos. *Only One Earth: The Care and Maintenance of a Small Planet.* New York: W. W. Norton & Company, 1972.

Wells, H. G. *The Open Conspiracy: Blue Prints for a World Revolution.* London: Victor Gollancz Ltd., 1928.

———. *World Brain.* London: Methuen Publishing Ltd., 1938.

Wiesner, Jerome. *Vannevar Bush, 1890–1974.* Washington, DC: National Academies Press, 1979.

Wijkman, Anders, and Johan Rockström. *Bankrupting Nature: Denying Our Planetary Boundaries.* UK: Routledge, 2012.

Winks, Robin W. *Laurance S. Rockefeller: Catalyst for Conservation*. Washington, DC: Island Press, 1997.

Winsemius, Pieter, and Ulrich Guntram. *A Thousand Shades of Green: Sustainable Strategies for Competitive Advantage*. London: Earthscan Publications Ltd., 2002.

Wood, Patrick M. *Technocracy: The Hard Road to World Order*. Mesa, AZ: Coherent Publishing, 2018.

———. *Technocracy Rising: The Trojan Horse of Global Transformation*. Mesa, AZ: Coherent Publishing, 2015.

World Economic Forum. *A Partner in Shaping History: The First 40 Years 1971–2010*. Geneva: World Economic Forum, 2009.

Wortman, Marc. *1941: Fighting the Shadow War: A Divided America in a World at War*. New York: Atlantic Monthly Press, 2016.

Weyler, Rex. *Greenpeace: How a Group of Ecologists, Journalists, and Visionaries Changed the World*. Vancouver, BC: Raincoast Books, 2004.

Reports

Aspen Institute. *Second Status Report to the Trustees of IFIAS, on the IFIAS Special Project, the Impact on Man of Climate Change*, October 8, 1976.

Barney, Gerald O., ed. *The Unfinished Agenda: The Citizen's Policy Guide to Environmental Issues: A Task Force Report Sponsored by the Rockefeller Brothers Fund*. New York: Thomas Y. Crowell Co., 1977.

———. *The Global 2000 Report to the President*. Washington, DC: US Government Printing Office, 1980.

Bergsten, C. Fred, et al. *The Reform of International Institutions: A Report of the Trilateral Task Force on International Institutions to the Trilateral Commission* (Triangle Papers). Washington, DC, 1976.

Bolin, Bert. International Council of Scientific Unions. *The Greenhouse Effect, Climatic Change, and Ecosystems*. Chichester, England. Published on behalf of the Scientific Committee on the Problems of the Environment of the International Council of Scientific Unions by Wiley, 1986.

Brandt Commission. *Common Crisis North-South: Cooperation for World Recovery*. Cambridge, MA: MIT Press, 1983.

Brookings Institution. *Managing Global Insecurity*. 2008. www.brookings.edu/author/managing-global-insecurity.

Bush, Vannevar. *Science the Endless Frontier: A Report to the President by Vannevar Bush, Director of the Office of Scientific Research and Development, July 1945*. Washington, DC: United States Government Printing Office, 1945.

California Environmental Associates. *Design to Win: Philanthropy's Role in the Fight against Global Warming.* San Francisco: California Environmental Associates, 2007.

Charney, Jule G., et al. *Carbon Dioxide and Climate: A Scientific Assessment: Report of an Ad Hoc Study Group on Carbon Dioxide and Climate.* Washington, DC: National Academy of Sciences, 1979.

Commission on Critical Choices for Americans. *Critical Choices for Americans*, Vol. 1–13. Lexington, MA, and Fallbrook, CA: Lexington Books and Aero Publishers, 1976–1977.

Conservation Foundation. *Implications of Rising Carbon Dioxide Content of the Atmosphere: A Statement of Trends and Implications of Carbon Dioxide Research Reviewed at a Conference of Scientists.* New York: Conservation Foundation, 1963.

Dodd, Norman. *The Dodd Report to the Reece Commission on Foundations.* New York: The Long House Inc., 1954.

Fay, Marianne, et al. *Decarbonizing Development: Three Steps to a Zero-Carbon Future.* Washington, DC: World Bank, 2015.

Richard N. Gardner, Saburo Okita, and B. J. Udin. *A Turning Point in North-South Relations.* Trilateral Commission Task Force Report 3, The Triangle Papers, 1974.

Gardner, Richard N. "The North-South Negotiations: Some Progress but Long Road Ahead," *Trialogue: A Bulletin of North American–European–Japanese Affairs,* "T10—Looking Back . . . and Forward." The Trilateral Commission, 1976. trilateral.org//download/doc/Looking_Back_And_Forward.pdf.

German Advisory Council on Global Change. *World in Transition: A Social Contract for Sustainabiliby,* 2011.

Global Transition Initiative. "Imagine All the People: Advancing a Global Citizens Movement." 2010. www.wideningcircle.org/documents/TWC%20Readings/GTI -Perspectives-Imagine_All_the_People.pdf.

Goodman, Gordon. "Responding to Climatic Change: Further Steps in Policy Development—A Proposal for the Continued Development in 1989 and 1990 of Work on the Possible Responses to Climatic Change." The Beijer Institute, February 1989.

Grand View Research. "Smart Cities Market Size, Share & Trends Analysis Report By Application (Education, Governance, Buildings, Mobility, Healthcare, Utilities), by Component (Services, Solutions) and Segment Forecasts 2018—2025," February 2018.

Greenpeace, *Annual Report 2001.*

International Task Force on Global Public Goods. "Meeting Global Challenges: International Cooperation in the National Interest," Final Report, Stockholm, Sweden, 2006. keionline.org/misc-docs/socialgoods/International-Task-Force-on -Global-Public-Goods_2006.pdf.

Internet of Things Council. "Europe's IoT." Pan European Networks. www. theinternetofthings.eu/sites/default/files/GOV18%20R%20van%20 Kranenburg%206007_ATL.pdf.

Jäeger, Jill. *Developing Policies for Responding to Climate Change: A Summary of the Discussions and Recommendations of the Workshops Held in Villach (28 September–2 October 1987) and Bellagio (9–13 November 1987), under the Auspices of the Beijer Institute, Stockholm.* World Climate Program, WMO, UNEP, April 1988.

Karas, J. H. W., and P. M. Kelly. *The Heat Trap: Threat Posed by Rising Levels of Greenhouse Gases.* Ottawa, Canada: Friends of the Earth, 1989.

Keeling, C. D. "The Influence of Mauna Loa Observatory on the Development of Atmospheric CO_2 Research," in *Mauna Loa Observatory: A 20th Anniversary Report*, edited by John Miller. Silver Spring, MD: Environmental Research Laboratories, Air Resources Laboratories, 1978.

Kellogg, William W., and Margaret Mead, eds. *The Atmosphere: Endangered and Endangering*, Fogarty International Center Proceedings No. 39, 1975. Washington, DC: US Government Printing Office, 1977.

King, Alexander, and Bertrand Schneider. *The First Global Revolution: A Report by the Council of the Club of Rome.* New York: Pantheon Books, 1991.

László, Ervin. *Goals for Mankind: A Report to the Club of Rome.* London: Hutchinson & Co., 1977.

Malasalaska, Pentti, and Matti Vapaavuori, eds. *The Club of Rome: "The Dossiers" 1965–1984.* Helsinki: Finnish Association for the Club of Rome, 2005. clubofrome.fi/wp-content/uploads/2014/10/Dossiers.pdf.

MacArthur, Dame Ellen, and Dominic Waughray. *Intelligent Assets: Unlocking the Circular Economy Potential.* Ellen MacArthur Foundation, 2016. www .ellenmacarthurfoundation.org/assets/downloads/publications/EllenMacArthur Foundation_Intelligent_Assets_080216.pdf.

McNeill, Jim, Pieter Winsemius, and Taizo Yakushiji. *Beyond Interdependence: The Meshing of the World's Economy and the Earth's Ecology*, a report to the Trilateral Commission. Oxford: Oxford University Press, 1991.

Meadows, Donella H., et al. *The Limits to Growth: A Report for the Club of Rome's Project on the Predicament of Mankind.* New York: Universe Books, 1972.

National Academy of Sciences. *The Biological Effects of Atomic Radiation: A Report to the Public.* Washington, DC: National Research Council, 1956.

Optimum Population Trust. *A Population-Based Climate Strategy: An Optimum Population Trust Briefing*, May 2007. populationmatters.org/documents/climate _strategy.pdf.

Pamlin, Dennis, and Stuart Armstrong. *Twelve Risks That Threaten Human Civilization: Executive Summary.* Global Challenges Foundation, January 2, 2015. g20ys.org/upload/auto/4eb9ce315f67d9d94aebf3d90522d5ee3d67a8bf.pdf.

Pasic, Amir. *Foundations in Security: An Overview of Foundation Visions, Programs, and Grantees.* Project on World Security, Rockefeller Brothers Fund. New York: Rockefeller Brothers Fund Inc., 1999.

Planet Under Pressure Conference. "State of the Planet Declaration," organized by IGBP, Diversitas, IHDP, WCRP, ICSU, London, March 26–29, 2012.

Project for a New American Century. *Rebuilding America's Defenses.* PNAC: Washington, DC, 2000.

Rajan, Sudhir Chella. *Global Politics and Institutions.* Tellus Institute, 2006.

Raskin, Paul, et al. *The Great Transition: The Promises and Lures of the Times Ahead,* a report of the Global Scenario Group. Boston: Stockholm Environment Institute-Boston, 2002.

Rockström, Johan, and Jeffrey Sachs. *Sustainable Development and Planetary Boundaries,* submitted to the High Level Panel on the Post-2015 Development Agenda, 2013.

Schwartz, Laura. *A History of Climate Action through Foundations,* Rockefeller Archives Center Research Report, May 29, 2018.

Siegel, Nicholas. *The German Marshall Fund of United States: A Brief History.* The German Marshall Fund, May 9, 2012.

Stavins, Robert N. *Project 88: Harnessing Market Forces to Protect the Environment.* John F. Kennedy School of Government, Harvard University, Cambridge, MA, 1988.

Stockholm Resilience Center. *Stockholm MISTRA Institute (SMI) on Sustainable Governance and Management of Social–Ecological Systems, Proposal to Mistra on a New Inter-disciplinary Research Centre on Sustainable Governance and Management of Social–Ecological Systems,* March 31, 2006.

Study of Critical Environmental Problems (SCEP). *Man's Impact on the Global Environment: Assessment and Recommendations for Action, Report of the Study of Critical Environmental Problems (SCEP),* Cambridge, MA: MIT Press, 1970.

Study of Man's Impact on Climate, ed. *Inadvertent Climate Modification: Report of the Study of Man's Impact on Climate* (SMIC), Cambridge, MA: MIT Press, 1971.

Sustainable Development Solutions Network. "The Structural Transformations towards Sustainable Development," Background paper for the High-Level Panel of Eminent Persons on the Post-2015 Development Agenda, 2013.

Trilateral Commission. *Commemorating 1989, the Year That Changed the Map of Europe (and Thereby the World),* November 2009. trilateral.org/file.view&fid=149.

Wasdell, David. *Beyond the Tipping Point.* Meridian Project, 2006. www.meridian.org .uk/_PDFs/BeyondTippingPoint.pdf.

———. *Global Warning.* Meridian Project, 2005.

———. *The Feedback Crisis in Climate Change,* Meridian Program, London, 2005.

———. *The Historical Background, Apollo-Gaia Project, Co-ordinated by David Wasdell as part of the Meridian Program.* 2007. www.apollo-gaia.org/A-GProject Development.pdf.

White House. *Restoring the Quality of Our Environment: Report of the Environmental Pollution Panel,* President's Science Advisory Committee, November 1965.

Wijkman, Anders, and Kristian Skånberg. *The Circular Economy and the Benefits for Society: Jobs and Clear Winners in an Economy Based on Renewable Energy and Resource Efficiency,* a study report at the request of the Club of Rome with support from the MAVA Foundation, 2016.

Wilson, Thomas W. Jr. *World Energy, the Environment and Political Action:* Summary of Second International Environmental Workshop, International Institute for Environmental Affairs, Washington, DC, 1972.

———. *World Population and a Global Emergency,* Aspen Institute for Humanistic Studies, Washington, DC, 1974.

Woods Hole Oceanographic Institution. *Annual Report 1970.*

World Climate Program. *Report from the International Conference on the Assessment of the Role of Carbon Dioxide and Other Greenhouse Gases in Climate Variations and Associated Impacts.* ICSU, UNEP, WMO, 1985.

World Economic Forum. *Accelerating Successful Smart Grid Pilots,* July 15, 2010. www3.weforum.org/docs/WEF_EN_SmartGrids_Pilots_Report_2010.pdf.

———. *Global Risks Report 2019,* 66, www3.weforum.org/docs/WEF_Global_Risks _Report_2019.pdf.

———. *Shaping the Future of Construction: A Breakthrough in Mindset and Technology.* WEF in collaboration with the Boston Consulting Group, May 2016.

Woodwell, G., G. J. MacDonald, R. Revelle, and C. D. Keeling. *The Carbon Dioxide Problem: Implications for Policy in the Management of Energy and Other Resources,* a Report to the Council on Environmental Quality, President's Council on Environmental Quality, Washington, DC, July 1979.

Scientific Works

Agrawala, Shardul. "Early Science-Policy Interactions in Climate Change: Lessons from the Advisory Group on Greenhouse Gases." *Global Environmental Change* 18, no. 2 (July 1999): 157–69.

———. "Explaining the Evolution of the IPCC Structure and Process." ENRP Discussion Paper E-97–05. Cambridge, MA: Harvard University, Kennedy School of Government, August 1997 (also as International Institute for Applied Systems Analysis Interim Report IR-97–032/August).

Ambirajan, S. "Malthusian Population Theory and Indian Famine Policy in the Nineteenth Century." *Population Studies* 30, no. 1 (1976): 5–14 (doi:10.2307/2173660).

Anderberg, Stefan. "Klimatfrågans utveckling ur ett svenskt perspektiv," in book, *Miljöhistoria över gränser*, edition *Skrifter med miljöhistoriska perspektiv*, Malmö: Malmö högskola, 2006, 160–178.

Batterbury, Simon. "The International Institute for Environment and Development: Notes on a Small Office." *Global Environmental Change* 14 (2004): 367–371.

Biermann, F., et al. "Navigating the Anthropocene: Improving Earth System Governance." *Science* 335, no. 6074 (March 16, 2012): 1306–1307.

Bolin, Bert. "Carl-Gustaf Rossby: The Stockholm Period 1947–1957." *Tellus A: Dynamic Meteorology and Oceanography* 51, no. 1 (1999): 4–12.

Bongaarts, John, and Brian C. O'Neill. "Global Warming Policy: Is Population Left Out in the Cold?" *Science* 17 (August 2018): 650–652.

Bostrom, Nick. "What is a Singleton?" *Linguistic and Philosophical Investigations* 5, no. 2 (2006): 48–54.

Bronk, Detlev. "Science Advice in the White House." *Science* 186, no. 4159 (October 11, 1974): 116–121.

Cartwright, Glenn F., and Adam Finkelstein. "Second Decade Symbionics and Beyond." Based on a paper presented at the Ninth General Assembly of the World Future Society, Washington, DC, July 31, 1999.

Ceballos, Gerardo, et al. "Accelerated Modern Human-induced Species Losses: Entering the Sixth Mass Extinction." *Science Advances* 1, no. 5 (June 19, 2015): e1400253.

Chamberlin, Shaun, Larch Maxey, and Victoria Hurth. "Reconciling Scientific Reality with Realpolitik: Moving Beyond Carbon Pricing to TEQs—An Integrated, Economy-wide Emissions Cap." *Carbon Management* 5, no. 4 (2014): 411–427, DOI: 10.1080/17583004.2015.1021563.

Cox, R. "Ideologies and the New International Economic Order: Reflections on Some Recent Literature." *International Organization* 33, no. 2 (1979): 257–302. doi:10.1017/S002081830003216

Daily, Gretchen C., Anne H. Ehrlich, and Paul R. Ehrlich. "Optimum Human Population Size." *Population and Environment* 15, no. 6 (July 1994): 469.

Franz, Wendy E. "The Development of an International Agenda for Climate Change: Connecting Science to Policy." Discussion Paper, E-97–07, Harvard University, Kennedy School of Government, July 31, 1997.

Foley, Jonathan A., et al. "Solutions for a Cultivated Planet." *Nature* 478 (October 20, 2011): 337–342.

Hamblin, J. "A Dispassionate and Objective Effort: Negotiating the First Study on the Biological Effects of Atomic Radiation." *Journal of the History of Biology* 40, no. 1 (2007): 147–177.

Hart, David. "Strategies of Research Policy Advocacy: Anthropogenic Climatic Change Research," 1957–1974, Center for Science and International Affairs at Harvard University, Kennedy School of Government (1992).

Hickey, Colin, Travis N. Rieder, and Jake Earl. "Population Engineering and the Fight against Climate Change." *Social Theory and Practice* 42, no. 4 (October 2016): 845–870.

Holling, C. S. "Resilience and Stability of Ecological Systems." *Annual Review of Ecology and Systematics* 4 (November 1973): 1–23.

Jantsch, Erich. *Perspectives of Planning. Proceedings of the OECD Working Symposium on Long-Range Forecasting and Planning, Bellagio, Italy, October 27–November 2, 1968.* Paris: Organization for Economic Cooperation and Development, 1969.

———. "The Bellagio Declaration on Planning." *Futures* 1, no. 3 (March 1969): 181–276.

Kellogg, W. W. "Mankind's Impact on Climate: The Evolution of an Awareness." *Climatic Change* 10, no. 2 (April 1987): 113–136, core.ac.uk/download/pdf/81712826.pdf.

Kirton, John. "The G20 System Still Works: Better Than Ever." *Caribbean Journal of International Relations & Diplomacy* 2, no. 3 (September 2014): 43–60.

Liao, S. Matthew, Anders Sandberg, and Rebecca Roache. "Human Engineering and Climate Change." *Ethics, Policy & Environment* 15, no. 2 (2012): 206–221.

Lybbert, Travis, and David Sumner. "Agricultural Biotechnology for Climate Change Mitigation and Adaptation." *Biores* 9, no. 3 (April 2015).

Malthus, Thomas. *An Essay on the Principle of Population.* London: J. Johnson, 1798. www.esp.org/books/malthus/population/malthus.pdf.

Matthews, M. A. "The Earth's Carbon Cycle." *New Scientist* 6 (October 8, 1959): 644–646.

McLoughlin, David. "The Third World Debt Crisis and the International Financial System." *Student Economic Review*, Trinity College Dublin (1989): 96–101.

Molitor, Graham T. T. "Molitor Forecasting Model: Key Dimensions for Plotting the 'Patterns of Change.'" *Journal of Futures Studies* 8, no. 1 (August 2003): 61–72.

Myers, Norman. "First Global Conference on the Future, Held in the Royal York Hotel and Harbour Hilton, Toronto, Canada, During 20–24 July 1980." *Environmental Conservation* 8, no. 1 (1981): 73–78. doi:10.1017/S0376892900026825.

National Academy of Sciences. *Biographical Memoirs Vol. 61.* Washington, DC: National Academies Press, 1992.

Nordangård, J. "Med brödfödan som drivkraft: En studie om att byta olja mot biodrivmedel i ett globalt perspektiv." (Dissertation.) Linköping University, 2007.

Nordangård, J. *ORDO AB CHAO: Den politiska historien om biodrivmedel i den Europeiska Unionen—Aktörer, nätverk och strategier.* (PhD dissertation.) Linköping University, 2012.

Pearce, Fred. "The Green Diplomat." *New Scientist*, March 21, 1992.

Persson, Ola. *What Is Circular Economy?—The Discourse of Circular Economy in the Swedish Public Sector.* (Dissertation.) Uppsala Universitet, 2015.

Phillips, Norman A. "Carl-Gustaf Rossby: His Times, Personality, and Actions." *Bulletin of the American Meteorological Society*, 79, no. 6 (June 1998): 1097–1112.

Plass, Gilbert. "The Carbon Dioxide Theory of Climatic Change." *Tellus* 8, no. 2 (May 1956): 140–154.

Revelle, Roger, and Hans E. Suess. "Carbon Dioxide Exchange between Atmosphere and Ocean and the Question of an Increase of Atmospheric CO_2 during the Past Decades." *Tellus* 9, no. 1 (February 1957): 18–27.

Schellnhuber, H. J., and J Kropp. "Geocybernetics: Controlling a Complex Dynamical System under Uncertainty." *Naturwissenschaften* 85, no. 9 (September 1998): 411–425.

The Pontifical Academy of Sciences. "Sustainable Humanity, Sustainable Nature: Our Responsibility." Proceedings of the Joint Workshop, 2–6 May 2014, Extra Series 41, Vatican City, 2015.

Weindling, Paul. "Julian Huxley and the Continuity of Eugenics in Twentieth-century Britain." *Journal of Modern European History* 10, no. 4 (2012): 480–499.

Articles

Afeez, Ahmed. "World Bank and UN Carbon Offset Scheme 'Complicit' in Fenocidal Land Grabs—NGOs." *The Guardian*, July 3, 2014.

AFP. "ExxonMobil, Rockefellers Face Off in Climate Battle." *The Express Tribune*, April 18, 2016.

Baraniuk, Chris. "Exclusive: UK Police Wants AI to Stop Violent Crime before It Happens." *New Scientist*, November 26, 2018.

Bastasch, Michael. "Pro-lifers: Pope Rejects Population Control, Abortion as Solutions to Global Warming." *Daily Caller*, June 17, 2015.

Bell, Larry. "Blood and Gore: Making a Killing on Anti-Carbon Investment Hype." *Forbes*, November 3, 2013.

Briggs, William M. "The Scientific Pantheist Who Advises Pope Francis." *The Stream*, June 22, 2015. https://stream.org/scientific-pantheist-who-advises-pope-francis/.

Bolling, Anders. "Johan Rockström är miljörörelsens egen Piketty." *Dagens Nyheter.* September 4, 2015.

Booker, Christopher. "Climate Change: This Is the Worst Scientific Scandal of Our Generation." *The Telegraph*, November 28, 2009.

Buchert, Peter. "Forskare sågar klimatavtalet." *Hufvudstadsbladet*, December 11, 2015.

"Building on the Combined Momentum of the U.S.–China Climate Accord and the Emerging U.S. ANSI LCA Standard, Winter 2015." *Climate Alert* 26, no. 4 (Winter 2015). https://climate.org/wp-content/uploads/2023/06/2015-winter.pdf.

Burdman, Mark. "World Futurists Turn Back the Clock." *Fusion*, November 1980.

"Can Rockefeller's Heirs Turn Exxon Greener?" *New York Times*, May 4, 2008.

Carney, Mathew. "Leave No Dark Corner." ABC News, Australia, September 17, 2018.

Carson, Biz. "Elon Musk Has Some Really Strange Ideas about Connecting Computers to Your Brain." *Business Insider*, September 15, 2016.

"China Has Highest Number of Smart City Pilot Projects: Report." *The Economic Times*, February 20, 2018.

Cornish, Edward. "The Search for Foresight—How the Futurist Were Born." *The Futurist*, November–December 2007.

Cuff, Madeline. "New Blockchain-Based Carbon Currency Aims to Make Carbon Pricing Mainstream." *Business Green*, September 19, 2017. https://www .businessgreen.com/news/3017564/new-blockchain-based-carbon-currency -aims-to-make-carbon-pricing-mainstream.

Daniel, Clifton. "Ford and 25th Amendment: Rockefeller Choice Held Free of Party Pressure." *New York Times*, December 23, 1974.

DeWitt, Karen. "Brzezinski, the Power and the Glory." *Washington Post*, February 4, 1977.

Dunlap, David W. "What David Rockefeller Wanted Built Got Built." *New York Times*, March 26, 2017.

Dutta, Anisha. "20 Smart Cities May Be Ready Only by 2021." *Hindustan Times*, November 12, 2018.

"Earth 'Entering New Extinction Phase'—US Study." June 20, 2015. www.bbc.com /news/science-environment-33209548.

Frank, Jeffrey. "Big Spender, Nelson Rockefeller's Grand Ambitions." *The New Yorker*, October 13, 2014.

Frisk, Martina. "Exxon Mobil utreds för dubbelspel om klimatrisker." *Aktuell Hållbarhet*, February 2, 2016.

Gardner, Richard N. "The Hard Road to World Order." *Foreign Affairs* 52, no. 3 (April 1, 1974): 556–76, doi:10.2307/20038069.

Garner, Rochelle. "Elon Musk, Stephen Hawking Win Luddite Award as AI 'Alarmists.'" *CNet*, January 19, 2016.

Genzlinger, Neil. "Barbara Marx Hubbard, 89, Futurist Who Saw 'Conscious Evolution,' Dies." *New York Times*, May 15, 2019.

"George H. W. Bush: Rockefeller Was a Valuable Adviser." Associated Press, March 20, 2017.

Glenn, Alan. "Turbulent Origins of Ann Arbor's First Earth Day." *Ann Arbor Chronicle*. April 22, 2009.

"Global Spirituality and Global Consciousness." *Kosmos Journal for Global Transformation*, 2003. www.kosmosjournal.org/article/global-spirituality-and -global-consciousness.

Gogman, Lars. "Hog farm satte alternativrörelsens verklighetsuppfattning i gungning." *Stockholm Fria*, May 9, 2008.

Goldenberg, Suzanne. "Doomsday Clock Stuck Near Midnight Due to Climate Change and Nuclear War." *The Guardian*, January 26, 2016.

Gorbachev, Mikhail. "What Role for the G20?" *New York Times*, April 27, 2009.

"Gordon Goodman." *Dagens Nyheter*, June 6, 2008.

"Gore Climate Film's Nine 'Errors.'" BBC News. October 11, 2007. news.bbc.co.uk/2/hi/7037671.stm.

Griffiths, Elle. "Paris Turned into a Sea of Shoes as Climate Change Campaigners Lay 20,000 Pairs in Symbolic Street Protest." *The Mirror*, November 29, 2015.

Hansell, Samuel. "Company News; Rockefeller Center Filing May Mean Big Tax Bill." *New York Times*, May 13, 1995.

Harris, Mark. "SpaceX Plans to Put More than 40,000 Satellites in Space." *New Scientist*, October 19, 2019.

Harris, Paul. "They're Called the Good Club—and They Want to Save the World." *The Guardian*, May 31, 2009.

Hickman, Leo. "James Lovelock on the Value of Sceptics and Why Copenhagen Was Doomed." *The Guardian*, March 29, 2010.

Hill, David. "How the Bauhaus Came to Aspen." *Custom Home*, April 4, 2019.

Hylton, Richard. "Rockefeller Family Tries to Keep a Vast Fortune from Dissipating." *New York Times*, February 16, 1992.

IESE Business School. "The Smartest Cities in the World in 2018." *Forbes*, July 13, 2018.

Frank, Jeffrey. "Big Spender, Nelson Rockefeller's Grand Ambitions." *The New Yorker*, October 13, 2014.

"From the Archives: "The Rising Threat of Carbon Dioxide." *New York Times*, October 28, 1956.

King, Gilbert. "A Halloween Massacre at the White House." *Smithsonian*, October 25, 2012. smithsonianmag.com/history/a-halloween-massacre-at-the-white-house -92668509.

Kissinger, Henry. "Henry Kissinger: My Friend David Rockefeller, a Man Who served the World." *Washington Post*, March 30, 2017.

———. "The Chance for a New World Order." *New York Times*, January 12, 2009.

Knapton, Sarah, "We Should Give Up Trying to Save the World from Climate Change, Says James Lovelock." *The Telegraph*, April 8, 2014.

Krause, Vivian. "Rockefellers Behind 'Scruffy Little Outfit.'" *Financial Post*, February 14, 2013.

Krauss, Clifford. "Rockefellers Seek Change at Exxon." *New York Times*, May 27, 2008.

Kreisberg, Jennifer Cobb. "A Globe, Clothing Itself with a Brain." *Wired*, June 1, 1995.

Krylmark, Viktor. "Rysk ansiktsigenkänning leder till fler gripanden." *NyTeknik*, September 29, 2017.

Kutz, Myer. "The Rockefeller Problem." *New York Times*, April 28, 1974.

Lavelle, Marianne. "A 50th Anniversary Few Remember: LBJ's Warning on Carbon Dioxide." *The Daily Climate*, February 2, 2015.

Lemann, Nick. "Poor Little Rich People, The Rockefellers: A Family Dynasty by Peter Collier and David Horowitz Holt." *Harvard Crimson*, April 22, 1976.

Ludden, Jennifer. "Should We Be Having Kids in the Age of Climate Change?" NPR, August 18, 2016.

Lynn, Frank. "Rockefeller Quits as Chairman of Critical Choices Commission." *New York Times*, March 1, 1975.

Markman, Jon. "Facial Recognition: A Force for Good . . . or Government?" *Forbes*, September 27, 2019.

McFadden, Robert D. "Rockefeller Gave Kissinger $50,000, Helped 2 Others." *New York Times*, October 6, 1974.

Marek, Kiersten. "Neva Rockefeller Goodwin and the Role of the Activist Investor in Steering Social Change." *Inside Philanthropy*, April 14, 2016.

Matai, D. K. "Preparing for the Super Convergence: Rise of the Bio-Info-Nano Singularity." *Business Insider*, April 22, 2011.

Maurk, Ben. "The Ludlow Massacre Still Matters," *The New Yorker*, April 18, 2014.

Mecklin, John. "It Is Still 3 Minutes to Midnight." 2016 Doomsday Clock Statement, Science and Security Board, *Bulletin of the Atomic Scientists*.

Mesmer, Philippe. "Songdo, Ghetto for the Affluent." *Le Monde*, May 29, 2017.

Morgan, Oliver, and Faisal Islam. "Saudi Dove in the Oil Slick." *The Guardian*, January 14, 2001.

Montgomery, Paul. "Business People; A Banking Star Moves to Société Générale." *New York Times,* June 23, 1988.

Mullen, Jethro. "Bitcoin Could 'Bring the Internet to a Halt,' Banking Group Warns." CNN Business, June 18, 2018.

Neate, Rupert. "Rockefeller Family Charity to Withdraw All Investments In Fossil Fuel Companies." *The Guardian*, March 23, 2016.

Obituaries: "José Argüelles." *The Telegraph*, April 5, 2011.

Olsson, Sarah. "Digital övervakning ska testas på förskolebarn." *Norrköpings Tidningar*, August 5, 2019.

———. "Digitala armbandsprojektet läggs ned." *Norrköpings Tidningar*, August 26, 2019.

Orlowski, Andrew. "Oops: Chief Climategate Investigator Failed to Declare Eco Directorship 'Dracula's In Charge of the Blood Bank.'" *The Register*, March 24, 2010.

Osborne, Alistair. "Rothschild and Rockefeller Families Team Up for Some Extra Wealth Creation." *The Telegraph*, January 23, 2012.

Palme, Olof. "Olof Palme: Förändrat klimat är största hotet." *Svenska Dagbladet*, November 27, 1974.

Pentin, Edward. "German Climatologist Refutes Claims He Promotes Population Control." *National Catholic Register*, June 19, 2015.

Revkin, Andrew. "Industry Ignored Its Scientists on Climate." *New York Times*, April 23, 2009.

Roberts, Sam. "Why Are Rockefellers Moving from 30 Rock? 'We Got a Deal.'" *New York Times*, November 23, 2014.

Rockström, Johan. "Common Boundaries." *Our Planet*, September 2011.

Sainato, Michael. "Stephen Hawking, Elon Musk, and Bill Gates Warn about Artificial Intelligence." *The Observer*, August 19, 2015.

Sanger, Margaret. "Cannibals." *The Woman Rebel*, May 1914. www.nyu.edu/projects /sanger/webedition/app/documents/show.php?sangerDoc=420013.xml.

Schwab, Klaus. "The Fourth Industrial Revolution: What It Means, How to Respond." *Foreign Affairs*, Dec 12, 2015, and World Economic Forum, January 24, 2016. www.weforum.org/agenda/2016/01/the-fourth-industrial-revolution-what-it -means-and-how-to-respond.

"Shell Boss "Fears for the Planet.'" *The Guardian*, June 17, 2004.

Smail, J. Kenneth. "Global Population Reduction: Confronting the Inevitable. *Worldwatch*, September/October, 2004.

Song, Lisa. "An Even More Inconvenient Truth: Why Carbon Credits or Forest Preservation May Be Worse than Nothing." *ProPublica*, May 22, 2019.

Such, Maria Iliana. "NOBLE in the 21st Century: Susan Rockefeller." *Huffington Post*, March 14, 2016.

Taibbi, Matt. "The Great American Bubble Machine." *Rolling Stone*, April 5, 2010.

"30 Years Ago Scientists Warned Congress on Global Warming—What They Said Sounds Eerily Familiar." *Washington Post,* November 30, 2016.

Thunberg, Greta. "'Our House Is on Fire': Greta Thunberg, 16, Urges Leaders to Act on Climate." *The Guardian*, January 25, 2019.

Timm, Trevor. "The Government Just Admitted It Will Use Smart Home Devices for Spying." *The Guardian*, February 9, 2016.

Tindera, Michela. "Inside Late Billionaire David Rockefeller's Will: Picassos, A Beetle Collection and a Maine Island." *Forbes*, April 20, 2017.

Tännsjö, Torbjörn. "Så kan klimatkrisen leda fram till en global despoti." *Dagens Nyheter,* December 5, 2018.

Visser, Nick. "The World Has Pledged to Divest $2.6 Trillion from Fossil Fuels." *Huffington Post,* September 22, 2015.

White, Chris. "South Korea's 'Smart City' Songdo: Not Quite Smart Enough?" *South China Morning Post,* March 25, 2018.

Wiedeman, Reeves. "The Rockefellers vs. the Company That Made Them Rockefellers." *New York*, January 8, 2018.

Wiseman, Virginia. "GLOBE World Summit of Legislators Resolution Urges Climate Action." IISD, June 11, 2014. climate-l.iisd.org/news/globe-world-summit-of -legislators-resolution-urges-climate-action.

Wortman, Marc. "Famed Architect Philip Johnson's Hidden Nazi Past." *Vanity Fair*, April 4, 2016.

Public Agency Documents

Council of European Communities. "Resolution of the Council and the Representatives of the Governments of the Member States, Meeting Within the Council of 1 February 1993 on a Community Programme of Policy and Action in Relation to the Environment and Sustainable Development—a European Community Programme of Policy and Action in Relation to the Environment and Sustainable Development." *Official Journal* C 138 (May 17, 1993): 0001–0004.

Die Bundesregierung. *Heiligendamm Process.* The Press and Information Office of the Federal German Government, 2008. www.g-8.de/nn_92160/Content/EN/ Artikel/_g8-summit/2007–06-08-heiligendamm-prozess_en.html.

European Commission. *A European Green Deal.* 2008. ec.europa.eu/info/strategy /priorities-2019–2024/european-green-deal_en.

European Parliament. *Report on Winning the Battle Against Global Climate Change, Committee on the Environment*, 2005/2049(INI), by Anders Wijkman. Brussels, October 20, 2005. tinyurl.com/y2ffnyyn.

European Parliament. *Digital Agenda for Europe.* April 2023. www.europarl.europa.eu /factsheets/en/sheet/64/digital-agenda-for-europe.

European Union. Acceptance Speech by Herman Van Rompuy Following His Nomination as First Permanent President of the European Council, November 19, 2009. www.consilium.europa.eu/media/25842/141246.pdf.

G20. *G20 Engagement Groups*. 2019. www.g20.org/en/engagementgroups.

———. *G20 Leaders' Communiqué*. Antalya Summit, November 15–16 2015, European Council, www.consilium.europa.eu/en/press/press-releases/2015/11/16/g20-summit-antalya-communique.

———. *G20 Leaders' Declaration*. Saint Petersburg Summit, September 5–6, 2013. www.g20.utoronto.ca/2013/Saint_Petersburg_Declaration_ENG.pdf.

———. G20 New Industrial Revolution Action Plan, www.mofa.go.jp/files/000185873.pdf.

Swedish Parliament. Riksdagens protokoll 1988/89:11. October 19, 1988. data.riksdagen.se/dokument/GC0911.

United Nations. *A More Secure World: Our Shared Responsibility*. Report of the High-level Panel on Threats, Challenges and Change, 2004. www.un.org/en/peacebuilding/pdf/historical/hlp_more_secure_world.pdf.

———. Address by Miguel d'Escoto Brockmann, President of the General Assembly, upon Adoption of the Outcome Document of the Conference on the World Financial and Economic Crisis and Its Impact on Development, June 26, 2009, New York. www.un.org/en/ga/president/63/pdf/statements/20090626-eccrisis-outcomedoc.pdf.

———. *An Action Agenda for Sustainable Development Report for the UN Secretary-General*. UN Sustainable Development Solutions Network. May 5, 2014.

———. "Concerts to Encourage Climate Action Announced at Davos." 2015. www.un.org/climatechange/blog/2015/01/concerts-encourage-climate-action-announced-davos.

———. *Open Working Group Proposal for Sustainable Development Goals*. sustainabledevelopment.un.org/owg.html.

———. *Report of the World Commission on Environment and Development: Our Common Future*. Oxford and New York: Oxford University Press, 1987.

———. *Rio Declaration on Environment and Development*. 1992. www.un.org/documents/ga/conf151/aconf15126–1annex1.htm.

———. "Secretary-General's Message to Globe International Second World Summit of Legislators [Delivered by Mr. Tomas Anker Christensen, Senior Partnerships Advisor, Executive Office of the Secretary-General]." June 6, 2014. www.un.org/sg/en/content/sg/statement/2014–06-06/secretary-generals-message-globe-international-second-world-summit.

———. "Speech to UN General Assembly by Prime Minister, Mr. Gordon Brown, 23 September 2009." www.un.org/en/ga/64/generaldebate/pdf/GB_en.pdf.

———. "Secretary-General, Marking Historic Donation to League of Nations Library, Hails Rockefeller Foundation's 'Global Philanthropy.'" September 10, 2012. www.un.org/press/en/2012/sgsm14498.doc.htm.

———. *Summary of the First Meeting of the High-Level Political Forum on Sustainable Development*. Note by the President of the General Assembly. November 13, 2013.

———. TST Issue Brief: Global Governance. https://sdgs.un.org/sites/default/files /documents/2429TST%2520Issues%2520Brief_Global%2520Governance _FINAL.pdf.

———. "UN Partners with the Rockefeller Foundation to Showcase Women's Role in Addressing Climate Change," press release, September 26, 2012. unfccc.int/files /secretariat/momentum_for_change/application/pdf/pr20120926_mfc_women _announce.pdf.

———. UN Secretary-General's Remarks at the United Nations Office at Geneva Library on the 85th Anniversary of the Donation by John D. Rockefeller to Endow the League of Nations Library." September 10, 2012. www.un.org/sg /en/content/sg/statement/2012–09-10/un-secretary-generals-remarks-united -nations-office-geneva-library.

———. "Transforming Our World: The 2030 Agenda for Sustainable Development." 2015. sustainabledevelopment.un.org/post2015/transformingourworld.

United Nations General Assembly. Resolution adopted by the General Assembly on 25 September 2015, 21 October 2015, www.un.org/ga/search/view_doc. asp?symbol=A/RES/70/1&Lang=E.

United Nations Archives. "Oscar Niemeyer and the United Nations Headquarters." archives.un.org/content/oscar-niemeyer-and-united-nations-headquarters.

United Nations Conference on Trade and Development (UNCTAD). *UNCTAD at 50: A Short History*. 2014. unctad.org/en/PublicationsLibrary/osg2014d1_en.pdf.

United Nations Framework Convention on Climate Change (UNFCCC). *Copenhagen Accord*. December 18, 2009. unfccc.int/resource/docs/2009/cop15/eng/l07.pdf.

United Nations General Assembly. "Resolution on the Report of the Ad Hoc Committee of the Sixth Special Session, United Nations. General Assembly Resolution 3201 (S-VI). May 1, 1974.

United Nations News Center. "At G20 Summit, Ban Says Response to Terrorism 'Needs to Be Robust, Always within Rule of Law.'" November 15, 2015. www.un.org/apps/news/story.asp?NewsID=52561#.V7x4APmLSUk.

United Nations Non-Governmental Liaison Service. "International Day of Climate Action." www.un-ngls.org/index.php/un-ngls_news_archives/2009/981 -international-day-of-climate-action.

United Nations Regional Information Centre for Western Europe. "Figueres: First Time the World Economy Is Transformed Intentionally." February 3, 2015. www

.unric.org/en/latest-un-buzz/29623-figueres-first-time-the-world-economy-is
-transformed-intentionally.

United States Department of State. "Memorandum to Richard McCormack from Frederick Bernthal." February 9, 1989.

United States House of Representatives. *Carbon Dioxide and Climate: The Greenhouse Effect*, Committee of Science and Technology, Hearing. Washington, DC: US Government Printing Office, 1981.

United States Senate. *How a Club of Billionaires and Their Foundations Control the Environmental Movement and Obama's EPA*. Committee on Environment and Public Works Minority Staff Report. July 30, 2014.

United States Senate Committee on Energy and Natural Resources. "Statement of Dr. James Hansen, Director, NASA Goddard Institute for Space Studies." June 23, 1988.

White House Office of the Press Secretary. "President Bush Discusses Global Climate Change." June 11, 2001.

White House Office of the Press Secretary. "Statement by the President on the Paris Climate Agreement." December 12, 2015. www.whitehouse.gov/the-press -office/2015/12/12/statement-president-paris-climate-agreement.

Internet Articles and Blogs

Art Story Foundation. "Museum of Modern Art." www.theartstory.org/museum-moma .htm.

Auken, Ida. "Welcome to 2030. I Own Nothing, Have No Privacy, and Life Has Never Been Better." World Economic Forum. November 11, 2016. www.weforum. org/agenda/2016/11/shopping-i-can-t-really-remember-what-that-is.

Beckow, Steve. "What Was the Harmonic Convergence?" Prepare for Change, 2017. prepareforchange.net/2017/08/20/what-was-the-harmonic-convergence.

Broom, Douglas. "The EU Wants to Create 10 Million Smart Lampposts." World Economic Forum. June 18, 2019. www.weforum.org/agenda/2019/06/the-eu-wants -to-create-10-million-smart-lampposts.

Carnegie Council. "Ricken Patel." www.carnegiecouncil.org/people/ricken-patel.

Center for Climate Strategies. "The Center for Climate Strategies Applauds U.S.–China Climate Agreement, Low Carbon Development Progress and Opportunities." November 12, 2014. www.climatestrategies.us/articles/articles /view/97.

Center for the Environment. "Leadership Fellowship Openings: Environmental Grantmakers Association, June 1–August 31." 2006. www.mtholyoke.edu/proj/cel /fellowship/envirograntmakers.shtml.

Christie's. "Rockefeller Sales Total $832.6 Million—Highest Ever for a Single
    Collection." May 11, 2018. www.christies.com/features/Rockefeller-sales-final
    -report-9206–3.aspx.

Club of Budapest. *The Manifesto on Planetary Consciousness*. www.clubofbudapest.org/
    clubofbudapest/index.php/en/about-us/the-manifesto-on-planetary-consciousness.

Council on Foundations. "History of the Council on Foundations." 2016. www.cof
    .org/sites/default/files/documents/files/History-Council-on-Foundations.pdf.

Earth Portals. "Planetary Pilgrims, José & Lloydine Argüelles." 2001. www
    .earthportals.com/Portal_Messenger/arguelles.html

Fitzgerald, John. "25th Amendment: The Architect." Watergate Amendment. www
    .watergatemendment.com/?page_id=3508.

Foundation for Conscious Evolution. "The Evolutionary Woman." barbaramarxhubbard
    .com/the-evolutionary-woman.

Foundation for the Law and Time. "Arcturus Remembered: A Crystal Earth Network
    Projection." www.lawoftime.org/timeshipearth/arcturus.html.

———. "Brief Biography of José Argüelles/Valum Votan." www.lawoftime.org/jose
    -arguelles-valum-votan.html?content=249.

———. "Holomind Perceiver: What It is and How to Use it." *Intergalactic Bulletin #2*.
    lawoftime.org/lawoftime/synchronotron-holomind-perceiver.html.

———. "Theory and History of the Noosphere." www.lawoftime.org/noosphere
    /theoryandhistory.html.

———. *Rinri Project Newsletter III: Mystery of the Stone Edition*. www.lawoftime.org
    /pdfs/Rinri-III-3.1.pdf.

———. Reviews of *Time, Synchronicity & Calendar Change*. 2011. www.lawoftime
    .org/time-sync/reviews.html.

Founders Daily. "Rockefeller Financial Services Buys $100,983,568 Stake in
    ExxonMobil Corporation." www.thefoundersdaily.com/rockefeller-financial-
    services-inc-buys-100983568-stake-in-exxon-mobil-corporation-xom/629884.

Friends of the Earth. "Climate Change and the Earth Summit." 2002. www.foe.
    co.uk/sites/default/files/downloads/climate_change_summit.pdf.

Global Challenges Foundation. "Earth Statement." www.globalchallenges.org/our
    -work/earth-statement-2015.

GLOBE International. "History." globelegislators.org/about-globe/24-history.

———. "The 1st Climate Legislation Summit." 2013. globelegislators.org/events
    /2013/1gcls-home.

———. "The 1st World Summit of Legislators." May 23, 2013. globelegislators.org
    /news/item/300-legislators-to-participate-in-1st-world-summit-of-legislators.

———. The 2nd Climate Legislation Summit. 2014. globelegislators.org/
    events/2014/2gcls-home.

———. "The 2nd World Summit of Legislators (WSL2014) Resolution." 2014. globelegislators.org/news/item/world-summit-of-legislators-resolution.

———. "Why GLOBE Catastrophic Risks Need Greater Awareness and Legislative Action. 2016. globelegislators.org/news/item/why-global-catastrophic-risks-need-greater-awareness-legislative-action.

———. "Towards Coherence & Impact: The Challenge of Paris and the 2030 Agenda for a Prosperous and Sustainable World." Communiqué, GLOBE COP21 Legislators Summit. National Assembly, 4–5 December 2015, Paris. globelegislators.org/images/PDF/EN_COP21_Summit_Communique.pdf.

Green Cross International. "Climate Change Taskforce Issues Urgent Appeal for Rio+20 to Take Global Warming Threat Seriously." 2011. www.gcint.org/climate-change-task-force-issues-urgent-appeal-for-rio20-conference-to-take-global-warming-threat-seriously.

Greenpeace. "From Hope to Despair." 2012. m.greenpeace.org/international/en/high/news/Blogs/makingwaves/from-hope-to-despair/blog/41051.

Helbing, Dirk. "A New, Global Fascism, Based on Mass Surveillance Is on the Rise." *FuturICT*. September 21, 2017. futurict.blogspot.com/2017/09/a-new-global-fascism-based-on-mass.html.

———. "The Birth of a Digital God." FuturICT. 2018. futurict.blogspot.com/2018/02/the-birth-of-digital-god_13.html.

Huxley, Julian. "UNESCO - It's Purpose and Philosophy." UNESCO. 1946. www.globalresearch.ca/the-task-of-unifying-the-world-mind/9593?.

Kabbalah. "Skapandet av en ny civilisation, World Wisdom Councils tredje möte." kabbalah.info/se/kabbala-runt-v%C3%A4rlden/skapandet-av-en-ny-civilisation.

Keeton, Rachel. "When Smart Cities Are Stupid." International New Town Institute. 2015. www.newtowninstitute.org/spip.php?article1078.

Koppes, Steve. "How the First Chain Reaction Changed Science." University of Chicago News. December 10, 2012. https://news.uchicago.edu/story/how-first-chain-reaction-changed-science.

Linux Information Project. "The Dismantling of The Standard Oil Trust." www.linfo.org/standardoil.html.

London School of Economics and Political Science. "LSE—Rockefeller's Baby?" June 24, 2015. blogs.lse.ac.uk/lsehistory/2015/06/24/lse-rockefellers-baby.

Love Song to the Earth. "A Song with the Power to Fight Climate Change and Maybe Even Change the World." 2015. lovesongtotheearth.org.

Matai, D. K. "Synopsis of the Philanthropia: The Philanthropia—Trinity Club, Syndicates and Ethical Investment Funds." 2006. www.mi2g.com/cgi/mi2g/press/philanthropia_trinity.pdf.

———. "Total Disruption by 2020 As Man-Machines Merge? 5 Billion Humans + 50 Bn IoT/Smart Devices + Q-BRAIN Singularity: Who Wins?" Mi2G Ltd. March 16, 2016. www.mi2g.com/cgi/mi2g/frameset.php?pageid=http%3A//www.mi2g .com/cgi/mi2g/press/150316.php.

Numbersusa.org. "Population Stabilization and the Modern Environmental Movement." www.numbersusa.org/pages/population-stabilization-and-modern-environmental -movement.

OECD Insights. "The Rise of the G20 and OECD's Role." November 17, 2015.

Oppenheimer, Michael. "How the IPCC Got Started." Environmental Defense Fund. November 1, 2017. blogs.edf.org/climate411/2007/11/01/ipcc_beginnings.

Rockström, Johan. "Bounding the Planetary Future: Why We Need a Great Transition." Essay, Great Transition Initiative. 2015. www.greattransition.org /publication/bounding-the-planetary-future-why-we-need-a-great-transition.

———. "Can This Revolution Save Our Warming Planet?" World Economic Forum. January 20, 2016. www.weforum.org/agenda/2016/01/revolution-warming-planet.

———. "How We Can Direct the Fourth Industrial Revolution towards a Zero Carbon Future—If We Act Now." World Economic Forum. January 19, 2018. www.weforum.org/agenda/2018/01/ how-we-can-direct-the-fourth-industrial-revolution-towards-a-zero-carbon-future.

Schellnhuber, Hans Joachim. "Expanding the Democracy Universe." Center for Humans and Nature. 2013. www.humansandnature.org/democracy-hans -joachim-schellnhuber.

Simcox, David. "NPG Forum Series: Nixon and American Population Policy." Floridians Sustainable Population. 1998. www.flsuspop.org/NixRockefeller.html.

Stockholm Resilience Center. "New Rockefeller Programme on Social Innovation." May 15, 2013. www.stockholmresilience.org/research/research-news/2013–05-15- new-rockefeller-programme-on-social-innovation.html.

Trilateral Commission. "The Trilateral Commission at 25: Between Past and future." 1998. trilateral.org/download/files/anniversary_evening.pdf.

UC San Diego, Revelle College. "Roger Randall Dougan Revelle, March 7, 1909— July 15, 1991." revelle.ucsd.edu/about/roger-revelle.html.

University of Chicago. "David Rockefeller, University Trustee and Descendent of UChicago's Philanthropic Founder, 1915–2017." 2017. news.uchicago.edu /article/2017/03/21/david-rockefeller-university-trustee-and-descendent-uchicagos -philanthropic.

University of Michigan. "ENACT Teach-In and Earth Day, Spring 1970." michiganintheworld.history.lsa.umich.edu/environmentalism/exhibits/show/main _exhibit/earthday.

Woeffrey, Oliver. "Could These 3 Ideas Reshape Governance?" World Economic
    Forum. February 26, 2016. www.weforum.org/agenda/2016/02/3-ideas-to
    -revive-global-governance.
Zubialde, Andrea. "Comparison Between the New Sustainable Development
    Goals and the Ethical Principles of the Earth Charter." *Bien Commun &
    Charte de la Terre* (blog), June 29, 2015. biencommuncharteldeterre.wordpress.
    com/2015/06/29/comparison-between-the-new-sustainable-development-goals
    -and-the-ethical-principles-of-the-earth-charter.

Websites

100 Resilient Cities. "About Us." www.100resilientcities.org/about-us#.
350.org. "History." 350.org/about.
AGRA. "Funding Partners." agra.org/funding-partners.
Ark of Hope. www.arkofhope.org.
Biohacker Summit. biohackersummit.com.
Biohax International. web.archive.org/web/20191124084123/https://www.biohax.tech.
Bulletin of the Atomic Scientists. "Timeline." thebulletin.org/timeline.
CAN Europe. "About Us." www.caneurope.org/about-us/learn-about-us.
The Climate Group. www.theclimategroup.org.
Club of Budapest Foundation. www.clubofbudapest.org.
Congressional Institute for the Future. "About." www.users.interport.net/f/u/future98
    /INST.html.
Democracy without Borders. "A World Parliament." www.democracywithoutborders
    .org/se/world-parliament-book-2.
Earth Charter Initiative. www.earthcharter.org.
Earth Day Network. "The History of Earth Day." www.earthday.org/about/the-history
    -of-earth-day.
Environmental Grantmakers Association. "Members." ega.org/about/members.
European Climate Foundation. "Funders." europeanclimate.org/people/funders.
Fleming Policy Centre. "TEQs in Summary." www.flemingpolicycentre.org.uk/teqs.
Foundation for Conscious Evolution. "Global Communication Hub."
    barbaramarxhubbard.com/global-communication-hub.
FuturICT. futurict.inn.ac.
George Marshall Institute. "About Us." marshall.org/about.
Global Challenges Foundation. "About Us." globalchallenges.org/en/about/about-us.
Global Philanthropy Forum. "Partners." philanthropyforum.org/community/members
    -partners.
Goi Peace Foundation. "Our Approach." www.goipeace.or.jp/en/about/approach.

HIVE EU. "Ethical Issues." hive-eu.org/about/ethical_issues.

London School of Economics and Political Science. Grantham Research Institute: Advisory Board. www.lse.ac.uk/GranthamInstitute/about/about-the-institute /advisory-board.

MacVie, Leah. "Erich Jantsch Biography." leahmacvie.com/erichjantsch.

Meridian Programme. "Introduction." www.meridian.org.uk/About/Origins/Pro -Origin-frameset.htm?p=1.

Neuralink. neuralink.com.

Noosphere Forum. "Mission Statement." www.noosphereforum.org/main.html.

———. "Resources, Networking and Partnerships." www.noosphereforum.org /collective/index.html.

Oxford Martin School. "Programmes: Mind & Machine." www.oxfordmartin.ox.ac .uk/research/programmes/mind-machine.

Paris Peace Forum. parispeaceforum.org.

Population Matters. "Patrons." www.populationmatters.org/about/people-and-story /patrons.

Poseidon Foundation. poseidon.eco.

SAS. "Emission Calculator and Carbon Offset." www.sasgroup.net/en/emission -calculator-and-carbon-offset.

Swedish Smartgrid. www.swedishsmartgrid.se.

Tellus Institute. "PoleStar Project." www.polestarproject.org.

Theosophy.net. "Noosphere–Global Mind–Ascension." theosophy.net/profiles/blogs /noosphere-global-mind.

Trilateral Commission. 1973 Members. swprs.files.wordpress.com/2017/07/trilateral -commission-members-1973.pdf.

Trilateral Commission. "Program of the 2002 Annual Meeting." trilateral.org/ File/151.

Technocracy Inc. www.technocracyinc.org.

Toronto Environmental Alliance. Campaigns: "Climate Change." www. torontoenvironment.org/campaigns/climate/climatechange.

Venrock. www.venrock.com.

Venus Project. www.thevenusproject.com.

Weatherhead Center for International Affairs. "In Theory and in Practice: Harvard's Center for International Affairs, 1958–1983." wcfia.harvard.edu/about/ theory-and-practice-harvards-center-international-affairs-1958%E2%80%931983.

Whitehead, Rennie. "The Club of Rome and CACOR: Recollections by J. Rennie Whitehead." www.whitehead-family.ca/drrennie/CACORhis.html."

Widening Circle. "Our Strategy." www.wideningcircle.org/keyIdeas/GCM.htm.

Winrock International. "Former Winrock Board Member: 'The Woman Who Could Stop Climate Change.'" 2015. winrock.org/former-winrock-board -member-the-woman-who-could-stop-climate-change.

Winrock International. "Winrock Board Member: Peter O'Neill." www.winrock.org /bio/peter-m-oneill.

Woods Hole Institute of Oceanography. "History and Legacy: Over 80 Years of Ocean Research, Education, and Exploration." www.whoi.edu/main/history-legacy.

World Economic Forum. "About." www.weforum.org/about/world-economic-forum.

———. "Leadership and Governance." www.weforum.org/about/leadership-and -governance.

———. "Platform for Accelerating the Circular Economy." www.weforum.org/projects /circular-economy.

———. "Timeline: The Davos Manifesto 1973." widgets.weforum.org/history/1973.html.

———. "World Economic Forum and UN Sign Strategic Partnership Framework." June 13, 2019. www.weforum.org/press/2019/06/world-economic-forum-and-un-sign -strategic-partnership-framework.

———. "The Great Reset." www.weforum.org/great-reset.

World Future Society. Conference of July 11–13, 2014. barbaramarxhubbard.com /ai1ec_event/world-future-society-july-2014-conscious-evolution-2–0/?instance_id=.

Interviews

Cheatham, Annie. Interview by Claire Wilson, Sophia Smith Collection, Smith College, Conway, MA, November 13, 2008. www.smith.edu/libraries/libs/ssc /activist/transcripts/Cheatham.pdf.

Hansen, James. "James Hansen on Nuclear Power," interview by Pandora's Promise, July 23, 2013. youtu.be/CZExWtXAZ7M.

Heintz, Stephen. "Interview with Stephen Heintz, CEO of the Rockefeller Brothers Fund," interview by Charles Keidan, *Alliance*, May 3, 2016.

Johnson, Philip. Interview by Sharon Zane, Museum of Modern Art Oral History Program, December 18, 1990. www.moma.org/momaorg/shared/pdfs/docs/learn /archives/transcript_johnson.pdf.

Marx Hubbard, Barbara. "An Evolutionary Conversation with Barbara Marx Hubbard," interview by Foundation for Conscious Evolution. barbaramarxhubbard.com/an -evolutionary-conversation-with-barbara-marx-hubbard.

———. "An Interview with Barbara Marx Hubbard," interview by Russ Volckmann, *Integral Review* 5, no.1 (June 2009).

Johnson, Philip. Interview by Sharon Zane, Museum of Modern Art Oral History Program, December 18, 1990. www.moma.org/momaorg/shared/pdfs/docs/learn /archives/transcript_johnson.pdf.

Nabarro, David. "INTERVIEW: World's Most Difficult Task—Ensuring UN Sustainable Development Agenda, Says Top Adviser," interview by UN News Service, January 26, 2016. news.un.org/en/story/2016/01/521002-interview-worlds-most-difficult-task-ensuring-un-sustainable-development-agenda.

Revelle, Roger Randall Dougan. "Roger Randall Dougan Revelle, Director of Scripps Institution of Oceanography, 1951–1964." Interview by Sarah L. Sharp in 1985, The Regents of the University of California SIO Reference Series No. 88–20, November 1988.

Rockefeller, David. "David Rockefeller, Philanthropist and Former C.E.O. of Chase Manhattan Bank, Shares His Autobiography, "Memoirs" Which Documents His Life and the History of His Legendary Family," interview by Charlie Rose, *Charlie Rose*, October 21, 2002. charlierose.com/videos/9606.

Snider, Dee. "Dee Snider on PMRC Hearing: 'I Was a Public Enemy,'" interview by Cory Grow, *Rolling Stone*, September 18, 2015.

Tickell, Sir Crispin. "Interview with Sir Crispin Tickell," inteview by Malcolm McBain, Churchill College, Cambridge, January 28, 1999. www.chu.cam.ac.uk/media/uploads/files/Tickell.pdf.

Wirth, Timothy. Interview by *PBS Frontline*, April 24, 2007. www.pbs.org/wgbh/pages/frontline/hotpolitics/interviews/wirth.html.

Audio and Visual Media

15 Minute History, episode 59, "John D. Rockefeller and the Standard Oil Company," (podcast), March 12, 2014.

An Inconvenient Truth, directed by Davis Guggenheim (2006; Paramount Classics).

Black Mirror, television series created by Charlie Brooker and executive produced by Annabel Jones (2011–2013; Zeppotron); House of Tomorrow (2014–19); Broke & Bones (2023–present).

BBC News. "The Women Too Scared of Climate Change to Have Children." March 4, 2019. www.bbc.com/news/av/uk-47442943/the-women-too-scared-of-climate-change-to-have-children

Columbia Journalism Review. "CJR Event on Covering Climate Change." April 30, 2019. www.cjr.org/watchdog/livestream-covering-climate-change.php.

C-SPAN. Annual Ambassadors' Dinner, September 14, 1994, with David Rockefeller's acceptance speech (video). www.c-span.org/video/?60201–1/annual-ambassadors-dinner.

David Rockefeller: Bridge Builder (A Tribute), produced, directed, and written by Roger Torda (2003; Synergos Institute, Balaton Film and Television). Rockefeller Archive Center. www.youtube.com/watch?v=y2cORk1ni10.

Future Shock, directed by Alexander Grasshof (1972; McGraw-Hill Films).

Global Brain/Awakening Earth, directed by Chris Hall, based on the book by Peter Russell (1982). youtu.be/s1fvEwzUovI.

Hidden Hands: A Different History of Modernism, episode 1, "Art and the CIA," directed by Tony Cash and produced by Frances Stonor Saunders (1995). youtu.be/k5YSikO6JRM.

In Their Own Words: Walter and Elizabeth Paepcke and the Birth of the Aspen Institute, produced and directed by Greg Poschman (2009; The Aspen Institute). youtu.be/7TOR1BVOzdA.

In Time, directed by Andrew Niccol (2011; 20th Century Studios).

Johns Hopkins Center for Health Security. Event 201 Pandemic Exercise. October 18, 2019. youtu.be/AoLw-Q8X174.

Merchants of Doubt, directed by Robert Kenner (2015; Sony Pictures Home Entertainment).

McCartney, Paul. "Love Song to the Earth." Music video. YouTube.

Neuralink. Neuralink Launch Event, July 17, 2019. youtu.be/r-vbh3t7WVI.

Obama, Barack. "Recorded Remarks to Global Climate Summit." November 18, 2008. AmericanRhetoric.com. www.americanrhetoric.com/speeches/barackobama/barackobamaglobalclimatesummit.htm.

Only One Earth: The Stockholm Conference, Peter Hollander, executive producer. (1972; UNEP documentary for the New York State Education Department). Part 1: youtu.be/mJUk70tfELA; Part 2: youtu.be/h3-TqHFkfy8.

Pandora's Promise. "James Hansen on Nuclear Power." July 23, 2013. youtu.be/CZExWtXAZ7M.

Progressives Today. "UN Climate Official: "We Should Make Every Effort to Decrease World Population" (speech). April 6, 2015. www.progressivestoday.com/un-climate-official-we-should-make-every-effort-to-decrease-world-population-video.

Survival of Spaceship Earth, directed by Dirk Wayne Summers (1972; United Productions Limited). youtu.be/KCHyP8sj92g.

The Day after Tomorrow, directed by Roland Emmerich (2004; 20th Century Fox).

The Unchained Goddess, directed by Richard Carlson and William T. Hurtz and produced by Frank Capra (1958; N.W. Ayer & Son Inc.)

White House. "The Vice President Delivers Remarks to the Council of the Americas Conference, the Loy Henderson Room, the Department of State, Washington, DC, May 6, 2002. youtu.be/NTVi5j1uiDA.

Year Million, television series created by Mark Elijah Rosenberg (2017; National Geographic).

Zeitgeist: Addendum, directed by Peter Joseph (2008; GMP LLC). www.zeitgeistmovie.com.

Miscellaneous

City of Stockholm. "Welcome to the World's Smartest City 2040," brochure, April 4, 2018.

Climate Action Network (2015). "Islamic Climate Declaration Calls for Fossil Fuel Phase Out," press release, August, 18, 2015.

Climate Reality Project. "24 Hours of Reality: The Dirty Weather Report to Spark Climate Change Call to Action," press release, November 14, 2012. www .climaterealityproject.org/press/24-hours-of-reality-the-dirty-weather-report-to -spark-climate-change-call-to-action.

Congressional Clearinghouse on the Future. Invitation to dinner discussion. "Role of Congress in the Third Wave" with Alvin Toffler, author of *Future Shock* and *The Third Wave*. John Heinz papers, Carnegie Mellon University Libraries Digital Collection. 1983. digitalcollections.library.cmu.edu.

———. Letter from Robert Edgar to the advisory group, John Heinz papers. Carnegie Mellon University Libraries Digital Collection. 1983. digitalcollections.library .cmu.edu.

Congressional Institute for the Future. "GLOBE—Global Legislators for a Balanced Environment, A Project of the Congressional Institute for the Future." John Heinz papers, Carnegie Mellon University Libraries Digital Collection. 1989. digitalcollections.library.cmu.edu.

GLOBE. "Global Legislators Offer Tough New Environmental Proposals," press release, November 16, 1990. digitalcollections.library.cmu.edu/awweb/awarchive ?type=file&item=638674.

Greenpeace. "Greenpeace Mourns Copenhagen Failure with 100 Crosses," press release, 2009. www.greenpeace.org/eastasia/press/releases/climate-energy/2009 /copenhagen-cross.

Hitchcock, Henry Russell, and Philip Johnson. *Modern Architecture: International Exhibition: New York, Feb. 10 to March 23, 1932*, exhibition catalogue, Museum of Modern Art, 1932. www.moma.org/documents/moma_catalogue_2044 _300061855.pdf.

ICA. Nytt verktyg hjälper kunderna att minska sin klimatpåverkan, press release, April 16, 2018. icagruppen.se.

Johnson, Philip, and Mark Wigley. *Deconstructivist Architecture*, exhibition catalogue, Museum of Modern Art, 1988. www.moma.org/documents/moma_catalogue _1813_300062863.pdf

March for Science. "March for Science—Frankfurt AM–Main—April 24th, 2018," poster, 2018. marchforscience.de/wp-content/uploads/2018/03/ScienceMarch FFM_PosterA1_2018_engl.pdf.

Thatcher, Margaret. "Speech to the Royal Society." September 27, 1988. www
.margaretthatcher.org/document/107346.

SEI. "Climate Calculator Tests Individual Impacts." 2017, updated February 1, 2019.
www.sei.org/projects-and-tools/tools/climate-calculator.

SEI International. "Möjligt att försörja 9 miljarder människor år 2050 inom hållbara
gränser," press release, October 12, 2011.

SLA. "Symbionese Liberation Army Declaration of Revolutionary War and the
Symbionese Program," press release, 1973. web.archive.org/web/20031020235849
/http://www.feastofhateandfear.com/archives/sla.htm.

Strong, Maurice. "Opening Statement to the Rio Summit (3 June 1992)." www
.mauricestrong.net/index.php/opening-statement.

———. "Remarks by Maurice Strong at Dinner Meeting of Energy Ministers of
OECD Countries at Aarhus, Denmark, Energy and the Environment. June 15,
1996. www.mauricestrong.net/index.php/speeches-remarks3/52-aarhus.

Trilateral Commission. *Democracies under Stress: Recreating the Trilateral Commission
to Revitalize Our Democracies to Uphold the Rules-Based International Order*,
Trilateral Commission, brochure, 1999.

von Moltke, Konrad. *Turning Up The Heat: Next Steps on Climate Change*, pamphlet,
Pocantico Conference Center of the Rockefeller Brothers Fund, 1995.

United Nations Climate Change Newsroom. "Six Oil Majors Say: We Will Act Faster
with Stronger Carbon Pricing," Open Letter to UN and Governments, June 1, 2015.
web.archive.org/web/20180330124154/http://newsroom.unfccc.int/unfccc
-newsroom/major-oil-companies-letter-to-un.

The Vatican. "Encyclical Letter Laudato Si' of the Holy Father Francis on Care for
Our Common Home." May 24, 2015. w2.vatican.va/content/francesco/en
/encyclicals/documents/papa-francesco_20150524_enciclica-laudato-si.html.

World Bank. "New Report Examines Risks of 4 Degree Hotter World by End of
Century," press release, November 18, 2012. www.worldbank.org/en/news/press
-release/2012/11/18/new-report-examines-risks-of-degree-hotter-world-by-end-of
-century.

World Future Society. *First Global Conference on the Future*, conference brochure,
1980.

WWF. "In Memoriam: Godfrey A. Rockefeller," press release, January 29, 2010. www
.worldwildlife.org/press-releases/in-memoriam-godfrey-a-rockefeller.

About the Author

J ACOB N ORDANGÅRD IS a Swedish researcher, author, and musician. PhD in Technology and Social Change at Linköping University; Master of Social Science in Geography; Master of Social Science in Culture and Media Production. Founding chairman of Stiftelsen Pharos (Pharos Foundation) and CEO of Pharos Media Productions. Formerly senior lecturer at the universities of Norrköping, Jönköping, and Stockholm.

Also the band leader, singer, and songwriter of the Swedish doom metal band Wardenclyffe, with lyrics inspired by his research and offered as soundtracks to his books. The latest project was the concept album *Temple of Solomon*, released one song per month (July 2021–February 2022) with a public lecture about the subject of each song.

This is his fourth of six books about the historical roots and development of the global management system that has emerged in recent years, including *An Inconvenient Journey*, *The Global Coup d'Etat*, and *The Digital World Brain*.

jacobnordangard.se
drjacobnordangard.substack.com

Image Credits

INTRODUCTION

Jacob Nordangård, 2023, photo by Stefan Ljung Photography

Taft oil well blow-out, c. 1920, photo by L.M. Clendenen, University of Southern California Libraries (Creative Commons, CC 0)

CHAPTER 1

John D. Rockefeller, Sr. & Jr., art by Kimmie Fransson (© Stiftelsen Pharos)

Rockefeller Center, New York, pxfuel.com (CC 0)

William Avery Rockefeller Jr., Library of Congress (CC 0)

John Davison Rockefeller, 1885 (CC 0)

Eliza Davison Rockefeller (CC 0)

William Rockefeller Sr. (CC 0)

University of Chicago, view from the Midway Plaisance park (CC 0)

Ida M. Tarbell, 1904, J. E. Purdy, Boston (CC 0)

CHAPTER 2

John D. Rockefeller III, art by Kimmie Fransson (© Stiftelsen Pharos)

United Nations Headquarters in New York, Veni Markovski, blog.veni.com (CC 4.0)

Pierre Teilhard de Chardin, 1955, Archives des Jésuites de France (CC 3.0)

The Abby Aldrich Rockefeller Sculpture Garden, MoMA, New York. Hibin, Flickr (CC 2.0)

Robert Moses with Battery Bridge model, 1939, Library of Congress (CC 0)

Empire State Plaza, Albany, New York, aereal photo by Jer21999 (CC 0)

David H. Koch Building at the Aspen Institute, photo by Lane Rasberry, Wikimedia (CC 4.0)

CHAPTER 3

Nelson Rockefeller, art by Kimmie Fransson (© Stiftelsen Pharos)

The Deluge, xylography by Gustave Doré, illustrated Bible, 1865 (CC 0)

Frederick Seitz, Ava Pauling, Detlev Bronk, Paul Weiss, 1963, photo by Fremont Davis (CC 0)

Carl-Gustaf Rossby on the cover of *Time Magazine* Dec 1956 (Fair Use)

Institute for Advanced Study, Princeton University. Smithsonian Institution (CC 0)

John von Neumann, Los Alamos National Laboratory (CC 0)

Henry Kissinger 1973, US Department of State (CC 0)

CHAPTER 4

Laurance Rockefeller, art by Kimmie Fransson (© Stiftelsen Pharos)

Oil money (Shutterstock license)

Lyndon B. Johnson, 1969, photo by Staffan Wennberg (with permission)
Buckminster Fuller, 1969, photo by Staffan Wennberg (with permission)
Rockefeller Foundation Conference Center in Bellagio, Italy, photo by Henry Kellner (CC 4.0)
Richard Nixon inauguration, January 20, 1969, photo by Staffan Wennberg (with permission)
Alvin Toffler, Beverly Hills, California, 2006, Vern Evans (CC 2.0)
Make love! Not Babies! poster, photo courtesy of Ewa Rudling, www.ewarudling.se
April 22. Earth Day ad in the *New York Times* 1970, *Give Earth a Chance: Environmental Activism in Michigan,* michiganintheworld.history.lsa.umich.edu/environmentalism/items/show/385
Maurice Strong in Skarpnäck, Sweden, 1972, photo by Staffan Wennberg (with permission)

CHAPTER 5

David Rockefeller, art by Kimmie Fransson (© Stiftelsen Pharos)
World Trade Center 1 & 2, New York City, aerial view, March 2001, Jeffmock (CC 0)
Nixon & Zhou Enlai, 25 Feb 1972, Richard Nixon Presidential Library and Museum (CC 0)
Henry Kissinger & Nelson Rockefeller, 3 Jan 1975, Rockefeller Archive Center (CC 0)
Walter Orr Roberts, c. 1960, NCAR Archives (CC 0)
Margaret Mead and John Lindsay, photo by Staffan Wennberg (with permission)
The Rockefeller Forest, Humboldt Redwood State Park. Photo by Jayson Sturner (CC 0)
Nelson Rockefeller & Jimmy Carter, 27 Oct 1977, US National Archives and Records Administration (CC 0)

CHAPTER 6

David Rockefeller Jr., art by Kimmie Fransson (© Stiftelsen Pharos)
Blue planet (cropped), everypixel.com (CC 0)
Willy Brandt, 13 May 1961, Library of Congress, photo by Marion S. Trikosko (CC 0)
Olof Palme at Schiphol Airport, 11 Sep 1974, Dutch National Archives (CC 0)
Gro Harlem Brundtland, 1974–1979, Teigens Fotoatelier, Norsk Teknisk Museum (CC 4.0)
North-South cover, Brandt Commission, 1980
Common Security cover, Palme Commission 1982
Our Common Future cover, Brundtland Commission 1987
Margaret Thatcher, 2005, Margaret Thatcher Foundation (CC 3.0)
Ronald Reagan, 1983, photo Michael Evans (CC 0)
George H. W. Bush, c. 1989 (CC 0)
Fabian Society/GLOBE entrance, London, July 2013, photo by SB (with permission)
The Shaw Window, London School of Economics, photo by Ruth Hartnup (CC 2.0)

CHAPTER 7

Abby Milton O'Neill, art by Kimmie Fransson (© Stiftelsen Pharos)
World Trade Center attack, photo by 9/11 Photos, Flickr (CC 2.0)
John D. "Jay" Rockefeller IV, 1968, photo by Staffan Wennberg (with permission)
Ark of Hope, 2001. Photo by courtesy of Don Shall

CHAPTER 8

Steven Rockefeller, art by Kimmie Fransson (© Stiftelsen Pharos)
Hurricane Katrina, 28 Aug 2005, NASA (CC 0)

Tony Blair at G8 Summit, Gleneagles, July 7, 2005, White House, photo by Eric Draper (CC 0)
New Orleans, Louisiana, after Hurricane Katrina, 2005, photo by NOAA (CC 0)

CHAPTER 9
Richard Rockefeller, art by Kimmie Fransson (© Stiftelsen Pharos)
One World Trade Center, by Olga1969 (CC 4.0)
Barack Obama, 6 Dec 2006, photo by Pete Souza, the White House (CC 0)
José Argüelles at the Babaji Ashram, Cisternino, Italy, 6 Sep 2009, Lawoftime (CC 0)

CHAPTER 10
Valerie Rockefeller, art by Kimmie Fransson (© Stiftelsen Pharos)
Wishing tree at COP21, 8 Dec 2015, photo by Inger Nordangård
Naomi Klein in Warsaw, 2008, photo by Mariusz Kubik (CC 3.0)
Susan Cohn Rockefeller, photo by Samira Bouaou (courtesy of *Epoch Times*)
Avaaz's shoe protest in Paris, December 2015, photo by John Englart, Flickr (CC 2.0)
Avaaz's WANTED posters, Paris, 11 Dec 2015, photo by Inger Nordangård
COP21 Climate Summit entrance, 8 Dec 2015, photo by Inger Nordangård
World leaders after signing the Paris Agreement, COP21, 12 Dec 2015, photo Flickr (CC 0)
CNN Weather Center on TV, 12 Dec 2015, photo by Inger Nordangård
Christiana Figueres, Bonn Climate Change Conference, May 2012, UNclimatechange, Flickr
    (CC 0)

CHAPTER 11
Henry Kissinger, art by Kimmie Fransson (© Stiftelsen Pharos)
Smart City (Shutterstock license)
Klaus Schwab at WEF, 2016, US Embassy, Bern, photo by Eric Bridiers (CC 0)
The Smart City Songdo, South Korea, pixabay.com (CC 0)

CHAPTER 12
Barbara Marx Hubbard, art by Kimmie Fransson (© Stiftelsen Pharos)
Homo noosphericus, art by Kimmie Fransson (© Stiftelsen Pharos)
Hans Joachim Schellnhuber & Johan Rockström, May 2016, photo by Inger Nordangård
Turning the internet of things into the internet of us, March 2017, photo by Inger Nordangård
Prometheus statue at Rockefeller Center, photo by Gigi Alt (CC 3.0)

CONCLUSION
Peter O'Neill & Neva Goodwin Rockefeller, art by Kimmie Fransson (@ Stiftelsen Pharos)
Ariana Rockefeller, 2018, photo by Chris Gabello (CC 0)

Endnotes

Prologue

1 "RFF's Decision to Divest," Rockefeller Family Fund, 2016. www.rffund.org/divestment.

2 Rupert Neate, "Rockefeller Family Charity to Withdraw All Investments in Fossil Fuel Companies, *The Guardian,* March 23, 2016.

3 "Can Rockefeller's Heirs Turn Exxon Greener?" *New York Times,* May 4, 2008.

4 Martina Frisk, "Exxon Mobil utreds för dubbelspel om klimatrisker, *Aktuell Hållbarhet,* 2 Feb 2016.

5 Rockefeller Brothers Fund, Grantees: Climate Accountability Institute. www.rbf.org/grantees /climate-accountability-institute.

6 Rockefeller Foundation. *Smart Globalization: Benefiting More People, More Fully, in More Places* —2007 Annual Report. www.rockefellerfoundation.org/app/uploads/Annual-Report-2007. pdf.

7 *World in Transition: A Social Contract for Sustainabiliby,* German Advisory Council on Global Change (WBGU), 2011.

Chapter 1

1 Ron Chernow, *Titan: The Life of John D. Rockefeller, Sr.* (New York: Random House, 1998), 556.

2 Chernow, *Titan,* 246–248.

3 Chernow, 6–7.

4 Chernow, 11.

5 Chernow, 24–25.

6 Chernow, 8–10.

7 Chernow, 43.

8 Chernow, 49.

9 Chernow, 64.

10 Chernow, 24–25.

11 Chernow, 54–55.

12 Robert Shaplen, *Toward the Well-Being of Mankind: Fifty Years of the Rockefeller Foundation* (Garden City, NY: Doubleday & Company, 1964).

13 Chernow, *Titan,* 328–329.

14 "David Rockefeller, University Trustee and Descendent of UChicago's Philanthropic Founder, 1915–2017," news.uchicago.edu/article/2017/03/21/david-rockefeller-university-trustee-and -descendent-uchicagos-philanthropic, accessed February 8, 2018.

15 John B. Boyer , *University of Chicago: A History* (Chicago: University of Chicago Press, 2015).

16 "John D. Rockefeller and the Standard Oil Company," Episode 59, *15 Minute History*, March 12, 2014, accessed January 24, 2018; "The Dismantling of the Standard Oil Trust," the Linux Information Project, May 21, 2004, www.linfo.org/standardoil.html.

17 "Evolution of a Foundation," Rockefeller Foundation Archives, rockfound.rockarch.org /evolution-of-a-foundation.

18 Gary Richardson and Jessie Romero, "The Meeting at Jekyll Island." Federal Reserve History, December 4, 2015, www.federalreservehistory.org/essays/jekyll-island-conference.

19 Chernow, 359.

20 "Free Tours by Foot, How to Get Top of the Rock Tickets," www.freetoursbyfoot.com/top-of -the-rock-ticket, accessed January 24, 2018.

21 Ben Maurk, "The Ludlow Massacre Still Matters," *The New Yorker*, April 18, 2014.

22 Margaret Sanger, "Cannibals," *The Woman Rebel*, May 1914, www.nyu.edu/projects/sanger /webedition/app/documents/show.php?sangerDoc=420013.xml.

23 Chernow, *Titan*, 613–614.

24 Donald Lett Jr., *Phoenix Rising: The Rise and Fall of the American Republic* (Bloomington, IN: AuthorHouse, 2008).

25 Stanley Baran and Dennis Davis, *Mass Communication Theory: Foundations, Ferment, and Future* (Boston: Cengage Learning, 2009).

26 "Finding a Footing," Rockefeller Foundation Archives, rockfound.rockarch.org/finding-a -footing, retrieved January 24, 2018.

27 "Simon Flexner," Rockefeller Foundation Archives, rockfound.rockarch.org/biographical/-/ asset_publisher/6ygcKECNI1nb/content/simon-flexner.

28 Samaan, A. E., *From a "Race of Masters" to a "Master Race": 1948 to 1848* (Charleston, SC: CreateSpace, 2013).

29 "Bureau of Social Hygiene," Rockefeller Foundation Archives, rockfound.rockarch.org/bureau -of-social-hygiene, accessed May 8, 2019.

30 "Family Planning," Rockefeller Foundation Archives, rockfound.rockarch.org/family-planning, accessed May 8, 2019.

31 "Kinsey Reports," Rockefeller Foundation Archives, rockfound.rockarch.org/kinsey-reports, accessed May 8, 2019.

32 Chernow, *Titan*, 377.

33 Rockefeller Panel Reports, *Prospect for America* (Garden City, NY: Doubleday, 1961), 88.

Chapter 2

1 Ibid.

2 "About," Rockefeller Brothers Fund, www.rbf.org/about, accessed January 24, 2018.

3 *1987 Annual Report,* Rockefeller Brothers Fund.

4 *1951–53 Annual Report*, Rockefeller Brothers Fund.

5 Robin W. Winks, *Laurance S. Rockefeller: Catalyst for Conservation* (Washington, DC: Island Press, 1997; Jennifer Cobb Kreisberg, "A Globe Clothing Itself in a Brain," *Wired*, January 6, 1995.

6 Winks, *Laurance S. Rockefeller.*

7 Julian Huxley, "UNESCO: Its Purpose and Its Philosophy," 1946, www.globalresearch.ca /the-task-of-unifying-the-world-mind/9593.

8 Paul Weindling, "Julian Huxley and the Continuity of Eugenics in Twentieth-Century Britain," *Journal of Modern European History*, vol. 10, no. 4 (2012), 480–499.

9   Angela Franks, *Margaret Sanger's Eugenic Legacy: The Control of Female Fertility* (Jefferson, NC: McFarland & Co., 2005).

10   Fred Pearce, *The Coming Population Crash: And Our Planet's Surprising Future* (Boston: Beacon Press, 2010).

11   Yu-ling Huang, "The Population Council, World Population Problem, and Contraceptive Studies during the Early Postwar Era," Rockefeller Archive Center, 2011, www.rockarch.org /publications/resrep/huang2.pdf, accessed January 24, 2018; Oscar Harkavy et al., *Curbing Population Growth: An Insiders Perspective on the Population Movement* (New York: Population Council, 1966).

12   "Agriculture," Rockefeller Foundation Archives, rockfound.rockarch.org/agriculture, accessed March 14, 2019.

13   "Agriculture."

14   Winks, *Laurance S. Rockefeller.*

15   Huxley, "UNESCO."

16   Pierre Teilhard de Chardin, Pierre, *Fenomenet människan* (Furulund, Sweden: Alhambra Förlag, 1955).

17   *1973 Annual Report*, Rockefeller Brothers Fund.

18   www.venrock.com.

19   Rockefeller Foundation Archives, "Computer Science," rockfound.rockarch.org/computer-science, accessed January 24, 2018.

20   Carroll Quigley, *Tragedy and Hope* (San Pedro, CA: GSG & Associates,1966).

21   "Secretary-General, Marking Historic Donation to League of Nations Library, Hails Rockefeller Foundation's 'Global Philanthropy,'" United Nations, September 10, 2012, www .un.org/press/en/2012/sgsm14498.doc.htm.

22   Dino Knudsen, *The Trilateral Commission and Global Governance* (London and New York: Routledge, 2013), 49; Council on Foreign Relations, *The War and Peace Studies of the Council on Foreign Relations 1939–1945* (New York: The Harold Pratt House, 1946).

23   Steven Rockefeller, "The Earth Charter at 15: A Spiritual Lens on Sustainability," interview by Allen White, Great Transition Initiative, December 2015, www.greattransition.org /publication/the-earth-charter-at-15.

24   Charlene Mires, *Absolutely Not New York: Finding a Home for the United Nations* (New York: NYU Press, 2014).

25   *1994 Annual Report*, Rockefeller Brothers Fund, 6.

26   "Oscar Niemeyer and the United Nations Headquarters," United Nations Archives, archives. un.org/content/oscar-niemeyer-and-united-nations-headquarters, accessed November 14, 2018.

27   "UN Secretary-General's Remarks at the United Nations Office at Geneva Library on the 85th Anniversary of the Donation by John D. Rockefeller to Endow the League of Nations Library," United Natioins, September 10, 2012, www.un.org/sg/en/content/sg /statement/2012–09-10/un-secretary-generals-remarks-united-nations-office-geneva-library.

28   The Dodd Report to the Reece Commission on Foundations, *Special Committee of the House of Representatives to Investigate Tax Exempt Foundations* (New York: The Long House Inc, 1954.)

29   "The Art Story," Museum of Modern Art, www.theartstory.org/museum-moma.htm, accessed January 24, 2018.

30   Michael FitzGerald, *Making Modernism: Picasso and the Creation of the Market for Twentieth-Century Art* (New York: Farrar, Straus and Giroux, 1995), 120.

31   Bernice Kert, *Abby Aldrich Rockefeller: The Woman in the Family* (New York: Random House, 1993).

32   "Rockefeller Sales Total $832.6 Million—Highest Ever for a Single Collection," Christie's, May 11, 2018, www.christies.com/features/Rockefeller-sales-final-report-9206–3.aspx; Michela Tindera, "Inside Late Billionaire David Rockefeller's Will: Picassos, a Beetle Collection and a Maine Island," *Forbes*, April 20, 2017.

33   Interview with Philip Johnson by Sharon Zane, Museum of Modern Art, December 18, 1990, www.moma.org/momaorg/shared/pdfs/docs/learn/archives/transcript_johnson.pdf.

34   In 1940, Nelson Rockefeller had been director of the Office of the Coordinator of Inter-American Affairs, a government agency that sponsored overseas art shows for propaganda purposes, with MoMA as contractor for these.

35   Frances Stonor Saunders, *Who Paid the Piper? The CIA and the Cultural Cold War* (London: Granta, 1999); Frances Stonor Saunders, *Hidden Hands: A Different History of Modernism*, episode 1, "Art and the CIA," directed by Tony Cash and produced by Frances Stonor Saunders (1995), youtu.be/k5YSikO6JRM.

36   Henry Russell Hitchcock, Philip Johnson, *Modern Architecture: International Exhibition, New York, Feb. 10 to March 23, 1932* (exhibition catalogue), Museum of Modern Art.

37   Marc Wortman, "Famed Architect Philip Johnson's Hidden Nazi Past," *Vanity Fair*, April 4, 2016; Wortman, *1941: Fighting the Shadow War: A Divided America in a World at War* (New York: Atlantic Monthly Press, 2016.)

38   Philip Johnson and Mark Wigley, *Deconstructivist Architecture* (exhibition catalogue), Museum of Modern Art, 1988.

39   Henry Hope Reed, *Rockefeller, New York: A Tour* (New York: Greensward Foundation, 1988).

40   David W. Dunlap, "What David Rockefeller Wanted Built Got Built," *New York Times*, March 26, 2017.

41   David Rockefeller, *Memoirs* (New York: Random House, 2003), 101.

42   Robert A. Caro, *The Power Broker: Robert Moses and the Fall of New York* (New York: Vintage Books, 1975).

43   Jane Jacobs, *The Death and Life of Great American Cities* (New York: Random House, 1961).

44   Peter Collier and David Horowitz, *The Rockefellers: An American Dynasty* (New York: Holt, Rinehart and Winston, 1976), 455–456.

45   Sidney Hyman, *The Aspen Idea* (Norman: University of Oklahoma Press, 1975), 36–42.

46   David Hill, "How the Bauhaus Came to Aspen, *Custom Home*, April 4, 2019.

47   *In Their Own Words: Walter and Elizabeth Paepcke and the Birth of the Aspen Institute*, produced and directed by Greg Poschman (2009; The Aspen Institute), www.youtube.com/watch?v=7TOR1BVOzdA.

48   Hyman, *Aspen Idea*, 88.

49   Hyman, 101.

50   Hyman, 16–18.

51   Mortimer J. Adler, *How to Read a Book: The Art of Getting a Liberal Education* (New York: Simon & Schuster, 1940).

52   Steve Koppes, "How the First Chain Reaction Changed Science," University of Chicago News, December 10, 2012, www.uchicago.edu/features/how_the_first_chain_reaction_changed_science.

53   Jeffrey Frank, "Big Spender, Nelson Rockefeller's Grand Ambitions," *The New Yorker*, October 13, 2014.

54   Suzanne Goldenberg, "Doomsday Clock Stuck Near Midnight Due to Climate Change and Nuclear War," *The Guardian*, January 26, 2016.

Chapter 3

1   Rockefeller Brothers Fund, *Prospect for America: The Rockefeller Panel Reports* (Garden City, NY: Doubleday, 1963), 197.

2   Vannevar Bush, *Science the Endless Frontier: A Report to the President by Vannevar Bush, Director of the Office of Scientific Research and Development*, July 1945 (Washington, DC: United States Government Printing Office, 1945).

3   Jerome Wiesner, *Vannevar Bush, 1890–1974* (Washington, DC: National Academies Press, 1979).

4   National Academy of Sciences, *The National Academy of Sciences: The First Hundred Years, 1863–1963* (Washington, DC: National Academies Press, 1978).

5   Detlev Bronk, "Science Advice in the White House," *Science* 186, no. 4159 (October 11, 1974): 116–121.

6   Donald Critchlow, *Intended Consequences: Birth Control, Abortion, and the Federal Government* (Oxford: Oxford University Press, 1999).

7   Thomas Robertson, *The Malthusian Moment: Global Population Growth and the Birth of American Environmentalism (Studies in Modern Science, Technology, and the Environment)* (New Brunswick, NJ: Rutgers University Press, 2012).

8   "Roger Randall Dougan Revelle, Director of Scripps Institution of Oceanography, 1951–1964," interview by Sarah Sharp in 1985, The Regents of the University of California SIO Reference Series No. 88–20, November 1988.

9   Thomas Malone, Edward Goldberg, and Walter Munk, *Roger Dougan Revelle 1909–1991: A Biographical Memoir* (Washington, DC: National Academy of Sciences, 1998).

10   Bert Bolin, *Carl-Gustaf Rossby: The Stockholm Period 1947–1957, Tellus A: Dynamic Meteorology and Oceanography* 51, no. 1 (1999).

11   Sverker Sörlin, *A Tribute to the Memory of Carl-Gustaf Rossby* (Stockholm: The Royal Swedish Academy of Engineering Sciences [IVA], 2015).

12   Horace Byers, *Carl-Gustaf Arvid Rossby* (Washington, DC: National Academy of Sciences, 1960).

13   "History and Legacy: Over 80 Years of Ocean Research, Education, and Exploration," Woods Hole Institute of Oceanography, www.whoi.edu/main/history-legacy, accessed January 24, 2018.

14   Lisa Rosner, *The Technological Fix: How People Use Technology to Create and Solve Problems* (New York: Routledge, 2004).

15   Rosner, *Technological Fix*.

16   Hermann Flohn, *Meterologie im Übergang: Erfahrungen und Erinnerungen (1931–1991)* (Bonn: Dümmler, 1992).

17   Sverker Sörlin, *Science, Geopolitics and Culture in the Polar Region: Norden beyond Borders* (Oxford: Taylor and Francis, 2016).

18   H Byers, *Carl-Gustaf Arvid Rossby*.

19   C. D. Keeling, "The Influence of Mauna Loa Observatory on the Development of Atmospheric CO_2 Research," in *Mauna Loa Observatory : A 20th Anniversary Report*, ed. John Miller (Silver Spring, MD: Environmental Research Laboratories, Air Resources Laboratories, 1978).

20   Bolin, *Carl-Gustaf Rossby*.

21 Carroll Quigley, *Tragedy and Hope* (San Pedro, CA: GSG, 1966).

22 Nick Begich, *Angels Don't Play This HAARP: Advances in Tesla Technology* (Anchorage: Earthpulse Press, 1995).

23 S Sörlin, *Tribute*.

24 John von Neumann, "Can We Survive Technology?" *Fortune*, June 1955, 106–108, 151–152, reprinted in *The Fabulous Future: America in 1980*, ed. David Sarnoff (New York: E. P. Dutton & Co., 1956), 41.

25 J. Hamblin, "'A Dispassionate and Objective Effort': Negotiating the First Study on the Biological Effects of Atomic Radiation," *Journal of the History of Biology* 40, no. 1 (2007): 147–177.

26 *1956 Annual Report*, Rockefeller Foundation.

27 National Academy of Sciences, "The Biological Effects of Atomic Radiation: A Report to the Public" (Washington, DC: National Academy of Sciences, National Research Council, 1956).

28 *1956 Annual Report*, Rockefeller Foundation.

29 Joshua P. Howe, *Behind the Curve: Science and the Politics of Global Warming* (Seattle: University of Washington Press, 2014).

30 Christopher Bailey, *US Climate Change Policy* (London: Routledge, 2015).

31 Roger Revelle and Hans Suess, "Carbon Dioxide Exchange between Atmosphere and Ocean and the Question of an Increase of Atmospheric CO_2 during the Past Decades," *Tellus* 9, no. 1 (1956): 18–27.

32 Gilbert Plass, "The Carbon Dioxide Theory of Climatic Change," *Tellus* 8, no. 2 (May 1956): 140–154.

33 "The Rising Threat of Carbon Dioxide, *New York Times*, October 28, 1956.

34 Norman A. Phillips, "Carl-Gustaf Rossby: His Times, Personality, and Actions," *Bulletin of the American Meteorological Society* 79, no. 6 (June 1998).

35 Bert Bolin, *The Atmosphere and the Sea in Motion: Scientific Contributions to the Rossby Memorial Volume* (New York: Rockefeller Institute Press, 1959).

36 National Academy of Sciences, *Biographical Memoirs, Vol. 61* (Washington, DC: National Academies Press, 1992).

37 Bronk, "Science Advice."

38 Bolin, *Atmosphere*.

39 Charles Campbell, *A Questing Life: The Search for Meaning* (Lincoln, Nebraska: iUniverse, 2006).

40 M. A. Matthews, "The Earth's Carbon Cycle," *New Scientist* 6 (October 8, 1959): 644–646.

41 *The Unchained Goddess*, directed by Richard Carlson and William T. Hurtz and produced by Frank Capra (1958, N. W. Ayer & Son).

42 Holly Sklar, *Trilateralism: The Trilateral Commission and Elite Planning for World Management* (Boston: South End Press, 1980).

43 Frank, "Big Spender."

44 Dino Knudsen, *The Trilateral Commission and Global Governance* (London and New York: Routledge Press, 2016).

45 Matthews, "Earth's Carbon Cycle."

46 "Subpanel II: International Security Objectives and Strategy, 1956–1964," Rockefeller Archives Center, dimes.rockarch.org, search: "International Security Objectives and Strategy."

47   Rockefeller Brothers Fund, *Prospect for America: The Rockefeller Panel Reports* (Garden City, NY: Doubleday, 1961): 196–199.

48   Bruce Grierson, *U-Turn: What If You Woke Up One Morning and Realized You Were Living the Wrong Life* (New York, NY: Bloomsbury Publishing, 2007), 53.

Chapter 4

1   WWF, "In Memoriam: Godfrey A. Rockefeller," press release, January 29, 2010, www.world wildlife.org/press-releases/in-memoriam-godfrey-a-rockefeller, accessed February 4, 2018.

2   Rockefeller, *Memoirs*, 208–209.

3   Conservation Foundation, *Implications of Rising Carbon Dioxide Content of the Atmosphere: a Statement of Trends and Implications of Carbon Dioxide Research Reviewed at a Conference of Scientists* (New York: Conservation Foundation, 1963).

4   Winks, *Laurance S. Rockefeller*.

5   Winks, 141.

6   The White House, *Restoring the Quality of Our Environment: Report of the Environmental Pollution Panel*, November 1965.

7   Marianne Lavelle, "A 50th Anniversary Few Remember: LBJ's Warning on Carbon Dioxide, *The Daily Climate*, February 2, 2015.

8   George Gonzales, *Corporate Power and the Environment* (Lanham, MD: Rowman & Littelfield, 2001).

9   Winks, *Laurance S. Rockefeller*.

10   Roger Randall Dougan Revelle, March 7, 1909–July 15, 1991, UC San Diego, Revelle College, revelle.ucsd.edu/about/roger-revelle.html, accessed January 24, 2018.

11   *1963 Annual Report*, Rockefeller Foundation, 74.

12   Edward Cornish, "The Search for Foresight: How THE Futurist Was Born," *The Futurist*, January–February 2007.

13   Lars-Göran Engfeldt, *From Stockholm to Johannesburg and Beyond: The Evolution of the International System for Sustainable Development Governance and Its Implications* (Stockholm: Swedish Ministry for Foreign Affairs, 2009).

14   Hyman, *Aspen Idea*, 288–299.

15   Hans Palmstierna and Lena Palmstierna, *Plundring, svält, förgiftning* (Stockholm: Rabén & Sjögren, 1968).

16   Hyman, *Aspen Idea*, 288.

17   Rennie Whitehead, "The Club of Rome and CACOR: Recollections by J. Rennie Whitehead," www.whitehead-family.ca/drrennie/CACORhis.html, accessed January 24, 2018.

18   Tapio Kanninen, *Crisis of Global Sustainability (Global Institutions)* (London: Routledge, 2012).

19   Leah MacVie, "Erich Jantsch Biography," leahmacvie.com/my-work/erichjantsch, accessed January 24, 2018.

20   MacVie, "Erich Jantsch."

21   Alexander Kristiakis, *How People Harness Their Collective Wisdom and Power to Construct the Future* (Charlotte: Information Age Publishing Inc., 2006).

22   Erich Jantsch, *Perspectives of Planning, Proceedings of the OECD Working Symposium on Long-Range Forecasting and Planning, Bellagio, Italy, October 27–November 2, 1968* (Paris: Organisation for Economic Cooperation and Development, 1969).

23 Eric Jantsch, "The Bellagio Declaration on Planning," *Futures* 1, no. 3 (March 1969): 181–276.

24 Knudsen, *Trilateral Commission*.

25 Pentti Malaska and Matti Vapaavuori, eds., *The Club of Rome: "The Dossiers" 1965–1984* (Helsinki: Finnish Association for the Club of Rome, 2005), clubofrome.fi/wp-content /uploads/2014/10/Dossiers.pdf.

26 Donella H. Meadows et al., *The Limits to Growth; A Report for the Club of Rome's Project on the Predicament of Mankind* (New York: Universe Books, 1972).

27 William W. Kellogg, "Mankind's Impact on Climate: The Evolution of an Awareness," *Climatic Change* 10, no. 2 (April 1987): 113–136, core.ac.uk/download/pdf/81712826.pdf.

28 *1970 Annual Report*, Woods Hole Oceanographic Institution, 16.

29 Study of Critical Environmental Problems (SCEP), *Man's Impact on the Global Environment: Assessment and Recommendations for Action, Report of the Study of Critical Environmental Problems (SCEP)* (Cambridge, MA: MIT Press, 1970), 45

30 Study of Man's Impact on Climate (SMIC), ed., *Inadvertent Climate Mmodification: Report of the Study of Man's Impact on Climate (SMIC)* (Cambridge, MA: MIT Press, 1971).

31 Winks, *Laurance S. Rockefeller*.

32 Jack Raymond, *Robert O. Anderson: Oil Man / Environmentalist and His Leading Role in the International Environmentalist Movement: a Monograph* (Aspen, CO: Aspen Institute for Humanistic Studies, 1988).

33 Dixy Lee Ray, *Trashing the Planet: How Science Can Help Us Deal with Acid Rain, Depletion of the Ozone, and the Soviet Threat among Other Things* (Washington, DC: Regnery Publishing, 1990).

34 Alvin Toffler, *Future Shock* (New York: Random House, 1970).

35 *Future Shock*, directed by Alexander Grasshof (1972; McGraw-Hill Films).

36 University of Michigan, "ENACT Teach-In and Earth Day, Spring 1970," michiganin-theworld.history.lsa.umich.edu/environmentalism/exhibits/show/mainexhibit/earthday, accessed May 21, 2019.

37 Hyman, *Aspen Idea*, 252.

38 Alan Glenn, "Turbulent Origins of Ann Arbor's First Earth Day, *Ann Arbor Chronicle*, April 22, 2009.

39 Glenn, "Turbulent Origins."

40 Ibid.

41 Numbersusa.org, "Population Stabilization and the Modern Environmental Movement," www .numbersusa.org/pages/population-stabilization-and-modern-environmental-movement, accessed January 28, 2018.

42 Earth Day Network, "The History of Earth Day," www.earthday.org/about/the-history-of -earth-day, accessed January 28, 2018.

43 Dan Toppin, *This Cybernetic Age* (New York: Human Development Corporation, 1969).

44 Kerry Bolton, *Revolution from Above* (Budapest, Arktos Media, 2011).

45 Rex Weyler, *Greenpeace: How a Group of Ecologists, Journalists, and Visionaries Changed the World* (Vancouver: Raincoast Books, 2004).

46 Alexis Schwarzenbach, *Saving the World's Wildlife: The WWF's First Fifty Years* (London: Profile Books, 2011).

47 *Only One Earth: The Stockholm Conference*, Peter Hollander, executive producer (1972; UNEP), UNEP documentary for the New York State Education Department, Part 1: youtu.be/mJUk70tfELA, Part 2: youtu.be/h3-TqHFkfy8.

48 Lars Gogman, "Rödgrönt samarbete med förhinder," *Arbetarhistoria* 2 (2012).

49 Gogman, "Hog farm satte alternativrörelsens verklighetsuppfattning i gungning," *Stockholms Fria*, March 9, 2008; through: Per Gudmundson, "Vem gör Hog Farm 2008?"

50 Declaration of the United Nations Conference on the Human Environment, Stockholm, June 16, 1972.

51 Ann Hironaka, *Greening the Globe: World Society and Environmental Change* (Oxford: Cambridge University Press, 2014).

52 *1972 Annual Report*, Rockefeller Foundation.

53 *1971 Annual Report*, Rockefeller Foundation.

54 Russell Train, *Politics, Pollution, and Pandas: An Environmental Memoir* (Washington, DC: Island Press, 2003), 49; Winks, 43–44.

55 Barbara Ward and René Jules Dubos, *Only One Earth: The Care and Maintenance of a Small Planet* (New York: W. W. Norton & Company, 1972).

56 David Satterthwaite, *Barbara Ward and the Origins of Sustainable Development* (London: International Institute for Environment and Development, 2006).

57 John D. Rockefeller III, *Population and the American Future: The Report of The Commission on Population Growth and the American Future*, Rockefeller Commission Report, The Center for Research on Population and Security, July 27, 1972, www.mnforsustain.org/pop_rockefeller_72.htm.

58 David Simcox, "NPG Forum Series: Nixon and American Population Policy," Floridians Sustainable Population, www.flsuspop.org/NixRockefeller.html, accessed January 24, 2018.

59 *Survival of Spaceship Earth*, directed by Dirk Wayne Summers (1972; United Productions Limited).

60 Bruce Grierson, *U-Turn: What If You Woke Up One Morning and Realized You Were Living the Wrong Life* (New York, NY: Bloomsbury Publishing, 2007), 53.

Chapter 5

1 World Economic Forum, Timeline: The Davos Manifesto, "1973," widgets.weforum.org/history/1973.html, accessed November 10, 2019.

2 World Economic Forum, *A Partner in Shaping History: The First 40 Years 1971–2010*, 2009.

3 The World Economic Forum, *A Partner*, 32.

4 Rockefeller, *Memoirs*.

5 Henry Kissinger, "Henry Kissinger: My Friend David Rockefeller, A Man Who Served the World," *Washington Post*, March 30, 2017.

6 Knudsen, *Trilateral Commission*.

7 Trilateral Commission, "The Trilateral Commission at 25: Between Past and Future," trilateral.org/download/files/anniversary_evening.pdf, accessed January 28, 2018.

8 Zbigniew Brzezinski, *Between Two Ages: America's Role in the Technetronic Era* (Santa Barbara, CA: Greenwood Press, 1970).

9 K DeWitt, "Brzezinski, the Power and the Glory," *Washington Post*, February 4, 1977.

10 Patrick M. Wood and Anthony C. Sutton, *Trilaterals over Washington* (Mesa, AZ: Coherent Publishing LCC, reprint, 2017).

11 Knudsen, *Trilateral Commission*.

12 *David Rockefeller: Bridge Builder (A Tribute)*, produced, directed, and written by Roger Torda (2003: Synergos Institute, Balaton Film and Television), Rockefeller Archive Center, youtu .be/y2cORk1ni10.

13 Rockefeller, *Memoirs*, 242–243.

14 David Rockefeller, "From a China Traveller," *New York Times*, August 10, 1973, www .nytimes.com/1973/08/10/archives/from-a-china-traveler.html.

15 Rockefeller, *Memoirs*, 258–260.

16 Oliver Morgan and Faisal Islam, "Saudi Dove in the Oil Slick, *The Guardian,* January 14, 2001.

17 Hyman, *Aspen Idea*, 293.

18 Thomas W. Wilson Jr., *World Energy, the Environment and Political Action: Summary of Second International Environmental Workshop* (Washington: International Institute for Environmental Affairs, 1972).

19 William Engdahl, *A Century of War* (London: Pluto Press, 2004).

20 Myer Kutz, "The Rockefeller Problem," *New York Times*, April 28, 1974.

21 Frank Lynn, "Rockefeller Quits as Chairman of Critical Choices Commission," *New York Times*, March 1, 1975.

22 Commission on Critical Choices for Americans, *Critical Choices for Americans , Volumes 1–13* (Lexington, MA: Lexington Books, 1976–77).

23 United Nations General Assembly, 3201 (S-VI) Declaration on the Establishment of a New International Economic Order, 6th special session, May 1, 1974.

24 R. Cox, "Ideologies and the New International Economic Order: Reflections on some Recent Literature," *International Organization* 33, no. 2 (1979): 257–302, doi:10.1017 /S002081830003216.

25 Trilateral Commission, "T10—Looking Back . . . and Forward," *Trialogue: A Bulletin of North American–European–Japanese Affairs*, 1976, trilateral.org/file.showdirectory&list =Trialogue-Series.

26 United Nations General Assembly, May 1, 1974.

27 Knudsen, *Trilateral Commission*, 114.

28 Richard N. Gardner, Saburo Okita, and B. J.Udink, *A Turning Point in North-South Relations*, Trilateral Commission Task Force Report 3, The Triangle Papers, 1974.

29 Knudsen, *Trilateral Commission*, 115.

30 Krundsen, 115.

31 Trilateral Commission, "T10."

32 United Nations Conference on Trade and Development, *UNCTAD at 50: A Short History*, 2014, unctad.org/en/PublicationsLibrary/osg2014d1_en.pdf.

33 J Colomer, *How Global Institutions Rule the World* (Basingstoke, UK: Palgrave Macmillan, 2014).

34 Knudsen, *Trilateral Commission*, 115.

35 Knudsen, 166–167.

36 Emmanue Mourlon-Druol and Federico Romero, eds., *International Summitry and Global Governance, The Rise of the G7 and the European Council, 1974–1991* (London: Routledge, 2014): 82–85.

37 David McLoughlin, "The Third World Debt Crisis and the International Financial System," *Student Economic Review*, Trinity College Dublin (1989): 96–101.

38   Walden Bello, *Dark Victory: The United States and Global Poverty* (2nd edition) (Oakland: Food First Books, 1999).

39   Clifton Daniel, "Ford and 25th Amendment: Rockefeller Choice Held Free of Party Pressure," *New York Times*, December 23, 1974.

40   Birch Bayh, *One Heartbeat Away: Presidential Disability and Succession* (Indianapolis: Bobbs-Merrill, 1968); John Fitzgerald, "25th Amendment: The Architect," Watergate Amendment, www.watergatemendment.com/?page_id=3508, accessed January 24, 2018.

41   Knudsen, *Trilateral Commission*, 133.

42   Mark O. Hatfield, "Nelson Rockefeller" in *Vice Presidents of the United States, 1789–1993* (Washington, DC: US Government Printing Office, 1997): 505–512.

43   Patrick M. Wood, *Technocracy Rising: The Trojan Horse of Global Transformation* (Mesa, AZ: Coherent Publishing, 2015): 60–66.

44   Hyman, *Aspen Idea*, 297.

45   Thomas W. Wilson Jr. *World Population and a Global Emergency* (Washington, DC: Aspen Institute for Humanistic Studies, 1974).

46   Hyman, *Aspen Idea*, 298.

47   Aspen Institute, *Second Status Report to the Trustees of IFIAS, on the IFIAS Special Project: The Impact on Man of Climate Change*, October 8, 1976.

48   Aspen Institute, *Second Status Report*.

49   David Hart, "Strategies of Research Policy Advocacy: Anthropogenic Climatic Change Research 1957–1974," Center for Science and International Affairs at Harvard University, Kennedy School of Government, 1992, belfercenter.ksg.harvard.edu/files/disc_paper_92_08 .pdf.

50   *1974 Annual Report*, Rockefeller Foundation.

51   Olof Palme, "Olof Palme: Förändrat klimat är största hotet," *Svenska Dagbladet*, November 27, 1974.

52   *1974 Annual Report*, Rockefeller Foundation, 64.

53   Kellogg, "Mankind's Impact."

54   *1974 Annual Report*, Rockefeller Brothers Fund, 25.

55   *1974 Annual Report*, 26.

56   Andrew Spekke, *The Next 25 Years: Crisis and Opportunity* (Washington, DC: World Future Society, 1975): III.

57   Spekke, *Next 25 Years*, 204–210.

58   Ervin László, *Goals for Mankind: A Report to the Club of Rome* (London: Hutchinson & Co., 1977).

59   Spekke, *Next 25 Years*, 57–59.

60   Ervin László, Robert Baker Jr., Elliott Eisenberg, and Venkata Raman, *The Objectives of the New International Economic Order* (Oxford: Pergamon Press, 1978).

61   Donald Keys, *Earth at Omega: Passage to Planetization* (Boston: The Branden Press, 1982).

62   Oliver Reiser, *Cosmic Humanism and World Unity* (New York: Gordon and Breach, 1975).

63   *Co-Creation* magazine, "A Tribute to Barbara Marx Hubbard 1929–2019," co-creationmaga-zine.com/a-tribute-to-barbara-marx-hubbard-1929–2019, accessed August 21, 2019.

64   Neil Genzlinger, "Barbara Marx Hubbard, 89, Futurist Who Saw 'Conscious Evolution,' Dies," *New York Times*, May 15, 2019.

65   Among the participants at Lindisfarne was chemist James Lovelock, who formulated the Gaia Hypothesis with clear influences from the teachings of Teilhard de Chardin.

66 Annie Cheatham interviewed by Claire Wilson, Sophia Smith Collection, Smith College, 2008, www.smith.edu/libraries/libs/ssc/activist/transcripts/Cheatham.pdf.

67 Congressional Clearinghouse on the Future, "Role of Congress in the Third Wave," 1983, digitalcollections.library.cmu.edu/portal/main.jsp?flag=browse&smd=1&awdid=5.

68 Alvin Toffler, *Creating a New Civilization: The Politics of the Third Wave* (Atlanta: Turner Publishing, 1995).

69 Congressional Clearinghouse on the Future, letter from Robert Edgar to the advisory group, 1983, digitalcollections.library.cmu.edu/awweb/awarchive?type=file&item=589201.

70 Congressional Institute for the Future, www.users.interport.net/f/u/future98/INST.html, accessed January 22, 2018.

71 Donald Gibson, *Environmentalism: Ideology and Power* (New York: Nova Science Publishers, 2002).

72 Gilbert King, "A Halloween Massacre at the White House," *Smithsonian*, October 25, 2012, smithsonianmag.com/history/a-halloween-massacre-at-the-white-house-92668509.

73 *1987 Annual Report*, Rockefeller Brothers Fund.

74 Sam Hodder, "Generations of Generosity: Remembering David Rockefeller," *Save the Redwood League* blog, March 22, 2017, www.savetheredwoods.org/blog/ generations-generosity -remembering-david-rockefeller.

75 "Symbionese Liberation Army Declaration of Revolutionary War and the Symbionese Program" (press release), 1973, web.archive.org/web/20031020235849/http://www.feastof hateandfear.com/archives/sla.html.

76 Gary Allen, *The Rockefeller File* (Cutchogue, NY: Buccaneer Books, 1976).

77 "David Rockefeller, Philanthropist and Former CEO of Chase Manhattan Bank, Shares His Autobiography, *Memoirs*, Which Documents His Life and the History of His Legendary Family," *Charlie Rose*, October 21, 2002, charlierose.com/videos/9606.

78 Peter Collier and David Horowitz Holt, *The Rockefellers*.

79 Nick Lemann, "Poor Little Rich People, the Rockefellers: A Family Dynasty by Peter Collier and David Horowitz Holt," *Harvard Crimson*, April 22, 1976.

80 Rockefeller, *Memoirs*, 322.

81 Rockefeller, 342–345.

82 Robert D. McFadden, "Rockefeller Gave Kissinger $50,000, Helped 2 Others," *New York Times*, October 6, 1974.

83 *1987 Annual Report*, Rockefeller Brothers Fund.

84 Frank, "Big Spender."

85 H Kissinger, "Nelson Rockefeller," February 2, 1979, www.henryakissinger.com/remembrances /nelson-rockefeller.

86 Rockefeller, *Memoirs*, 322.

87 Gerald O. Barney, *The Global 2000 Report to the President* (Washington, DC: US Government Printing Office, 1980).

88 Barney, *The Unfinished Agenda: The Citizen's Policy Guide to Environmental Issues: A Task Force Report Sponsored by the Rockefeller Brothers Fund* (New York: Thomas Y. Crowell Co., 1977).

89 *1980 Annual Report*, Rockefeller Foundation.

90 Mark Burdman, "World Futurists Turn Back the Clock," *Fusion*, November 1980.

91 World Future Society, *First Global Conference on the Future* (conference brochure), 1980.

92   Warren W. Wagar, "Technocracy as the Highest State of Capitalism," in Frank Feather, *Through the '80s: Thinking Globally, Acting Locally* (Washington, DC: World Future Society, 1980).

93   Wagar, *The City of Man: Prophecies of a World Civilization in Twentieth-Century Thought* (Boston: Houghton Mifflin, 1963).

94   Glenn F. Cartwright and Adam Finkelstein, "Second Decade Symbionics and Beyond," based on a paper presented at the Ninth General Assembly of the World Future Society, Washington DC, July 31, 1999.

95   Norman Myers, "First Global Conference on the Future, Held in the Royal York Hotel and Harbour Hilton, Toronto, Canada, During 20–24 July 1980," *Environmental Conservation* 8, no. 1 (1981): 73–78, doi:10.1017/S0376892900026825.

Chapter 6

1   *1989 Annual Report*, Rockefeller Brothers Fund.

2   Rockefeller, *Memoirs*, 322.

3   Rockefeller Brothers Fund, *Sustainable Development Program Review 2005–2010*, November 2010.

4   Disclosure Project, www.disclosureproject.org.

5   Rockefeller Brothers Fund, "75 Years of Engaged Philanthropy Timeline: The One World Program 1983," 2015, 75.rbf.org/#!trigger=one-world-program.

6   Jim McNeill, Pieter Winsemius, and Taizo Yakushiji, *Beyond Interdependence: The Meshing of the World's Economy and the Earth's Ecology*, a report for Trilateral Commission (Oxford: Oxford University Press, 1991).

7   Laura Schwartz, *A History of Climate Action through Foundations*, Rockefeller Archives Center Research Report, May 28, 2018, rockarch.issuelab.org/resources/30792/30792.pdf.

8   "1973 Members," Trilateral Commission, swprs.files.wordpress.com/2017/07/rilateral -commission-members-1973.pdf.

9   Weatherhead Center for International Affairs, "In Theory and in Practice: Harvard's Center for International Affairs, 1958–1983," wcfia.harvard.edu/about/theory-and-practice-harvards -center-international-affairs-1958%E2%80%931983, accessed January 20, 2018; Trilateral Commission, "The Trilateral Commission at 25: Between Past and Future," 1998, trilateral. org/download/files/anniversary_evening.pdf, accessed January 28, 2018.

10   Laura Mazur and Louella Miles, eds., "Sir Crispin Tickell," in *Conversations with Green Gurus: The Collective Wisdom of Environmental Movers and Shakers* (Hoboken, NJ: John Wiley & Sons, Inc., 2012).

11   Fred Pearce, "The Green Diplomat," *New Scientist*, March 21, 1992, 38.

12   "Patrons," Population Matters, www.populationmatters.org/about/people-and-story/patrons, accessed January 20, 2018.

13   G. Woodwell, G. J. MacDonald, R. Revelle, and C. D. Keeling, *The Carbon Dioxide Problem: Implications for Policy in the Management of Energy and Other Resources*, A Report to the Council on Environmental Quality, July 1979, President's Council on Environmental Quality, Washington, DC, July 1979.

14   *Carbon Dioxide and Climate: A Scientific Assessment*, Report of an Ad Hoc Study Group on Carbon Dioxide and Climate, Climate Research Board Assembly of Mathematical and Physical Sciences National Research Council, July 23–27, 1979.

15 Shardul Agrawala, "Early Science-Policy Interactions in Climate Change: Lessons from the Advisory Group on Greenhouse Gases," *Global Environmental Change* 9, no. 2 (1999): 157–169.

16 Schwartz, *Climate Action*.

17 Ibid.

18 *1986 Annual Report*, Rockefeller Brothers Fund, 24.

19 Bert Bolin, International Council of Scientific Unions (1986), *The Greenhouse Effect, Climatic Change, and Ecosystems*, Chichester, West Sussex, published on behalf of the Scientific Committee on the Problems of the Environment of the International Council of Scientific Unions by Wiley, 1986.

20 World Climate Program, *Report from the International Conference on the Assessment of the role of Carbon Dioxide and other Greenhouse Gases in Climate Variations and Associated Impacts*, ICSU, UNEP, WMO, 1985.

21 Agrawala, "Early Science-Policy."

22 "Memorandum from Thomas Wahman to Russell Phillips, September 23, 1985," Rockefeller Brothers Fund Records, RG3 Project Files, Subgroup 2, Box 1163, Folder 7175, FA005.

23 Schwartz, *Climate Action*.

24 "30 Years Ago Scientists Warned Congress on Global Warming—What They Said Sounds Eerily Familiar," *Washington Post*, November 30, 2016.

25 Commission of the European Communities, *The Greenhouse Effect and the Community. Communication to the Council. Commission Work Programme Concerning the Evaluation of Policy Options to Deal with the "Greenhouse Effect."* Draft Council Resolution on the Greenhouse Effect and the Community, COM (88) 656 final, Brussels, November 16, 1988, 57–58.

26 *1987 Annual Report*, Rockefeller Foundation.

27 Wendy E. Franz, "The Development of an International Agenda for Climate Change: Connecting Science to Policy," Discussion Paper, E-97–07, Harvard University, Kennedy School of Government, July 31, 1997.

28 Jill Jäger, *Developing Policies for Responding to Climate Change: A Summary of the Discussions and Recommendations of the Workshops Held in Villach (28 September–2 October 1987) and Bellagio (9–13 November 1987), Under the Auspices of the Beijer Institute, Stockholm*, World Climate Program, WMO, UNEP, 1988.

29 Franz, "Developoment," 26.

30 Schwarz, *Climate Action*, 143.

31 United Nations, *Report of the World Commission on Environment and Development: Our Common Future* (Oxford: Oxford University Press, 1987).

32 Simon Batterbury, "The International Institute for Environment and Development: Notes on a Small Office," *Global Environmental Change* 14 (2004): 367–371.

33 Schwartz, *Climate Action*.

34 Toronto Environmental Alliance, Campaigns: "Climate Change," www.torontoenvironment .org/campaigns/climate/climatechange, accessed January 28, 2018.

35 United States Senate Committee on Energy and Natural Resources, "Statement of Dr. James Hansen, Director, NASA Goddard Institute for Space Studies," June 23, 1988.

36 Interview with Timothy Wirth on *PBS Frontline*, April 24, 2007, www.pbs.org/wgbh/pages /frontline/hotpolitics/interviews/wirth.html.

37   Bert Bolin, *A History of the Science and Politics of Climate Change: The Role of the Intergovernmental Panel on Climate Change* (Cambridge, UK: Cambridge University Press, 2007): 49.

38   Stefan Anderberg, "Klimatfrågans utveckling ur ett svenskt perspektiv" in book *Miljöhistoria över gränser,* edition *Skrifter med miljöhistoriska perspektiv* (Malmö: Malmö Högskola, 2006): 160–178.

39   Margaret Thatcher, "Speech to the Royal Society," September 27, 1988, www.margaretthatcher .org/document/107346.

40   Swedish Parliament, Riksdagens protokoll 1988/89:11, October 19, 1988, data.riksdagen.se /dokument/GC0911.

41   Leigh Glover, *Postmodern Climate Change* (London: Routledge, 2006): 119.

42   Shaedul Agrawala, "Explaining the Evolution of the IPCC Structure and Process," ENRP Discussion Paper E-97–05, Kennedy School of Government, Harvard University, August 1997, and also as International Institute for Applied Systems Analysis Interim Report IR-97–032/August.

43   Michael Oppenheimer, "How the IPCC Got Started," Environmental Defense Fund, November 1, 2007, blogs.edf.org/climate411/2007/11/01/ipcc_beginnings.

44   Agrawala, "Early Science-Policy," 612.

45   "George H. W. Bush: Rockefeller Was a Valuable Adviser," Associated Press, March 20, 2017.

46   United States Department of State, "Memorandum to Richard McCormack from Frederick Bernthal," February 9, 1989.

47   Trilateral Commission, *Commemorating 1989, the Year That Changed the Map of Europe (and Thereby the World)*, November 2009, trilateral.org/file.view&fid=149.

48   Congressional Institute for the Future, GLOBE—Global Legislators for a Balanced Environment, A Project of the Congressional Institute for the Future, John Heinz papers, Carnegie Mellon University Libraries Digital Collection, digitalcollections.library.cmu.edu.

49   GLOBE International, "Global Legislators Offer Tough New Environmental Proposals," (press release), November 16, 1990, digitalcollections.library.cmu.edu/awweb/awarchive?type =file&item=638674.

50   GLOBE International, "History," globelegislators.org/about-globe/24-history, accessed January 21, 2018.

51   Sudhir Chella Rajan, *Global Politics and Institutions*, Tellus Institute, 2006, greattransition. org/archives/papers/Global_Politics_and_Institutions.pdf.

52   GLOBE International, "Global Legislators."

53   Robert N. Stavins, *Project 88: Harnessing Market Forces to Protect the Environment*, John F. Kennedy School of Government, Harvard University, Cambridge, MA, 1988.

54   London School of Economics and Political Science, "Our History," www.lse.ac.uk/about-lse /our-history, accessed January 23, 2018.

55   London School of Economics and Political Science, "LSE—Rockefeller's Baby?" June 24, 2015, blogs.lse.ac.uk/lsehistory/2015/06/24/lse-rockefellers-baby.

56   Graham T. T. Molitor, "Molitor Forecasting Model: Key Dimensions for Plotting the 'Patterns of Change'," *Journal of Futures Studies* 8, no. 1 (August 2003), www.metafuture.org/articles bycolleagues/graham%20mollitor/Molitor%20forecasting%20model%202003.pdf.

57   Rockefeller Brothers Fund, *Sustainable Development Program Review*.

58   Gordon Goodman, "Responding to Climatic Change: Further Steps in Policy Development—A Proposal for the Continued Development in 1989 and 1990 of Work on

the Possible Responses to Climatic Change," The Beijer Institute, February 1989, in Schwartz, *Climate Action.*

59   Schwartz, *Climate Action.*

60   J. H. W Karas and P. M., *Heat Trap: Threat Posed by Rising Levels of Greenhouse Gases* (Ottawa, Canada: Friends of the Earth, 1989).

61   R Lamb, *Promising the Earth* (London: Routledge, 1996).

62   Jeremy Legget, *Carbon War: Global Warming and the End of the Oil Era* (London: Routledge, 2011).

63   CAN Europe, "About Us," www.caneurope.org/about-us/learn-about-us, accessed January 20, 2018.

64   Andrew Revkin, "Industry Ignored Its Scientists on Climate," *New York Times*, April 23, 2009.

65   George Marshall Institute, "About Us," marshall.org/about, accessed January 23, 2018.

66   *1977 Annual Report*, Rockefeller Foundation.

67   Charles Slichter, *Frederick Seitz 1911–2008* (Washington, DC: National Academy of Sciences, 2010).

68   *1989 Annual Report*, Rockefeller Brothers Fund.

Chapter 7

1   *1989 Annual Report*, Rockefeller Brothers Fund.

2   McNeill, Winsemius, and Yakushiji, *Beyond Interdependence*, 124–125.

3   Daniel Bodansky, "The History of the Global Climate Change Regime," chapter in Urs Luterbacher and Detlef F. Sprinz, *International Relations and Global Climate Change* (Cambridge, MA: MIT Press, 2001).

4   McNeill, Winsemius, and Yakushiji, *Beyond Interdependence.*

5   Peter Winsemius and Ulrich Guntram, *A Thousand Shades of Green: Sustainable Strategies for Competitive Advantage* (London: Earthscan Publications Ltd, 2002).

6   *1991 Annual Report*, Rockefeller Brothers Fund.

7   Alexander King and Bertrand Schneider, *The First Global Revolution: A Report by the Council of The Club of Rome* (New York: Pantheon Books, 1991).

8   McNeill, Winsemius, and Yakushiji, *Beyond Interdependence.*

9   Maurice Strong, "Opening Statement to the Rio Summit (3 June 1992)," www.mauricestrong .net/index.php/opening-statement6.

10   United Nations, *Rio Declaration on Environment and Development*, 1992, www.un.org /documents/ga/conf151/aconf15126–1annex1.htm.

11   The UNFCCC Secretariat was located in Bonn and the convention held annual meetings from 1995.

12   Bill Clinton, *My Life* (New York: Knopf Publishing Group, 2004).

13   Clinton, *My Life*, 453.

14   Clinton, 742.

15   *1993 Annual Report*, Rockefeller Brothers Fund, 10–11.

16   *1993 Annual Report.*

17   *1993 Annual Report.*

18   K von Moltke, *Turning Up The Heat: Next Steps on Climate Change* (pamphlet), Pocantico Conference Center of the Rockefeller Brothers Fund, 1995.

19   *1994 Annual Report*, Rockefeller Brothers Fund.

20 The Club of Budapest, "The Club of Budapest Foundation," www.clubofbudapest.org /clubofbudapest/attachments/article/72/BP-Club_A_2014_nyomda.pdf, accessed January 22, 2018.

21 Earth Charter Initiative, "History of the Earth Charter," www.earthcharter.org/discover /history-of-the-earth-charter, accessed January 23, 2018.

22 Steven Rockefeller, "The Legitimacy of the Earth Charter. A Lecture Presented at Exeter College, Oxford University February 9, 2007," earthcharter.org/virtuallibrary2/images /uploads/Oxford-Legitimacy%20of%20the%20Earth%20Charter%20rev%209%2019%20 2008%20(2).pdf.

23 The Council of European Communities, "Resolution of the Council and the Representatives of the Governments of the Member States, Meeting Within the Council of 1 February 1993 on a Community Programme of Policy and Action in Relation to the Environment and Sustainable Development—a European Community Programme of Policy and Action in Relation to the Environment and Sustainable Development," *Official Journal* C 138, May 17, 1993, 0001–0004.

24 McNeill, Winsemius, and Yakushiji, *Beyond Interdependence*, 123.

25 Earth Charter Initiative, "A Short History of the Earth Charter Initiative," www.earthcharter china.org/eng/history2.html, accessed January 21, 2018.

26 Steven Rockefeller, "Teilhard's Vision and the Earth Charter," chapter 5 in Thierry Meynard, ed., *Teilhard and the Future of Humanity* (New York: Fordham University Press, 2006, https: //earthcharter.org/library/teilhards-vision-and-the-earth-charter/.

27 Earth Charter Initiative, "The Earth Charter," earthcharter.org/discover/the-earth-charter, accessed January 23, 2018.

28 Earth Charter Initiative, "Earth Charter."

29 Ark of Hope, www.arkofhope.org, accessed January 21, 2018.

30 Earth Charter Initiative, "Short History."

31 Steven Rockefeller, "Commentary on Meaning, Religion, and a Great Transition," Great Transition Initiative, December 2014, www.greattransition.org/commentary/steven-rockefeller -meaning-religion-and-a-great-transition-michael-karlberg.

32 White House, "The Vice President Delivers Remarks to the Council of the Americas Conference," The Loy Henderson Room, The Department of State, Washington, DC, May 6, 2002, youtu.be/NTVi5j1uiDA.

33 Trilateral Commission, Program of the 2002 Annual Meeting, trilateral.org/File/151.

34 Project for a New American Century, *Rebuilding America's Defenses*, PNAC, Washington, DC, 2002.

35 Paul Raskin et al., *Great Transition, The Promise and Lure of the Times Ahead* (Boston: Stockholm Environment Institute-Boston, 2002).

36 Tellus Institute, PoleStar Project, 2017, www.polestarproject.org.

37 *1994 Annual Report*, Rockefeller Foundation.

38 Raskin et al., *Great Transition*.

39 Oliver Reiser, *Cosmic Humanism*, 23.

40 Great Transition Initiative (GTI), "Imagine All the People: Advancing a Global Citizens Movement, 2010, www.wideningcircle.org/documents/TWC%20Readings/GTI-Perspectives -Imagine_All_the_People.pdf.

41 The Widening Circle, "About," www.wideningcircle.org/about/index.htm, accessed September 6, 2018.

42 The Widening Circle, "Our Strategy," www.wideningcircle.org/keyIdeas/GCM.htm, accessed September 6, 2018.

43 Goi Peace Foundation, "Our Approach," www.goipeace.or.jp/en/about/approach, accessed January 17, 2017.

44 *1997 Annual Report*, Rockefeller Brothers Fund.

45 Rockefeller Brother Fund, *Sustainable Development*.

46 White House, "President Bush Discusses Global Climate Change," Office of the Press Secretary, June 11, 2001.

47 *1998 Annual Report*, Rockefeller Brothers Fund.

48 Rockefeller Brothers Fund, *Sustainable Development*.

49 *2001 Annual Report*, Rockefeller Brothers Fund.

50 *2001 Annual Report*, Greenpeace.

51 Friends of the Earth, "Climate Change and the Earth Summit," 2002, www.foe.co.uk/sites /default/files/downloads/climate_change_summit.pdf, accessed January 21, 2018.

52 Rockefeller Brothers Fund, *Sustainable Development*.

53 *2004 Annual Review*, Rockefeller Brother Fund.

54 The Climate Group, www.theclimategroup.org, accessed January 21, 2018.

Chapter 8

1 European Commission, "Winning the Battle against Global Climate Change," memo, 2005, ec.europa.eu/commission/presscorner/detail/en/MEMO_05_42.

2 Rockefeller Brothers Fund, *Sustainable Development*.

3 Rockefeller Brothers Fund, "Solutions to Global Warming: A National Conversation We Desperately Need to Have," 2006, www.rbf.org/news/solutions-global-warming-national -conversation-we-desperately-need-have.

4 Rockefeller Brothers Fund, "Solutions."

5 Rockefeller Brothers Fund.

6 Rockefeller Brothers Fund.

7 European Parliament, *Report on Winning the Battle against Global Climate Change*, Committee on the Environment, Anders Wijkman, reporter, Public Health and Food Safety, Brussels, October 20, 2005, tinyurl.com/y2ffnyyn.

8 Malcolm McBain, interview with Sir Crispin Tickell, January 28, 1999, www.chu.cam.ac.uk /media/uploads/files/Tickell.pdf.

9 David Wasdell, *Global Warning*, Meridian Project, 2005, www.meridian.org.uk/_PDFs /GlobalWarning1.pdf.

10 Wasdell, *Global Warning*.

11 G8, "British Prime Minister Tony Blair Reflects on 'Significant Progress' of G8 Summit," July 8, 2005, www.g8.utoronto.ca/summit/2005gleneagles/blair_050708.html.

12 David Wasdell, *Apollo-Gaia Project: The Historical Background*, 2007, www.apollo-gaia.org /A-GProjectDevelopment.pdf.

13 Wasdell, *Apollo-Gaia Project*.

14 David Wasdell, David, *The Feedback Crisis in Climate Change*, Meridian Programme, London, 2005, www.meridian.org.uk/Resources/Global%20Dynamics/Feedback%20Crisis/index.htm.

15 David Wasdell, *Beyond the Tipping Point*, Meridian Project, 2006, www.meridian.org.uk /_PDFs/BeyondTippingPoint.pdf.

16 Wasdell, *Apollo-Gaia Project*.

17 Meridian Program, "About," www.meridian.org.uk/About/Origins/Pro-Origin-frameset. htm?p=1, accessed January 21, 2018.

18 US House of Representatives, *Carbon Dioxide and Climate: The Greenhouse Effect*, Committee of Science and Technology, Hearing (Washington, DC: US Government Printing Office, 1981).

19 Bill Turque, *Inventing Al Gore: A Biography* (Boston: Houghton Mifflin Co., 2000).

20 C Grow, "Dee Snider on PMRC Hearing: 'I Was a Public Enemy,'" *Rolling Stone*, September 18, 2015.

21 Edward Cornish, "The Search for Foresight—How the Futurist Were Born," *The Futurist*, November-December 2007; David Livingstone, *Transhumanism: The History of a Dangerous Idea*, Sabilillah Publications, 2005).

22 Jennifer Cobb Kreisberg, "A Globe, Clothing Itself with a Brain," *Wired*, June 1, 1995.

23 Kosmos Associates Inc., "Global Spirituality and Global Consciousness," *Kosmos Journal for Global Transformation*, 2003, www.kosmosjournal.org/article/global-spirituality-and-global -consciousness.

24 Al Gore, *Earth in the Balance: Ecology and the Human Spirit* (New York: Earthscan, 2000).

25 "Gore Climate Film's Nine 'Errors,'" BBC News, October 11, 2007.

26 Matthew Bishop and Michael Green, *Philanthrocapitalism: How the Rich Can Save the World* (New York: Bloomsbury Press, 2008), 136.

27 Climate Reality Project, "24 Hours of Reality: The Dirty Weather Report," November 14, 2012, www.climaterealityproject.org/press/24-hours-of-reality-the-dirty-weather-report -to-spark-climate-change-call-to-action.

28 Rockefeller Brothers Fund, "Grantees: Rockefeller Philanthropy Advisors," www.rbf.org /grantees/rockefeller-philanthropy-advisors, accessed January 21, 2018.

29 Rockefeller Archive Center, "Rockefeller Philanthropy: A Selected Guide," www.rockarch .org/philanthropy/pdf/RockefellerPhilanthropy.pdf, accessed January 21, 2018.

30 Bishop and Green, *Philanthrocapitalism*, 233.

31 Environmental Grantmakers Association, "Members," ega.org/about/members, accessed January 22, 2018; Center for the Environment, "Leadership Fellowship Openings: Environmental Grantmakers Association, June 1–August 31, 2006, New York City," www .mtholyoke.edu/proj/cel/fellowship/envirograntmakers.shtml.

32 United States Senate, Committee on Environment and Public Works, Minority Staff Report: *"How a Club of Billionaires and Their Foundations Control the Environmental Movement and Obama's EPA,"* July 30, 2014.

33 Global Philanthropy Forum, "Partners," philanthropyforum.org/community/members -partners, accessed November 14, 2018.

34 Council on Foundations, History of the Council on Foundations, 2016, www.cof.org/sites / default/files/documents/files/History-Council-on-Foundations.pdf.

35 Amir Pasic, *Foundations in Security: An Overview of Foundation Visions, Programs, and Grantees*, Project on World Security, Rockefeller Brothers Fund, 1999, 7.

36 *Bulletin of the Atomic Scientists*, "Timeline," thebulletin.org/timeline, accessed January 21, 2018.

37 California Environmental Associates, *Design to Win: Philanthropy's Role in the Fight against Global Warming* (San Francisco: California Environmental Associates, 2007), www.climate works.org/wp-content/uploads/2015/02/design_to_win_final_8_31_07.pdf.

38 European Climate Foundation, "Funders," europeanclimate.org/people/funders, accessed January 21, 2018.

39 Die Bundesregierung, "Heiligendamm Process," press release, The Press and Information Office of the Federal German Government, 2008, www.g-8.de/nn_92160/Content/EN /Article/__g8-summit/2007–06-08-heiligendamm-prozess__en.html.

40 *2007 Annual Review*, Rockefeller Foundation.

41 *2008 Annual Review*, Rockefeller Brothers Fund.

Chapter 9

1   Henry Kissinger, "The Chance for a New World Order," *New York Times*, January 12, 2009.

2   Clifford Krauss, "Rockefellers Seek Change at Exxon," *New York Times*, May 27, 2008.

3   Krauss. "Rockefellers."

4   Richard Hylton, "Rockefeller Family Tries to Keep a Vast Fortune from Dissipation," *New York Times*, February 16, 1992.

5   Rockefeller Brothers Fund, *Sustainable Development*.

6   Rockefeller Brothers Fund, "Grantees: 350.org," www.rbf.org, accessed January 22, 2018.

7   350.org, "History," 350.org/about, accessed September 19, 2019.

8   Bill McKibben, *The End of Nature* (New York: Anchor/Random House, 1989).

9   Vivian Krause, "Rockefellers Behind 'Scruffy Little Outfit,'" *Financial Post*, February 14, 2013.

10 Krause, "Rockefellers."

11 Board members of TechRocks included Peter O'Neill, Wendy Rockefeller, Stuart Rockefeller, chairman Richard Rockefeller, and Lee Wasserman (president of theRockefeller Family Fund).

12 Barack Obama, "Recorded Remarks to Global Climate Summit," November 18, 2008, www .americanrhetoric.com/speeches/barackobama/barackobamaglobalclimatesummit.htm.

13 Rockefeller Brothers Fund, "Grantees: Presidential Climate Action Project," www.rbf.org /grantees/natural-capitalism-solutions-inc, accessed January 21, 2018.

14 Funding also came from the Denver Foundation, Aspen Community Foundation, and Prentice Foundation.

15 Henry Kissinger, "Chance."

16 Paul Harris, "They're Called the Good Club—and They Want to Save the World," *The Guardian*, May 31, 2009.

17 United Nations, *A More Secure World: Our Shared Responsibility*, Report of the High-Level Panel on Threats, Challenges, and Change, 2004, www.un.org/en/peacebuilding/pdf /historical/hlp_more_secure_world.pdf.

18 International Task Force on Global Public Goods, *Meeting Global Challenges: International Cooperation in the National Interest, Final Report*, Stockholm, Sweden, 2006, keionline.org /misc-docs/socialgoods/International-Task-Force-on-Global-Public-Goods_2006.pdf.

19 Brookings Institution, "Managing Global Insecurity," 2008, www.brookings.edu/author /managing-global-insecurity, accessed August 29, 2019.

20 OECD Insights, "The Rise of the G20 and OECD's Role," November 17, 2015, oecdinsights .org/2015/11/17/the-rise-of-the-g20-and-oecds-role.

21 European Union, "Acceptance Speech by Herman Van Rompuy Following His Nomination as First Permanent President of the European Council," November 19, 2009, www.consilium .europa.eu/media/25842/141246.pdf.

22   United Nations, "Speech to UN General Assembly by Prime Minister, Mr. Gordon Brown, 23 September 2009," gadebate.un.org/sites/default/files/gastatements/64/64_GB_en.pdf.

23   United Nations, "Speech."

24   The UN Non-Governmental Liaison Service, "International Day of Climate Action," www .un-ngls.org/index.php/un-ngls_news_archives/2009/981-international-day-of-climate -action, accessed January 21, 2018.

25   United Nations Framework Convention on Climate Change (UNFCCC), "Copenhagen Accord," December 18, 2009, unfccc.int/resource/docs/2009/cop15/eng/l07.pdf.

26   Greenpeace, "Greenpeace Mourns Copenhagen Failure with 100 Crosses," press release, 2009, www.greenpeace.org/eastasia/press/releases/climate-energy/2009/copenhagen-cross.

27   *2009 Annual Review*, Rockefeller Brothers Fund.

28   Christopher Booker, "Climate Change: This Is the Worst Scientific Scandal of Our Generation," *The Telegraph*, November 28, 2009.

29   "Shell Boss 'Fears for the Planet,'" *The Guardian*, June 17, 2004; Andrew Orlowski, "Oops: Chief Climategate Investigator Failed to Declare Eco Directorship 'Dracula's in Charge of the Blood Bank,'" *The Register*, March 24, 2010.

30   Boris Petrovic, "Noosphere—Global Mind—Ascension," May 20, 2009, theosophy.net/profiles /blogs/noosphere-global-mind.

31   Noosphere Forum, "Mission Statement," www.noosphereforum.org/main.html, accessed January 20, 2018.

32   Foundation for the Law and Time, Reviews of *Time, Synchronicity & Calendar Change*, 2011, www.lawoftime.org/time-sync/reviews.htm.

33   Foundation for Conscious Evolution, Global Communication Hub, barbaramarxhubbard. com/global-communication-hub, accessed January 22, 2018.

34   *Global Brain*/Awakening Earth, directed by Chris Hall, based on the book by Peter Russell, 1982, youtu.be/s1fvEwzUovI.

35   Foundation for the Law and Time, "Theory and History of the Noosphere," www.lawoftime. org/noosphere/theoryandhistory.html, accessed January 22, 2018.

36   Noosphere Forum, "Mission Statement."

37   Earth Portals, "Planetary Pilgrims, José & Lloydine Argüelles," 2001, www.earthportals.com /Portal_Messenger/arguelles.html.

38   Steve Beckow, "What Was the Harmonic Convergence?" Prepare for Change, 2017, prepare forchange.net/2017/08/20/what-was-the-harmonic-convergence.

39   Noosphere Forum, "Resources, Networking and Partnerships," www.noosphereforum.org /collective/index.html, accessed January 22, 2018.

40   Foundation for the Law and Time, "Brief Biography of José Argüelles/Valum Votan," www .lawoftime.org/jose-arguelles-valum-votan.html?content=249, accessed January 22, 2018.

41   Oliver Reiser, *The World Sensorium: The Social Embryology of World Federation* (Whitefish, MT: Kessinger Publishing, LLC, 1946).

42   Foundation for the Law and Time, "Arcturus Remembered: A Crystal Earth Network Projection," www.lawoftime.org/timeshipearth/arcturus.html, accessed January 22, 2018.

43   Foundation for the Law and Time, *Rinri Project Newsletter III: Mystery of the Stone Edition*, www.lawoftime.org/pdfs/Rinri-III-3.1.pdf, accessed January 22, 2018.

44   Foundation for the Law and Time, "Holomind Perceiver: What It Is and How to Use It," *Intergalactic Bulletin* #2, lawoftime.org/lawoftime/synchronotron-holomind-perceiver.html, accessed January 22, 2018.

45   United Nations, "Address by Miguel d'Escoto Brockmann, President of the General Assembly, upon Adoption of the Outcome Document of the Conference on the World Financial and Economic Crisis and Its Impact on Development," June 26, 2009, New York, www.un.org/en /ga/president/63/pdf/statements/20090626-eccrisis-outcomedoc.pdf.

46   Barbara Marx Hubbard, *The Revelation: Our Crisis is a Birth (The Book of Co-Creation)* (Sonoma, CA: Foundation for Conscious Evolution, 1993), 147.

47   Barbara Marx Hubbard, *The Book of Co-Creation: An Evolutionary Interpretation of the New Testament* (New Visions, 1980).

48   Neale Donald Walsch, *The Mother of Invention: The Legacy of Barbara Marx Hubbard and the Future of YOU* (New York: Hay House, 2011).

49   Barbara Marx Hubbard, *Conscious Evolution: Awakening the Power of Our Social Potential* (Novato, CA: New World Library, 2010), viii.

50   Foundation for Conscious Evolution, "About Barbara Marx Hubbard," web.archive.org /web/20160129090833/http://barbaramarxhubbard.com/barbara-marx-hubbard, accessed January 29, 2016.

51   Kiersten Marek, "Neva Rockefeller Goodwin and the Role of the Activist Investor in Steering Social Change," *Inside Philanthropy*, April 14, 2016, insidephilanthropy.com/home /2016/4/15/neva-rockefeller-goodwin-and-the-role-of-the-activist-invest.html.

52   H. G. Wells, *The Open Conspiracy: Blue Prints for a World Revolution* (London: Victor Gollancz Ltd., 1928).

53   H. G. Wells, *World Brain* (London: Methuen Publishing Ltd., 1938).

54   Willis Harman and O. W. Markley, eds., *Changing Images of Man* (Pergamon Press 1982).

55   Club of Budapest, "Club of Budapest Foundation," www.clubofbudapest.org/clubofbudapest /attachments/article/72/BP-Club_A_2014_nyomda.pdf.

56   Club of Budapest, "The Manifesto on Planetary Consciousness," www.clubofbudapest.org /clubofbudapest/index.php/en/about-us/the-manifesto-on-planetary-consciousness, accessed January 22, 2018.

57   Goi Peace Foundation, "Our Approach," www.goipeace.or.jp/en/about/approach, acessed October 18, 2018.

58   "Obituaries: 'José Argüelles'," *The Telegraph*, April 5, 2011.

59   Robert Constanza and Ida Kubiszewski, eds., *Creating a Sustainable and Desirable Future: Insights from 45 Global Thought Leaders* (Singapore: World Scientific Publishing, 2014).

60   Green Cross International, "Climate Change Taskforce Issues Urgent Appeal for Rio+20 to Take Global Warming Threat Seriously," 2011, www.gcint.org/climate-change-task-force -issues-urgent-appeal-for-rio20-conference-to-take-global-warming-threat-seriously.

61   Planet under Pressure Conference, "State of the Planet Declaration," organized by IGBP, Diversitas, IHDP, WCRP, ICSU, London, March 26–29, 2012.

62   Anders Wijkman and Johan Rockström, *Bankrupting Nature: Denying Our Planetary Boundaries* (UK: Routledge, 2012).

63   F Biermann et al., "Navigating the Anthropocene: Improving Earth System Governance," *Science* 335, no. 6074 (March 16, 2012): 1306–1307.

64   Biermann et al., "Navigating."

65   John Kirton, "The G20 System Still Works: Better Than Ever," *Caribbean Journal of International Relations & Diplomacy* 2, no. 3 (September 2014): 43–60.

66   Globe International, "300 Legislators, 38 Presidents of Congress to Participate in 1st World Summit of Legislators," May 23, 2013, globelegislators.org/news/item/300-legislators-to -participate-in-1st-world-summit-of-legislators.

67   Greenpeace, "From Hope to Despair," 2002, m.greenpeace.org/international/en/high/news /Blogs/makingwaves/from-hope-to-despair/blog/41051.

68   United Nations, "UN Secretary-General's Remarks at the United Nations Office at Geneva Library on the 85th Anniversary of the Donation by John D. Rockefeller to Endow the League of Nations Library," September 10, 2012, www.un.org/sg/en/content/sg/statement /2012–09-10/ un-secretary-generals-remarks-united-nations-office-geneva-library.

69   Alistair Osborne, "Rothschild and Rockefeller Families Team Up for Some Extra Wealth Creation," *The Telegraph*, January 23, 2012.

70   Rockefeller Financial, "RIT Capital Partners plc ('RIT'), Chaired by Lord (Jacob) Rothschild and Rockefeller & Co. Announce Strategic Partnership," press release, May 30, 2012, www .ritcap.com/sites/default/files/RIT_-_Rockefeller_Press_Release_FINAL_5–29-12.pdf.

Chapter 10

1   Charles Keidan, "Interview with Stephen Heintz, CEO of the Rockefeller Brothers Fund," Alliance, May 3, 2016.

2   Stockholm Resilience Center, "New Rockefeller Programme on Social Innovation," May 15, 2013, www.stockholmresilience.org/research/research-news/2013–05-15-new-rockefeller -programme-on-social-innovation.html.

3   Stockholm Resilience Center, "Stockholm MISTRA Institute on Sustainable Governance and Management of Social-Ecological System: Proposal to Mistra on New Inter-disciplinary Research Center on Sustainable Governance and Management of Social-Ecological Systems," Stockholms Universitet, March 31, 2006, www.stockholmresilience.org/download18. aeea46911a31274279800035608.

4   C. S. Holling, "Resilience and Stability of Ecological Systems," *Annual Review of Ecology and Systematics* 4 (November 1973):1–23.

5   Lance Gunderson and C. S. Holling, *Panarchy: Understanding Transformations in Human and Natural Systems* (Washington, DC: Island Press, 2001).

6   Joseph Tainter, *The Collapse of Complex Societies* (Cambridge, UK: Cambridge University Press, 1998).

7   Thomas Heinze and Richard Münch, *Innovation in Science and Organizational Renewal: Historical and Sociological Perspectives* (Berlin: Springer, 2016).

8   100 Resilient Cities, "About Us," www.100resilientcities.org/about–us#, accessed January 23, 2018.

9   László Szombatfalvy, *Vår tids största utmaningar* (Stockholm: Ekerlids, 2009).

10   The World Bank, "New Report Examines Risks of 4 Degree Hotter World by End of Century." press release, November 18, 2012, www.worldbank.org/en/news/press-release/2012/11/18 /new-report-examines-risks-of-degree-hotter-world-by-end-of-century.

11   H. J. Schellnhuber and J. Kropp, "Geocybernetics: Controlling a Complex Dynamical System under Uncertainty," *Naturwissenschaften* 85, no. 9 (September 1998): 411–425.

12   Mikhail Gorbachev, "What Role for the G20?," *New York Times*, April 27, 2009.

13   G20, "G20 Leaders' Declaration," September 6, 2013, St. Petersburg, www.g20.utoronto .ca/2013/Saint_Petersburg_Declaration_ENG.pdf.

14 United Nations, "Open Working Group on Sustainable Development Goals," 2013, sustainable development.un.org/owg.html.

15 United Nations Sustainable Development Solutions Network, *An Action Agenda for Sustainable Development Report for the UN Secretary-General*, May 5, 2014.

16 United Nations, Summary of the first meeting of the high-level political forum on sustainable development, Note by the President of the General Assembly, November 13, 2013, www .un.org/ga/search/view_doc.asp?symbol=A/68/588&Lang=E.

17 United Nations, "TST Issue Brief: Global Governance," 2013, sustainabledevelopment. un.org/content/documents/2429TST%20Issues%20Brief_Global%20Governance_FINAL .pdf.

18 Globe International, "The 1st Climate Legislation Summit," 2013, globelegislators.org/events /2013/1gcls-home.

19 Globe International, "The 2nd Climate Legislation Summit," 2014, globelegislators.org/events /2014/2gcls-home.

20 Virginia Wiseman, "GLOBE World Summit of Legislators Resolution Urges Climate Action," IISD, June 11, 2014, climate-l.iisd.org/news/globe-world-summit-of-legislators -resolution-urges-climate-action.

21 Globe International, "2nd World Summit of Legislators (WSL2014) Resolution," 2014, globe legislators.org/news/item/world-summit-of-legislators-resolution.

22 United Nations Secretary General, "Secretary-General's message to Globe International Second World Summit of Legislators (Delivered by Mr. Tomas Anker Christensen, Senior Partnerships Advisor, Executive Office of the Secretary-General)," June 6, 2014, www.un.org/sg/en/content/sg/statement/2014–06-06/secretary-generals-message-globe -international-second-world-summit.

23 *2014 Annual Review*, Rockefeller Brothers Fund.

24 Rockefeller Brothers Fund, "Divestment Statement," 2017, www.rbf.org/sites/default/files /rbf-divestment_statement-2017-oct.pdf.

25 John C. Topping Jr., "Building on the Combined Momentum of the U.S.–China Climate Accord and the Emerging U.S. ANSI LCA Standard," *Climate Alert* 26, no. 4 (Winter 2015): 2–3.

26 Judith Rodin, President, Rockefeller Foundation, "Addressing the UN Climate Summit," September 24, 2014, www.rockefellerfoundation.org/blog/addressing-un-climate-summit.

27 United States Senate, *Club of Billionaires*, www.wrongkindofgreen.org/wp-content/uploads /2015/10/chainofenvironmentalcommand.pdf.

28 The Center for Climate Strategies, "The Center for Climate Strategies Applauds U.S. China Climate Agreement, Low Carbon Development Progress and Opportunities," November 12, 2014, www.climatestrategies.us/articles/articles/view/97.

29 Sam Roberts, "Why Are Rockefellers Moving from 30 Rock? 'We Got a Deal," *New York Times*, November 23, 2014.

30 United Nations, "Concerts to Encourage Climate Action Announced at Davos," 2015, www .un.org/climatechange/blog/2015/01/concerts-encourage-climate-action-announced-davos.

31 Global Challenges Foundation, "Earth Statement," www.globalchallenges.org/our-work /earth-statement-2015, accessed January 23, 2018.

32 Rockefeller Brothers Fund, *Sustainable Development*.

33 The Vatican, "Encyclical Letter Laudato Si' of the Holy Father Francis on Care for Our Common Home," May 24, 2015, w2.vatican.va/content/francesco/en/encyclicals/documents /papa-francesco_20150524_enciclica-laudato-si.html.

34 The Pontifical Academy of Sciences, "Sustainable Humanity Sustainable Nature Our Responsibility," Proceedings of the Joint Workshop, 2–6 May 2014, Extra Series 41, Vatican City, 2015.

35 Climate Action Network, "Islamic Climate Declaration Calls for Fossil Fuel Phase Out," press release, August 18, 2015, www.climatenetwork.org/press-release/islamic-climate-declaration -calls-fossil-fuel-phase-out.

36 United Nations Climate Change Newsroom, "Six Oil Majors Say: We Will Act Faster with Stronger Carbon Pricing, Open Letter to UN and Governments," June 1, 2015, web.archive.org/web/20180330124154/http://newsroom.unfccc.int/unfccc-newsroom /major-oil-companies-letter-to-n.

37 GLOBE International, "Towards Coherence & Impact: The Challenge of Paris and the 2030 Agenda for a Prosperous and Sustainable World," communique, GLOBE COP21 Legislators Summit, National Assembly, 4–5 December 2015, Paris, globelegislators.org/images/PDF /EN_COP21_Summit_Communique.pdf.

38 Anders Wijkman and Kristian Skånberg, *The Circular Economy and the Benefits for Society*, A Study Report at the Request of the Club of Rome with Support from the MAVA Foundation, 2016, clubofrome.org/wp-content/uploads/2016/03/The-Circular-Economy-and-Benefits-for -Society.pdf.

39 Paul McCartney, "Love Song to the Earth," YouTube, September 15, 2015.

40 "'Love Song to the Earth,': A Song with the Power to Fight Climate Change and Maybe Even Change the World," lovesongtotheearth.org, accessed January 23, 2018.

41 United Nations General Assembly, "Resolution Adopted by the General Assembly on 25 September 2015," October 21, 2015, www.un.org/ga/search/view_doc.asp?symbol=A/RES /70/1&Lang=E.

42 Andrea Zubialde, "Comparison between the New Sustainable Development Goals and the Ethical Principles of the Earth Charter," *Bien Commun & Charte de la Terre*, blog, June 29, 2015, biencommunchartedelaterre.wordpress.com/2015/06/29/comparison-between -the-new-sustainable-development-goals-and-the-ethical-principles-of-the-earth-charter.

43 UN News, "INTERVIEW: World's Most Difficult Task—Ensuring UN Sustainable Development Agenda, Says Top Adviser," January 27, 2016, news.un.org/en/story/2016/01/521002-interview -worlds-most-difficult-task-ensuring-un-sustainable-development-agenda.

44 UN News Center, "At G20 Summit, Ban Says Response to Terrorism 'Needs to Be Robust, Always within Rule of Law,'" November 15, 2015, www.un.org/apps/news/story.asp?NewsID =52561#.V7x4APmLSUk.

45 European Council, "G20 Leaders' Communiqué, Antalya Summit, 15–16 November 2015," press release, www.consilium.europa.eu/en/press/press-releases/2015/11/16/ g20-summit -Antalya-communique.

46 Elle Griffiths, "Paris Turned into a Sea of Shoes as Climate Change Campaigners Lay 20,000 Pairs in Symbolic Street Protest," *The Mirror*, November 29, 2015.

47 Bjørn Lomborg, *Cool It: The Skeptical Environmentalist's Guide to Global Warming* (New York: Knopf Publishing Group, 2007).

48 Peter Buchert, "Forskare sågar klimatavtalet," *Hufvudstadsbladet*, December 11, 2015.

49   Janice Sinclaire, "Doomsday Clock Hands Remain Unchanged, Despite Iran Deal and Paris Talks," *The Bulletin of the Atomic Scientists*, January 26, 2016.

50   Charles Keidan, "Stephen Heintz."

51   Rockefeller Brothers Fund (2015), Stephen Heintz Praises the Agreement Reached at the Paris Climate Conference, www.rbf.org/news/stephen-heintz-praises-agreement-reached-paris-climate-conference (retrieved 23 Jan 2018).

52   Charles Keidan, "Stephen Heintz."

53   White House, "Statement by the President on the Paris Climate Agreement," December 12, 2015, www.whitehouse.gov/the-press-office/2015/12/12/statement-president-paris-climate-agreement.

54   UNRIC, "Figueres: First Time the World Economy Is Transformed Intentionally," February 3, 2015, www.unric.org/en/latest-un-buzz/29623-figueres-first-time-the-world-economy-is-transformed-intentionally.

55   United Nations, "UN Partners with the Rockefeller Foundation to Showcase Women's Role in Addressing Climate Change," press release, September 26, 2012, unfccc.int/files/secretariat/momentum_for_change/application/pdf/pr20120926_mfc_women_announce.pdf.

56   Winrock International (2015), "Former Winrock Board Member: 'The Woman Who Could Stop Climate Change,'" winrock.org/former-winrock-board-member-the-woman-who-could-stop-climate-change.

Chapter 11

1   Klaus Schwab, "The Fourth Industrial Revolution: What It Means, How to Respond," Foreign Affairs, December 12, 2015 and World Economic Forum, January 14, 2016, www.weforum.org/agenda/2016/01/the-fourth-industrial-revolution-what-it-means-and-how-to-respond.

2   Johan Rockström, "Can This Revolution Save Our Warming Planet?" World Economic Forum, January 20, 2016, www.weforum.org/agenda/2016/01/revolution-warming-planet.

3   G20, "G20 New Industrial Revolution Action Plan," www.mofa.go.jp/files/000185873.pdf.

4   Johan Rockström, "How We Can Direct the Fourth Industrial Revolution towards a Zero Carbon Future—If We Act Now," World Economic Forum, January 19, 2018, www.weforum.org/agenda/2018/01/how-we-can-direct-the-fourth-industrial-revolution-towards-a-zero-carbon-future.

5   Klaus Schwab, *Shaping the Fourth Industrial Revolution* (Geneva: World Economic Forum, 2018).

6   Rockefeller Foundation, Innovative Frontiers of Development Conference, 5–9 June 2019, www.rockefellerfoundation.org/bellagio60/innovative-frontiers-development.

7   Rachel Keeton, "When Smart Cities Are Stupid," International New Town Institute, 2015.

8   Philippe Mesmer, "Songdo, Ghetto for the Affluent," *Le Monde*, May 29, 2017.

9   Chris White, "South Korea's 'Smart City' Songdo: Not Quite Smart Enough?" *South China Morning Post*, March 25, 2018.

10   Anisha Dutta, "20 Smart Cities May Be Ready Only by 2021," *Hindustan Times*, November 12, 2018.

11   "China Has Highest Number of Smart City Pilot Projects: Report," *The Economic Times*, February 20, 2018.

12   Grand View Research, "Smart Cities Market Size, Share & Trends Analysis Report by Application (Education, Governance, Buildings, Mobility, Healthcare, Utilities), by Component (Services, Solutions), and Segment Forecasts 2018–2025," February 2018.

13   The Internet of Things Council, "Europe's IoT," Pan European Network, www.theinternet ofthings.eu/sites/default/files/GOV18%20R%20van%20Kranenburg%206007_ATL.pdf, accessed August 14, 2018.

14   Douglas Broom, "The EU Wants to Create 10 Million Smart Lampposts," World Economic Forum, June 19, 2019, www.weforum.org/agenda/2019/06/the-eu-wants-to-create-10-million -smart-lampposts.

15   Chris Baraniuk, "Exclusive: UK Police Wants AI to Stop Violent Crime before It Happens," *New Scientist*, November 26, 2018.

16   Jon Markman, "Facial Recognition: A Force For Good . . . Or Government?" *Forbes*, September 27, 2019.

17   World Economic Forum, *Shaping the Future of Construction: A Breakthrough in Mindset and Technology*, WEF in collaboration with The Boston Consulting Group, May 2016.

18   Jethro Mullen, "Bitcoin Could 'Bring the Internet to a Halt,' Banking Group Warns," *CNN Business*, June 18, 2018.

19   Schwab, *Shaping the Fourth Industrial Revolution*, 123.

20   Rockefeller Foundation, "Meet Africa's Inclusive AI Community Solving Global Problems," blog post, August 26, 2019, www.rockefellerfoundation.org/blog/meet-africas -inclusive-ai- community-solving-global-problems.

21   J. Nordangård, "Med brödfödan som drivkraft: En studie om att byta olja mot biodrivmedel i ett globalt perspektiv," dissertation, Department of Geography, Linköping University, 2007, urn.kb.se/resolve?urn=urn:nbn:se:liu:diva-8116.

22   J. Nordangård, *ORDO AB CHAO: Den politiska historien om biodrivmedel i den Europeiska Unionen—Aktörer, nätverk och strategier*, PhD dissertation, Linköping University, 2012, urn. kb.se/resolve?urn=urn:nbn:se:liu:diva-85821.

23   Pandora's Promise, "James Hansen on Nuclear Power," July 23, 2013, youtu.be/CZEx WtXAZ7M.

24   *Global Risks Report* 2019, World Economic Forum, 66.

25   Schellnhuber and Kropp, "Geocybernetics," 411–425.

26   Rockefeller Panel Reports, *Prospect for America* (Garden City, NY: Doubleday, 1961).

27   Maurice Strong, "Remarks by Maurice Strong at Dinner Meeting of Energy Ministers of OECD Countries at Aarhus, Denmark, Energy and the Environment," June 15, 1996, www .mauricestrong.net/index.php/speeches-remarks3/52-aarhus.

28   Marianne Fay et al., *Decarbonizing Development: Three Steps to a Zero-Carbon Future* (Washington, DC: World Bank Group, 2015).

29   William W. Kellogg and Margaret Mead, eds., *The Atmosphere: Endangered and Endangering*, Fogarty International Center Proceedings No. 39 (Washington, DC: US Government Printing Office, 1977).

30   Matt Taibbi, "The Great American Bubble Machine," *Rolling Stone*, April 5, 2010.

31   Technocracy Inc., Transition Plan 2016, www.technocracyinc.org/student-intern-video-2016.

32   Effekt Magasin, Effekt i riksdagen, May 20, 2010.

33   Shaun Chamberlin, Larch Maxey, and Victoria Hurth, "Reconciling Scientific Reality with Realpolitik: Moving Beyond Carbon Pricing to TEQs—An Integrated, Economy-wide Emissions Cap," *Carbon Management* 5, no. 4 (2014): 411–427, DOI: 10.1080 /17583004.2015.1021563.

34   Fleming Policy Centre, "TEQs in summary," www.flemingpolicycentre.org.uk/teqs.

35   Ibid.

36   Madeleine, "New Blockchain-Based Carbon Currency Aims to Make Carbon Pricing Mainstream," *BusinessGreen*, September 19, 2017, https://www.businessgreen.com/news /3017564/new-blockchain-based-carbon-currency-aims-to-make-carbon-pricing-mainstream.

37   Swedish Smartgrid, www.swedishsmartgrid.se.

38   Greta Thunberg, "'Our House Is on Fire': Greta Thunberg, 16, Urges Leaders to Act on Climate," *The Guardian*, January 25, 2019.

39   SEI, "Climate Calculator Tests Individual Impacts," 2017, updated February 1, 2019, www .sei.org/projects-and-tools/tools/climate-calculator.

40   ICA, "Nytt verktyg hjälper kunderna att minska sin klimatpåverkan," press release, April 16, 2018, icagruppen.se.

41   MyClimate.org, www.myclimate.org.

42   SAS, "Emission Calculator and Carbon Offset," www.sasgroup.net/en/emission-calculator -and-carbon-offset, accessed March 10, 2019.

43   Poseidon Foundation, poseidon.eco.

44   Lisa Song, "An Even More Inconvenient Truth: Why Carbon Credits or Forest Preservation May Be Worse Than Nothing," *ProPublica*, May 22, 2019.

45   Ahmed Afeez, "World Bank and UN Carbon Offset Scheme 'Complicit' in Genocidal Land Grabs—NGOs," *The Guardian*, July 3, 2014.

46   World Economic Forum, "Platform for Accelerating the Circular Economy," www.weforum .org/projects/circular-economy, accessed January 23, 2018.

47   Dame Ellen MacArthur and Dominic Waughray, *Intelligent Assets: Unlocking the Circular Economy Potential*, Ellen MacArthur Foundation, 2016.

48   Barbara Ward, *Spaceship Earth* (New York: Columbia University Press, 1966).

49   Kenneth E. Boulding, "The Economics of the Coming Spaceship Earth," in Henry Jarrett, ed., *Environmental Quality in a Growing Economy: Essays from the Sixth RFF Forum* (Baltimore: Johns Hopkins University Press, 1966): 3–14.

50   Buckminster Fuller, *Operating Manual for Spaceship Earth* (Carbondale, IL: Southern Illinois University Press, 1969).

51   Wijkman and Skånberg, *Circular Economy*.

52   Maria Iliana Such, "NOBLE in the 21st Century: Susan Rockefeller," *Huffington Post*, March 14, 2016.

53   Tom Slee, *Whats Yours Is Mine: Against the Sharing Economy* (Berkeley: OR Books, 2017).

54   Ida Auken, "Welcome to 2030. I Own Nothing, Have No Privacy, and Life Has Never Been Better," World Economic Forum, 2016, www.weforum.org/agenda/2016/11/ shopping -i-can-t-really-remember-what-that-is.

55   Patrick M. Wood, *Technocracy Rising*.

56   Zbigniew Brzezinski, *Between Two Ages*.

57   Beverly Burris *Technocracy at Work* (Albany, NY: SUNY Press, 1993); Deborah Harkness, *John Dee's Conversation with Angels* (Cambridge, UK: Cambridge University Press, 1999).

58   The Venus Project, www.thevenusproject.com; *Zeitgeist: Addendum*, directed by Peter Joseph, 2008, GMP LLC, film, www.zeitgeistmovie.com.

59   Mathew Carney, "Leave No Dark Corner" ABC News, Australia, September 17, 2018.

60   City of Stockholm, "Welcome to the World's Smartest City 2040," April 4, 2018, international.stockholm.se/governance/smart-and-connected-city/welcome-to-the-worlds -smartest-city-2040.

61 Viktor Krylmark, "Rysk ansiktsigenkänning leder till fler gripanden," *NyTeknik*, September 29, 2017.

62 Trevor Timm, "The Government Just Admitted It Will Use Smart Home Devices for Spying," *The Guardian*, February 9, 2016.

63 Sarah Olsson, "Digital övervakning ska testas på förskolebarn," *Norrköpings Tidningar*, August 5, 2019.

64 Olsson, "Digitala armbandsprojektet läggs ned,"*Norrköpings Tidningar*, August 26, 2019.

65 *2016 Annual Review*, Rockefeller Brothers Fund.

66 Trilateral Commission, *Democracies under Stress: Recreating the Trilateral Commission to Revitalize Our Democracies to Uphold the Rules-Based International Order*, brochure, Summer 2019.

67 Columbia Journalism Review, "CJR Event on Covering Climate Change," April 30, 2019, www.cjr.org/watchdog/livestream-covering-climate-change.php.

68 D Helbing, "A New, Global Fascism, Based on Mass Surveillance Is on the Rise," *FuturICT*, September 21, 2017, futurict.blogspot.com/2017/09/a-new-global-fascism-based-on-mass .html.

69 *2016 Annual Review*, Rockefeller Brothers Fund.

70 Oliver Woeffrey, "Could These 3 Ideas Reshape Governance?" World Economic Forum, February 26, 2016, www.weforum.org/agenda/2016/02/3-ideas-to-revive-global-governance.

71 Dennis Pamlin and Stuart Armstrong, *Twelve Risks That Threaten Human Civilisation: Executive Summary,* Global Challenges Foundation, January 2, 2015, g20ys.org/upload/auto /4eb9ce315f67d9d94aebf3d90522d5ee3d67a8bf.pdf.

72 Globe International, "Why GLOBE Catastrophic Risks Need Greater Awareness and Legislative Action," 2016, globelegislators.org/news/item/why-global-catastrophic-risks-need -greater- awareness-legislative-action.

73 Leo Hickman, "James Lovelock on the Value of Sceptics and Why Copenhagen Was Doomed," *The Guardian*, March 29, 2010.

74 Andreas Bummel and Jo Leinen, *A World Parliament: Governance and Democracy in the 21st Century* (Berlin: Democracy Without Borders, 2018), www.democracywithoutborders.org /se/world-parliament-book-2/.

75 Torbjörn Tännsjö, "Så kan klimatkrisen leda fram till en global despoti," *Dagens Nyheter*, December 5, 2018.

76 Anders Bolling, "Johan Rockström är miljörörelsens egen Piketty," *Dagens Nyheter*, September 4, 2015.

77 Reiser, *World Sensorium*.

78 Nick Bostrom, "What Is a Singleton?" *Linguistic and Philosophical Investigations* 5, no. 2 (2006): 48–54.

79 G20, "G20 Engagement Groups," www.g20.org/en/engagementgroups, accessed September 28, 2019.

80 Rockefeller Brothers Fund, *Sustainable Development*.

81 Rockefeller Foundation, "Bellagio 60, Innovative Frontiers of Development, Upgrading the System—Rethinking Capitalism, Digital State, Strengthening Multilateralism," June 5–9, 2019, rockefellerfoundation.org/bellagio60/innovative-frontiers-development.

Chapter 12

1 Russ Volckmann, "An Interview with Barbara Marx Hubbard," *Integral Review* 5, no. 1 (June 2009).

2 D. K. Matai, "Preparing for The Super Convergence: Rise of the Bio-Info-Nano Singularity," *Business Insider*, April 22, 2011.

3 Paul Montgomery, "Business People; A Banking Star Moves to Société Générale," *New York Times*, June 23, 1988.

4 D. K. Matai, "Synopsis of the Philanthropia: The Philanthropia—Trinity Club, Syndicates and Ethical Investment Funds," 2006, www.mi2g.com/cgi/mi2g/press/philanthropia_trinity .pdf.

5 BBC News, "Earth 'Entering New Extinction Phase'—US Study," June 20, 2015.

6 Gretchen C. Ehrlich, Anne H Ehrlich, and Paul R. Ehrlich, "Optimum Human Population Size," *Population and Environment* 15, no. 6 (July 1994): 469, doi:10.1007/BF02211719.

7 Schellnhuber and Kropp, "Geocybernetics," 411–425.

8 Michael Bastasch, "Pro-Lifers: Pope Rejects Population Control, Abortion as Solutions to Global Warming," *Daily Caller*, June 17, 2015.

9 Edward Pentin, "German Climatologist Refutes Claims He Promotes Population Control," *National Catholic Register*, June 19, 2015.

10 Global Challenges Foundation, "About Us," globalchallenges.org/en/about/about-us, accessed January 23, 2018.

11 C-SPAN, "Annual Ambassadors' Dinner," video, September 14, 1994, www.c-span.org /video/?60201-1/annual-ambassadors-dinner.

12 AGRA sponsors include MasterCard, Norad, SIDA, and UNEP; AGRA, "Funding Partners," agra.org/funding-partners, accessed May 11, 2019.

13 Anuradha Mittal and Melissa Moore, *African Farmers and Environmentalists Speak Out Against a New Green Revolution in Africa* (Oakland, CA: The Oakland Institute, 2009).

14 Johan Rockström, "Bounding the Planetary Future: Why We Need a Great Transition," Great Transition Initiative, essay, 2015, www.greattransition.org/publication/bounding -the-planetary-future-why-we-need-a-great-transition.

15 Jonathan A. Foley et al., "Solutions for a Cultivated Planet," *Nature* 478 (October 20, 2011): 337–342, doi:10.1038/nature10452.

16 SEI International, "Möjligt att försörja 9 miljarder människor år 2050 inom hållbara gränser," press release, October 12, 2011.

17 Travis Lybbert and David Sumner, "Agricultural Biotechnology for Climate Change Mitigation and Adaptation," *Biores* 9, no. 3 (2015).

18 Progressives Today, "UN Climate Official: 'We Should Make Every Effort to Decrease World Population' (Video)," 2015, www.progressivestoday.com/un-climate-official-we-should -make-every-effort-to-decrease-world-population-video.

19 J. Kenneth Smail, "Global Population Reduction: Confronting the Inevitable," *Worldwatch* magazine, September/October 2004.

20 Gibson, *Environmentalism*.

21 S. Ambirajan, "Malthusian Population Theory and Indian Famine Policy in the Nineteenth Century," *Population Studies* 30, no. 1 (1976): 5–14, doi:10.2307/2173660.

22 Optimum Population Trust, "A Population-Based Climate Strategy—An Optimum Population Trust Briefing," May 2007, web.archive.org/web/20160616222500population-matters.org/documents/climate_strategy.pdf.

23 Johan Rockström and Jeffrey Sachs, "Sustainable Development and Planetary Boundaries," submitted to the High Level Panel on the Post-2015 Development Agenda, 2013, www.eesc.europa.eu/resources/docs/sustainable-development-and-planetary-boundaries.pdf.

24 John Bongaarts and Brian C. O'Neill, "Global Warming Policy: Is Population Left Out in the Cold?" *Science* 17 (August 2018): 650–652.

25 Colin Hickey, Travis N. Rieder, and Jake Earl, "Population Engineering and the Fight against Climate Change," *Social Theory and Practice* 42, no. 4 (October 2016): 845–870, www.npr.org/documents/2016/jun/population_engineering.pdf.

26 Jennifer Ludden, "Should We Be Having Kids in the Age of Climate Change?" NPR, August 18, 2016.

27 BBC News, "The Women Too Scared of Climate Change to Have Children," March 4, 2019, www.bbc.com/news/av/uk-47442943/the-women-too-scared-of-climate-change-to-have-children.

28 Pamlin and Armstrong, *Twelve Risks*.

29 D Helbing, "The Birth of a Digital God," *FuturICT*, February 13, 2018, futurict.blogspot.com/2018/02/the-birth-of-digital-god_13.html.

30 Michael Sainato, "Stephen Hawking, Elon Musk, and Bill Gates Warn about Artificial Intelligence," *The Observer*, August 19, 2015.

31 Rochelle Garner, "Elon Musk, Stephen Hawking Win Luddite Award as AI 'Alarmists,'" *CNet*, January 19, 2016.

32 S. Matthew Liao, Anders Sandberg, and Rebecca Roache, "Human Engineering and Climate Change," *Ethics, Policy & Environment* 15, no. 2 (2012).

33 Sarah Knapton, "We Should Give Up Trying to Save the World from Climate Change, Says James Lovelock," *The Telegraph*, April 8, 2014.

34 Knapton, "Climate Change."

35 Mi2g Limited, "Total Disruption by 2020 as Man-Machines Merge? 5 Billion Humans + 50 Bn IoT/Smart Devices + Q-BRAIN Singularity: Who Wins?," March 15, 2016, www.mi2g.com/cgi/mi2g/frameset.php?pageid=http%3A//www.mi2g.com/cgi/mi2g/press/150316.php.

36 K Schwab, *Shaping the Fourth Industrial Revolution*, 169.

37 HIVE EU, "Ethical Issues," hive-eu.org/about/ethical_issues, accessed January 23, 2018.

38 Oxford Martin School, "Programmes: Mind & Machine," www.oxfordmartin.ox.ac.uk/research/programmes/mind-machine, accessed January 23, 2018.

39 Biohax International, www.biohax.tech.

40 Biohacker Summit, biohackersummit.com.

41 D Broderick, *Year Million: Science at the Far Edge of Knowledge* (New York: Atlas & Co., 2008).

42 SpaceX, www.spacex.com.

43 M Harris, "SpaceX Plans to Put More Than 40,000 Satellites in Space," *New Scientist*, October 19, 2019.

44 Neuralink, neuralink.com; Neuralink, Launch Event, July 17, 2019, youtu.be/r-vbh3t7WVI.

Epilogue

1   Rockefeller, *Memoirs*.

2   AFP, "ExxonMobil, Rockefellers Face Off in Climate battle," *The Express Tribune*, April 18, 2016.

3   Winrock International, "Winrock Board Member: Peter O'Neill," www.winrock.org/bio /peter-m-oneill, accessed January 23, 2018.

4   Nick Visser, "The World Has Pledged to Divest $2.6 Trillion from Fossil Fuels," *Huffington Post*, September 22, 2015.

5   Reeves Wiedeman, "The Rockefellers vs. the Company That Made Them Rockefellers," *New York* magazine, January 8, 2018.

6   Samuel Hansell, "Company News; Rockefeller Center Filing May Mean Big Tax Bill," *New York Times*, May 13, 1995.

7   Rockefeller, *Memoirs*, 474–481.

8   Rockefeller Brothers Fund, "On the Loss of David Rockefeller, 1915–2017," March 20, 2017, www.rbf.org/news/on-the-loss-of-David-Rockefeller.

9   Rockefeller, *Memoirs*.

10 Divest-Invest, www.divestinvest.org.

11 World Economic Forum, "World Economic Forum and UN Sign Strategic Partnership Framework," June 13, 2019, www.weforum.org/press/2019/06/world-economic-forum-and -un-sign-strategic-partnership-framework.

12 World Economic Forum, *Unlocking Technology for the Global Goals*, WEF with PwC, January 2020, www3.weforum.org/docs/Unlocking_Technology_for_the_Global_Goals.pdf.

13 Selam Gebrekidan, "For Autocrats, and Others, Coronavirus Is a Chance to Grab Even More Power," *New York Times*, March 30, 2020.

14 Rockefeller Foundation, *Scenarios for the Future of Technology and International Development*, The Rockefeller Foundation and Global Business Alliance, May 2010.

15 Johns Hopkins Center for Health Security, "Event 201 Pandemic Exercise," YouTube, October 18, 2019, youtu.be/AoLw-Q8X174.

16 ID2020, "Alliance," id2020.org/alliance, accessed March 20, 2020.

17 Sandrine Dixson-Declève et al. "Could COVID-19 Give Rise to a Greener Global Future?" World Economic Forum, March 25, 2020.

18 Dixy Lee Ray, *Environmental Overkill: Whatever Happened to Common Sense?* (New York: Harper Perennial, 1994).

19 World Economic Forum, "The Great Reset," July 17, 2020, www.weforum.org/great-reset.

Conclusions

1   Harris, "SpaceX"; Neuralink, neuralink.com.

Index of Persons

Index of Organizations, Projects, and Misc.